Inorganic Chemistry of Main Group Elements

OTHER RELATED BOOKS BY VCH

N. Auner and J. Weis (editors)
Organosilicon Chemistry: From Molecules to Materials

James A. Cowan
Inorganic Biochemistry: An Introduction

A. Dedieu (editor)
Transition Metal Hydrides

Christoph Elschenbroich and Albrecht Salzer
Organometallics: A Concise Introduction
Second Edition

J. Nentwig, M. Kreuder, and K. Morgenstern
General and Inorganic Chemistry Made Easy

T. A. O'Donnell
Superacids and Acidic Melts as Inorganic Chemical Reaction Media

C. N. R. Rao and Bernard Raveau
Transition Metal Oxides

G. H. Robinson (editor)
Coordination Chemistry of Aluminum

U. Schwertmann and R. M. Cornell
Iron Oxides in the Laboratory: Preparation and Characterization

Inorganic Chemistry of Main Group Elements

R. Bruce King

VCH

R. Bruce King
Department of Chemistry
University of Georgia
Athens, Georgia 30602

This book is printed on acid-free paper. ⊗

Library of Congress Cataloging-in-Publication Data

King, R. Bruce.
Inorganic chemistry of main group elements / R. Bruce King.
    p.   cm.
Includes bibliographical references and index.
ISBN 1-56081-679-1 (acid-free)
1. Chemistry, Inorganic.   I. Title.
QD151.2.K5   1994   94-35433
546—dc20   CIP

Printed in the United States of America

ISBN 1-56081-679-1 VCH Publishers, Inc.

Printing History:
10 9 8 7 6 5 4 3 2 1

Published jointly by

VCH Publishers, Inc.	VCH Verlagsgesellschaft mbH	VCH Publishers (UK) Ltd.
220 East 23rd Street	P.O. Box 10 11 61	8 Wellington Court
New York, New York 10010	69451 Weinheim, Germany	Cambridge CB1 1HZ
		` Kingdom

Preface

Some of the most important applications of inorganic chemistry involve the chemistry of the main group elements, which are defined here to include all the elements except for the d-block transition metals. Examples of well-known applications of the main group elements in other areas of science and technology are the use of silicon, germanium, and arsenic in the fabrication of semiconductors, the importance of aluminum and silicon chemistry in zeolite technology, the importance of phosphates in biological systems, and the use of reagents containing diverse main group elements such as phosphorus, selenium, boron, lithium, and magnesium in organic synthesis. For this reason there is a clear need for an up-to-date book presenting the highlights of the descriptive chemistry of the main group elements in a concise manner.

The objective of this book is to provide a summary of the most important aspects of the descriptive inorganic chemistry of the main group elements with liberal references to important books, review articles, and seminal papers. The book is designed to be a text for a one-quarter or one-semester graduate level course as well as a first source of general information on the chemistry of main group elements for research workers in other fields. The reader is assumed to be acquainted with the fundamental ideas of chemical structure and bonding; this book is devoted entirely to descriptive chemistry and organized according to the periodic table. The contents of the book are based on a graduate level course given by the author at the University of Georgia during the spring quarters of 1989, 1991, and 1992 and the winter quarter of 1993.

R. Bruce King
Athens, Georgia
July 1994

Contents

12. Lanthanides and Actinides   289

Introduction and General Organization

The organization of this book is based on the periodic table (see Table I.1, p. xx). In each cell of the periodic table in Table I.1 the atomic number is given in the first line, the chemical symbol in the second line, and the atomic weight in the third line. The d-block transition metals excluded from consideration in this book (i.e., elements 22–29, 40–47, and 72–79) are enclosed in double lines in this periodic table. The f-block lanthanides (elements 57–71) and actinides (elements 89–103) are included in this book, since the f electrons do not have a major effect on the chemistry of most of these elements, particularly the lanthanides. Zinc, cadmium, and mercury (group 12) are included in this book, since the filled d^{10} shell is retained in all their stable compounds.

The *groups* in the periodic table are represented by the columns, which are numbered from 1 for the alkali metals (Li, Na, K, Rb, Cs, Fr) to 18 for the noble gases (He, Ne, Ar, Kr, Xe, Rn). These groups correspond to the number of valence electrons for the rows K–Kr and Rb–Xe. Elements below a given element in the same column (hence in the same group) of the periodic table are called its *congeners*. For example, the congeners of carbon in group 14 are silicon, germanium, tin, and lead.

The noble gases (group 18) are of significance because their electronic configurations represent stable closed shells, which all elements try to attain in their chemical combinations; most chemistry can be rationalized on this basis leading to the "octet rule" for light elements using only s and p orbitals and the "18-electron rule" for transition metals using s, p, and d orbitals. Correspondingly, the noble gases, because of their stable electronic configurations, are very unreactive, and stable compounds of helium, neon, and

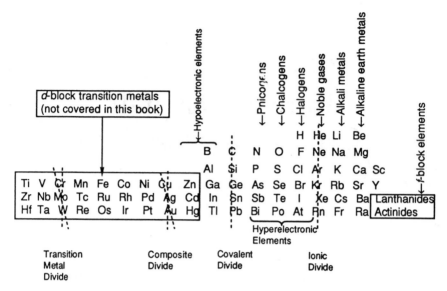

Figure I.1. Metallurgist's interpretation of the periodic table showing the four "divides." The *d*-block and *f*-block elements, hypo- and hyperelectronic elements, and the special families of elements are also shown. (From Ref. 1.)

argon are unknown. The noble gases form the so-called ionic divide in the metallurgists' interpretation of the periodic table (Figure I.1).[1] Elements in the columns to the immediate left of the noble gases in the periodic table tend to acquire electrons to form anions. Thus the halogens (group 17) acquire a single electron to form halides X^- (X = F, Cl, Br, I) and the chalcogens (group 16) acquire two electrons to form the anions E^{2-} (E = O, S, Se, Te). All these anions have the favored electronic configuration of the nearest noble gas and are very stable. Conversely, elements in the columns to the immediate right of the noble gases in the periodic table tend to lose electrons to form cations. Thus the alkali metals (group 1) lose a single electron to form the monocations M^+ (M = Li, Na, K, Rb, Cs), and the alkaline earth metals (group 2) lose two electrons to form the dications M^{2+} (M = Be, Mg, Ca, Sr, Ba). All these cations have the favored electronic configuration of the nearest noble gas and are very stable.

The lightest elements (lithium through fluorine) always use a four-orbital manifold consisting of one spherically symmetric *s* orbital and three orthogonal uninodal *p* orbitals (Figure I.2). These elements thus have a maximum coordination number of 4 as long as all their bonds are the usual type of two-electron, two-center bonds. The sp^3 hybrids formed by this four-orbital manifold have the well-known geometry of the tetrahedron (Figure I.2). The central element of the Li–F row, namely carbon, has four valence electrons of its own and acquires the favored eight-electron noble gas configuration if its sp^3 manifold is used to form four two-electron, two-center bonds directed

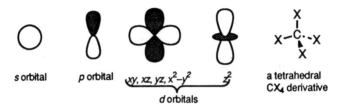

s orbital   p orbital   xy, xz, yz, x²-y²   z²   a tetrahedral
                                                                CX₄ derivative
                              d orbitals

Figure I.2. Shapes of the atomic orbitals involved in main group element bonding; a tetrahedral CX_4 derivative.

toward the vertices of a tetrahedron in a CX_4 derivative (e.g., carbon tetra-chloride) so that the central carbon atom acquires one electron from each of the four atoms to which it is bonded. The vast series of saturated organic compounds are constructed from such carbon tetrahedra with extensive linkage ("catenation") through carbon–carbon bonding. Such compounds are the province of the organic chemist and are thus beyond the scope of this book. The heavier congeners of carbon in group 14, namely silicon, germanium, tin, and lead, also form large numbers of analogous tetrahedral derivatives (Figure I.2) through similar hybridization. Group 14 forms the so-called covalent divide in the metallurgist's interpretation of the periodic table (Figure I.1).[1] Silicon and germanium, two of the group 14 elements, as well as isoelectronic compounds such as gallium arsenide (GaAs), are key materials in semiconductor technology.

Elements to the right of the covalent divide, namely the pnicogens (group 15), chalcogens (group 16), and the halogens (group 17), with five, six, and seven valence electrons, respectively, need lone pairs in one or more of their valence orbitals to attain the eight-electron favored noble gas electronic configuration, assuming that they use a four-orbital sp^3 manifold. Such elements may be called hyperelectronic elements, where a *hyperelectronic element* is defined as an element with more than four valence electrons for a four-orbital sp^3 manifold. Doping group 14 semiconductors by addition of appropriately chosen hyperelectronic elements such as arsenic adds electrons as carriers of electric current leading to so-called *n-type* semiconductors. Elements to the left of the covalent divide having fewer than four valence electrons, such as group 13 elements (B, Al, Ga, In, Tl) with only three valence electrons and the alkaline earth metals and group 12 elements (Zn, Cd, Hg) with only two valence electrons, are not able to attain the eight-electron noble gas configuration even if they use all four sp^3 hybrids to form four tetrahedrally disposed two-electron, two-center bonds. Such elements may be called *hypoelectronic elements*, that is, having fewer than four valence electrons for a four-orbital sp^3 manifold. Hypoelectronic elements tend to acquire the noble gas electronic configuration by forming multicenter bonds; the two-electron, three-center B—H—B bonds in diborane, B_2H_6 (Figure I.3) provide a simple example of chemical bonding of this type. Compounds with such multicenter bonds are frequently called

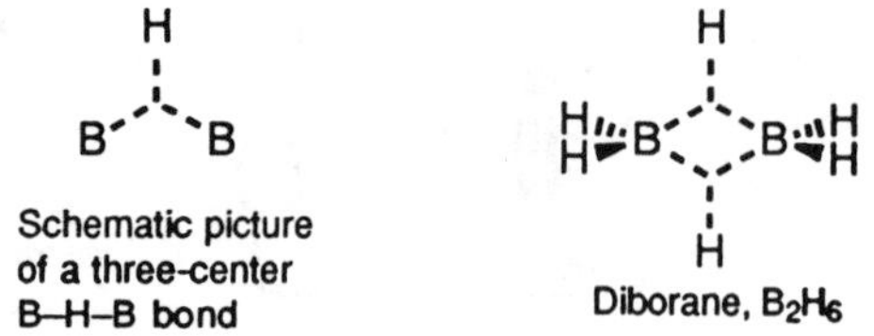

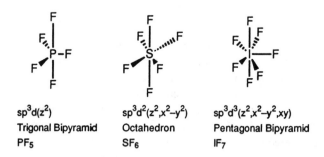

Figure I.3. A three-center B—H—B bond and its occurrence in diborane.

Figure I.4. Role of d orbitals in hypervalent main group element fluorides.

electron-deficient, since an electron pair is not available for each pair of atomic orbitals to form only two-electron, two-center bonds. Doping group 14 semiconductors by addition of appropriately chosen hypoelectronic elements such as boron adds holes (i.e., absence of electrons) as carriers of electric current leading to so-called *p-type* semiconductors.

The foregoing discussion assumes that only s and p orbitals are involved in the chemical bonding of main group elements. However, in some cases the d orbitals, which play key roles in the chemistry of the d-block transition metals not discussed in this book, can become involved in the chemical bonding of main group elements, leading to the so-called *hypervalent* compounds. In hypervalent compounds the central main group element may have a co-ordination number larger that 4, such as 5 in trigonal bipyramidal PF_5, 6 in octahedral SF_6, or 7 in pentagonal bipyramidal IF_7 (Figure I.4). In such hypervalent compounds the central main group element has an electronic configuration in excess of that of the next noble gas.

Many hypervalent compounds are formed by the hyperelectronic elements to the right of the covalent divide (group 14) in the periodic table (Table I.1 and Figure I.1). The central hyperelectronic atoms in such hypervalent compounds frequently have lone pairs, which are often, although not always, stereochemically active. Examples of hypervalent fluorides of hyperelectronic elements with stereochemically active lone pairs are given in Figure I.5. This figure indicates the geometry of the fluorine atoms as well as the underlying coordination polyhedron (i.e., the "pseudopolyhedron") formed by both the

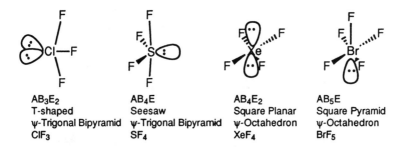

AB₃E₂	AB₄E	AB₄E₂	AB₅E
T-shaped	Seesaw	Square Planar	Square Pyramid
ψ-Trigonal Bipyramid	ψ-Trigonal Bipyramid	ψ-Octahedron	ψ-Octahedron
ClF₃	SF₄	XeF₄	BrF₅

Figure I.5. Examples of hypervalent hyperelectronic element fluorides with stereochemically active lone electron pairs (E = electron pair).

fluorine atoms and the stereochemically active lone pairs. In this sense a *pseudo-polyhedron* (or *ψ-polyhedron*) is defined as the polyhedron formed by the central atom including both its chemical bonds *and* stereochemically active lone pairs.

Another important aspect of the chemistry of some main group elements is the formation of multiple $p\pi–p\pi$ bonds similar to the C=C double bond in ethylene or the C≡C triple bond in acetylene. Compounds containing such multiple $p\pi–p\pi$ bonds can be very stable if both atoms forming the bond are carbon or the lightest hyperelectronic elements such as nitrogen and oxygen. However, until the last 15 years similar $p\pi–p\pi$ multiple bonding involving third-row elements such as phosphorus or silicon was not believed to give stable compounds despite some long-known examples of stable sulfur compounds with C=S $p\pi–p\pi$ double bonds such as carbon disulfide, :S̈=C=S̈:, and thiourea, :S̈=C(NH₂)₂. However, within the last 15 years a number of kinetically stable compounds containing $p\pi–p\pi$ double and triple bonds involving heavier elements such as silicon, germanium, phosphorus, and arsenic have been prepared using bulky organic alkyl and aryl groups such as *tert*-butyl, mesityl (2,4,6-trimethylphenyl), "supermesityl" (2,4,6-tri-*tert*-butylphenyl), and bis(trimethylsilyl)methyl to block sterically the otherwise reactive multiple bonds. In fact, the use of sterically demanding terminal groups has often been a very useful method for kinetically stabilizing unusual multiple bonds or oxidation states, as indicated by the examples depicted in Figure I.6.

The chapters in this book present the highlights of the descriptive chemistry of the main group elements in the following order:

1. Hydrogen, the lightest element with many unique properties.
2. Carbon and the other elements in the covalent divide (silicon, germanium, tin, and lead).
3. The hyperelectronic elements, starting with the nitrogen column (group 15) and moving to the right as far as the noble gases (i.e., the ionic divide).
4. The hypoelectronic element boron and its heavier congeners in group 13 (aluminum, gallium, indium, and thallium).
5. The highly electropositive alkali and alkaline earth metals.

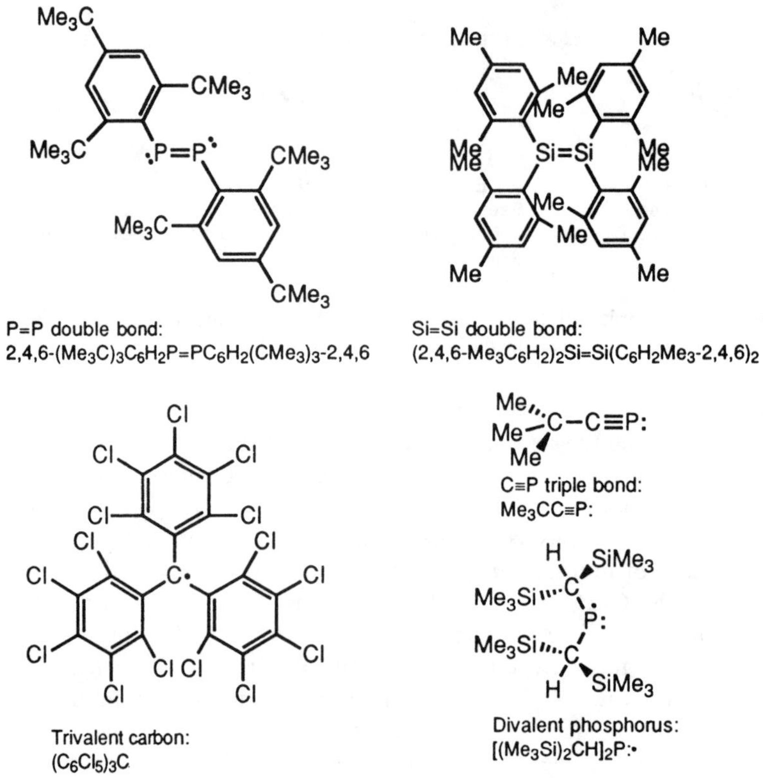

P=P double bond:
2,4,6-(Me₃C)₃C₆H₂P=PC₆H₂(CMe₃)₃-2,4,6

Si=Si double bond:
(2,4,6-Me₃C₆H₂)₂Si=Si(C₆H₂Me₃-2,4,6)₂

C≡P triple bond:
Me₃CC≡P:

Trivalent carbon:
(C₆Cl₅)₃C

Divalent phosphorus:
[(Me₃Si)₂CH]₂P:·

Figure I.6. Examples of the kinetic stabilization of reactive main group element units (multiple bonds and free radicals) by the use of bulky substituents.

6. Zinc, cadmium, and mercury.
7. The *f*-block elements, namely the lanthanides and actinides.

For further general details on either descriptive chemistry or background information the reader is referred to a number of comprehensive and/or general inorganic textbooks.[2-8] For more specialized information on the nonmetals up to the mid-1970s, a book by Steudel[9] is useful. Smith[10] has written a useful prelude to the study of descriptive inorganic chemistry, such as that present in this book.

References

1. Stone, H. E. N., *Acta Metall.*, 1979, **27**, 259–264.

2. Cotton, F. A.; Wilkinson, G., *Advanced Inorganic Chemistry*, 5th ed., Wiley, New York, 1988.

3. Greenwood, N. N.; Earnshaw, A., *Chemistry of the Elements*, Pergamon, Oxford, 1984.

4. Lagowski, J. J., *Modern Inorganic Chemistry*, Dekker, New York, 1973.

5. Heslop, R. B.; Jones, K., *Inorganic Chemistry: A Guide to Advanced Study*, Elsevier, Amsterdam, 1976.

6. Huheey, J. E., *Inorganic Chemistry*, 3rd ed., Harper & Row, New York, 1983.

7. Purcell, K. F.; Kotz, J. C., *An Introduction to Inorganic Chemistry*, Saunders, Philadelphia, 1980.

8. Douglas, B. E.; McDaniel, D. H.; Alexander, J. J., *Concepts and Models of Inorganic Chemistry*, 2nd ed., Wiley, New York, 1983.

9. Steudel, R., *Chemistry of the Non-Metals*, de Gruyter, Berlin, 1977.

10. Smith, D. W., *Inorganic Substances*, Cambridge University Press, Cambridge, 1990.

Table I.1   The Periodic Table of the Elements

Group 1 (except H) = alkali metals; group 2 = alkaline earth metals; group 15 = pnicogens; group 16 = chalcogens; group 17 = halogens.

Group 0 = noble gases; *(elements 57–71) = lanthanides; †(elements 89–103) = actinides.

The d-block transition metals not covered in this book (elements 22–29, 40–47, and 72–79) are enclosed in double lines.

1

Hydrogen

1.1 General Aspects of Hydrogen Chemistry

Hydrogen is the most abundant element in the universe. In addition, hydrogen is the third most abundant element in the earth's surface (after oxygen and silicon). Thus hydrogen in combined form accounts for 15.4% of the atoms in the earth's crust and oceans and is the ninth element in order of abundance (0.9%).

The chemistry of hydrogen is related to three electronic processes, which are discussed in turn.

1.1.1 Loss of an Electron to Form a Proton

The loss of an electron from a hydrogen atom to form a bare proton is an endothermic process:

$$H(g) \rightarrow H^+(g) + e^- \qquad \Delta H = 569 \text{ kJ/mol} = 13.59 \text{ eV} \qquad (1.1)$$

Because of its small size and very high charge density, however, the proton is always associated with other atoms or molecules in condensed phases. Because of its large charge-to-radius ratio, the total hydration enthalpy of the proton is highly exothermic and much larger than that of other singly charged cations:

$$H^+(g) + nH_2O(l) \rightarrow H_3O^+(aq) \qquad \Delta H^\circ = -1090 \text{ kJ/mol} \qquad (1.2)$$

This highly exothermic proton hydration enthalpy makes the conversion of hydrogen to solvated protons an exothermic process even though the ionization of hydrogen (equation 1.1) is an endothermic process.

1.1.2 Gain of an Electron to Form a Hydride Ion

Hydrogen can gain an electron to form the hydride ion with the favored helium electronic configuration:

$$H + e^- \rightarrow H^- \qquad \Delta H^\circ = 72 \text{ kJ/mol} \qquad (1.3)$$

The hydride ion is found in the saline hydrides of the most electropositive metals (e.g., NaH, KH, CaH_2).

1.1.3 Formation of a Covalent Electron Pair Bond

Covalent element–hydrogen bonds are found in the diverse binary molecules FH (acidic), OH_2 (neutral), NH_3 (basic), CH_4 (neutral, inert), and BH_3 (dimerizes to B_2H_6 through B—H—B three-center bonding—see Figure I.3). In addition, hydrogen exhibits the following special bonding features:

1. Formation of nonstoichiometric hydrides rather than saline hydrides with transition metals and other less electropositive metals;
2. Formation of multicenter hydrogen bridge bonds (e.g., three-center B—H—B and M—H—B bonds) in boranes and transition metal hydrides;
3. Involvement of "hydrogen bonding" through bridges of the type X—H$\cdots$Y in the chemistry of water, aqueous solutions, hydroxylic species, carboxylic acids, and biological systems including linking of polypeptide chains in proteins and base pairs in nucleic acids.

1.2 Hydrogen Isotopes and Elemental Hydrogen

1.2.1 Hydrogen Isotopes

Table 1.1 summarizes some important properties of the three hydrogen isotopes. The mass differences of the hydrogen isotopes are the largest of any set of isotopes, leading to the largest isotope effects. However, the three hydrogen isotopes are essentially identical in chemistry except for differences in rate and equilibrium constants.

Deuterium[1] is separated from water in ton quantities by fractional

Table 1.1 Selected Properties of the Hydrogen Isotopes

Isotope	Natural Abundance (%)	Half-life	Spin	NMR Sensitivity*
Protium, ^{1}H	99.985%	Stable	$\frac{1}{2}$	1.000
Deuterium, ^{2}H, D	0.015%	Stable	1	0.0097
Tritium, ^{3}H, T	10^{-17}	12.4 years	$\frac{1}{2}$	1.21

* For equal numbers of nuclei.

distillation, electrolysis, and/or isotopic exchange.[2] In this connection the Girdler sulfide reaction

$$HOH(l) + HSD(g) \rightleftharpoons HOD(l) + HSH(g) \qquad K \sim 1.01 \text{ at } 25°C \qquad (1.4)$$

is used to enrich natural water to $\sim 15\%$ deuterium followed by further enrichment by fractional electrolysis. Deuterium oxide is used as a moderator for nuclear reactors for two reasons: it is effective at reducing the energies of fast fission neutrons to thermal energies and, since it has a lower capture cross section for neutrons than does protium, the neutron flux is not reduced significantly. Deuterium compounds are widely used as solvents in proton nuclear magnetic resonance (NMR) work to avoid interference from solvent protons. In addition, deuterium compounds are widely used for mechanistic and spectroscopic studies.

Tritium, the only radioactive isotope of hydrogen, is a low energy beta emitter with a half-life of 12.4 years.[3] It is the most sensitive NMR nucleus known, although because of its radioactivity tritium is not normally used in high concentrations. Naturally occurring tritium is formed from neutron capture by nitrogen of cosmic irradiation according to the following equation:

$$^{14}_{7}N + ^{1}_{0}n \rightarrow ^{3}_{1}T + ^{12}_{6}C \qquad (1.5)$$

Tritium is produced artificially by neutron irradiation of enriched lithium-6 in a nuclear reactor:

$$^{6}_{3}Li + ^{1}_{0}n \rightarrow ^{4}_{2}He + ^{3}_{1}T \qquad (1.6)$$

The lithium is introduced in the form of an alloy with magnesium or aluminum; this alloy retains much of the tritium until treatment with acid. Tritium is a source of energy from fusion reactions[4]:

$$^{3}_{1}T + ^{2}_{1}D \rightarrow ^{4}_{2}He \ (3.5 \text{ MeV}) + ^{1}_{0}n \ (14.1 \text{ eV}) \qquad (1.7)$$

1.2.2 Elemental Hydrogen, H_2, "Dihydrogen"

Dihydrogen, H_2, is a colorless, odorless, tasteless, highly flammable gas (bp $-252.5°C = 20.3$ K), which is obtained by electrolysis of water or by treatment of metals such as zinc or iron with dilute nonoxidizing acids. On the industrial scale, H_2 can be obtained from coke and steam by the following sequence of reactions:

$$C + H_2O \xrightarrow{1000°C} CO + H_2 \qquad \text{(water gas reaction)} \qquad (1.8a)$$

$$CO + H_2O \xrightarrow{catalyst} CO_2 + H_2 \qquad \text{(water gas shift reaction)} \qquad (1.8b)$$

The CO_2 by-product is removed by scrubbing with K_2CO_3 solution or ethanolamine. The 1:1 CO/H_2 mixture obtained by reaction 1.8a is sometimes called *synthesis gas*. An alternative approach to the preparation of H_2 uses the

1:3 CO/H_2 mixture obtained by the steam re-forming of methane over a promoted nickel catalyst at 750°C according to the equation:

$$CH_4 + H_2O \xrightarrow{\text{catalyst}} CO + 3H_2 \tag{1.9}$$

Molecular H_2 exists as a mixture of ortho and para H_2 with aligned (parallel) and opposed (antiparallel) nuclear spins, respectively.[5] This is the best example of *nuclear spin isomerism*. Although, the o-H_2 → p-H_2 conversion is exothermic, it is a forbidden triplet–singlet transition, which cannot be accomplished thermally without complete dissociation and recombination of the H_2. For this reason the o-H_2 → p-H_2 conversion is catalyzed by interaction with species that break the H—H bond, such as palladium, platinum, active Fe_2O_3, and nitric oxide. Concentration of o-H_2 to greater than 75% is impossible but 3:1 *ortho/para* mixtures are in metastable equilibrium.

Dihydrogen is cleaved by many transition metal complexes with complete rupture of its H—H bond, for example,

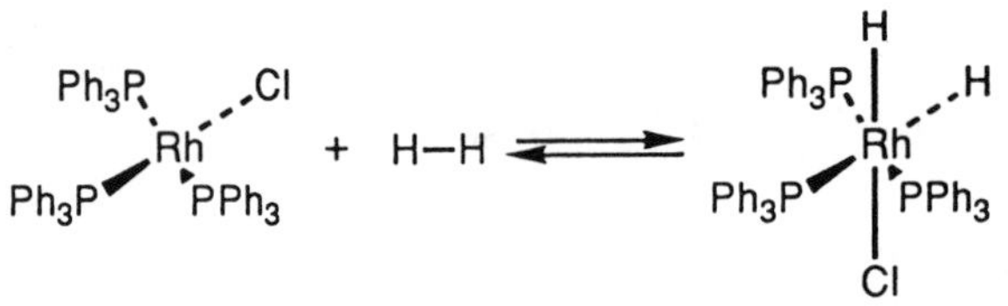

$$\tag{1.10}$$

1.1

The resulting product (e.g., **1.1**) is a transition metal dihydride in which the coordination number of the transition metal has increased by 2 (in this case from 4 to 6) in a so-called *oxidative addition* reaction. Other transition metal complexes react with dihydrogen with some retention of H—H bonding to form a molecular hydrogen complex,[6] for example,

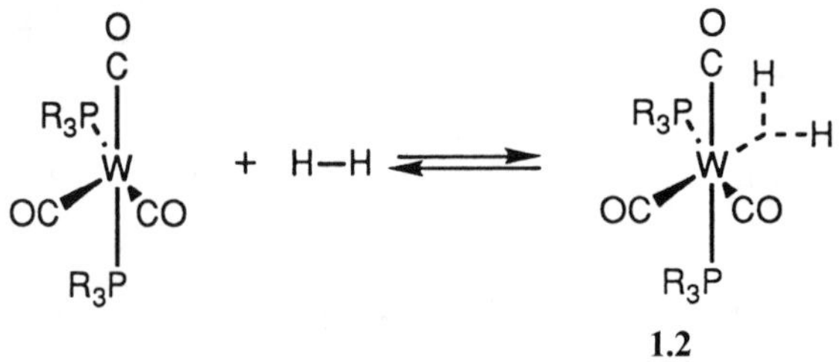

$$\tag{1.11}$$

1.2

In molecular hydrogen (H_2) complexes such as **1.2**, the electron pair in the σ bond of H_2 can be considered to coordinate to the transition metal. Alternatively, but equivalently, the complex can be formulated with a three-center H—M—H bond (compare Figure I.3).

1.2.3 Atomic Hydrogen, H

Atomic hydrogen is produced from molecular hydrogen by the endothermic process

$$H_2 \rightarrow 2H \qquad \Delta H^\circ = 434.1 \text{ kJ/mol} \qquad (1.12)$$

using ultraviolet irradiation or discharge arcs at high current density in discharge tubes at low H_2 pressure. The half-life of atomic hydrogen is relatively short ($t_{1/2} = 0.3$ s). The heat of recombination of atomic hydrogen to H_2 is high enough to produce high temperatures, leading to the use of atomic hydrogen in metal welding. Atomic hydrogen is a much more reactive reducing agent than molecular hydrogen and combines directly with elemental Ge, Sn, As, Sb, and Te to form the corresponding hydrides.

1.3 Hydrogen Bonding

Hydrogen bonding is used to describe the relatively weak secondary interaction between molecules with partially positive hydrogen atoms and electronegative atoms possessing lone electron pairs, such as F, O, and N, and to a lesser extent Cl, S, and P.[7-9] A hydrogen bond can be represented by structure **1.3**, in which X—H is a *proton donor* and Y is a *proton acceptor*. Hydrogen bonds are considered to be a special bond type with energies between those of van der Waals interactions and normal covalent bonds. Hydrogen bonding interactions thus have typical energies of 4 to 40 kJ/mol but occasionally may have larger energies up to 110 kJ/mol.

$$X—H \cdots Y$$
1.3

The concept of hydrogen bonding was suggested initially by the boiling points and enthalpies of vaporization of the binary hydrogen compounds HX (X = F, Cl), H_2X' (X' = O, S), and H_3X'' (X" = N, P) (Table 1.2). In all cases the hydrogen compound of the lighter element is found to have a much higher boiling point than the corresponding hydrogen compound of the heavier element despite the much higher molecular weight of the latter compound. Subsequently hydrogen bonding (**1.3**) was found in the structures of crystalline solids by neutron diffraction. For such studies requiring the location of hydrogen atoms, X-ray diffraction is less suitable than neutron diffraction. This is because the dependence on electron density of X-ray scattering makes hydrogen atoms, which have no core electrons, very difficult to observe. Such structural studies show that in hydrogen-bonded structures X—H$\cdots$Y, the X$\cdots$Y distances are significantly shorter (e.g., by ≥ 0.2 Å) than the van der Waals contacts. Thus for O—H$\cdots$O structures, O$\cdots$O distances below 3 Å indicate hydrogen bonding. For example, in solid $NaHCO_3$ there are four different O$\cdots$O distances between oxygen atoms of the bicarbonate ions,

Table 1.2   Boiling Points of Nonmetal Hydrides

NH_3,	$-33.5°C$	H_2O,	$+100.0°C$	HF,	$+19.5°C$
PH_3,	$-87.7°C$	H_2S,	$-60.7°C$	HCl,	$-84.9°C$

namely 2.61, 3.12, 3.15, and 3.19 Å. The first $O \cdots O$ distance of 2.26 Å corresponds to the $O—H \cdots O$ hydrogen bond, whereas the remaining three $O \cdots O$ distances correspond only to van der Waals contacts.

A number of spectroscopic methods can be used to detect hydrogen bonding in solution. Thus hydrogen bonding $X—H \cdots Y$ has the following effects on infrared and Raman frequencies:

1. The $v(X—H)$ stretching frequency shifts to lower wavenumbers but increases in width and intensity often more than tenfold. This change of frequency is less than 1000 cm^{-1} for $N—H \cdots F$ hydrogen bonds and is in the range of $1500–2000 \text{ cm}^{-1}$ for $O—H \cdots O$ and $F—H \cdots F$ hydrogen bonds.
2. The $\delta(X—H)$ bending frequency shifts to higher wavenumbers.

Proton NMR spectra indicate hydrogen bonding by a shift to lower fields. This low field shift is generally interpreted, at least qualitatively, in terms of a decrease in the diamagnetic shielding of the proton.[10]

Hydrogen bonds have the following general properties:

1. Most hydrogen bonds $X—H \cdots Y$ are unsymmetrical; that is, the hydrogen atom is much closer to X than to Y. Only the strongest hydrogen bonds are symmetrical, such as the hydrogen bonds in HF_2^-, as discussed in connection with Figure 1.1.
2. Hydrogen bonds $X—H \cdots Y$ are linear or only slightly bent, thus maximizing attraction between H and Y and minimizing $X \cdots Y$ repulsion, so that the bond energy is maximized.
3. The valence angle θ, formed between the hydrogen bond and the $Y—R$ bond (1.4), usually varies between 100 and 140°.[11]

1.4

4. Normally the hydrogen atoms in hydrogen bonds are two-coordinate, but there are examples of hydrogen bonds with three-coordinate and four-coordinate hydrogen atoms.[12,13] Thus in a survey of 1509 $N—H \cdots O{=}C$ hydrogen bonds observed by X-ray or neutron diffraction in 889 organic crystal structures, about one-fifth (i.e., 304) were "bifurcated" with three-coordinate hydrogen and only six "trifurcated" hydrogen bonds with four-coordinate hydrogen were found.

5. In most cases only one hydrogen bond is directed toward each lone pair of Y, but in crystalline ammonia, three hydrogen bonds extend from each nitrogen atom.

The strength of a hydrogen bond corresponds to the enthalpy of its dissociation by the reaction

$$X—H \cdots Y \rightarrow X—H + Y \begin{cases} \text{weak: } \Delta H < 15 \text{ kJ/mol} \\ \text{intermediate: } \Delta H = 15\text{–}40 \text{ kJ/mol} \\ \text{strong } \Delta H > 40 \text{ kJ/mol} \end{cases} \quad (1.13)$$

The enthalpy of the hydrogen bond is zero if the $X \cdots Y$ distance is close to the van der Waals distance. The enthalpy is greater than 100 kJ/mol for a few very short hydrogen bonds such as those in KHF_2 ($\Delta H \approx 212$ kJ/mol; $F—H \cdots F = 2.26$ Å). The hydrogen bond in KHF_2 can be interpreted as an F—H—F three-center, two-electron bond. The most common hydrogen bonds are $O—H \cdots O$. If the $O \cdots O$ distance is in the 3.0–2.8 Å range, the hydrogen bond is relatively weak, whereas if the $O \cdots O$ distance is in the 2.8–2.6 Å range, the hydrogen bond is much stronger, with ΔH in the range of 15–40 kJ/mol.

Hydrogen bonding is important in the structure of ice (solid water), of which there are nine known modifications.[14] At 0°C and 1 atmosphere pressure, water solidifies to ice I, which has an open structure constructed from puckered six-membered rings. Each oxygen atom is surrounded tetrahedrally by four other oxygen atoms through unsymmetrical $O—H \cdots O$ hydrogen bonds with an $O \cdots O$ distance of 2.75 Å and the hydrogen atom lying 1.01 Å from one oxygen atom and 1.74 Å from the other. Each oxygen atom has two near and two far hydrogen atoms leading to six possible arrangements, two of which are illustrated in Figure 1.1. The presence of these unsymmetrical hydrogen bonds in ice I leads to a zero-point entropy of 3.4 J mol^{-1}deg^{-1}. By contrast, KHF_2, with symmetrical hydrogen bonds in its HF_2^- ion, has no zero-point entropy.

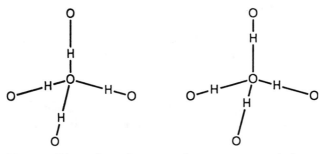

Figure 1.1. Two possible configurations around an oxygen atom in ice.

Hydrogen bonding is also found in species of the following types.

1. *Hydrates of metal salts.* The oxygen atom of water is normally coordinated to the metal atom. However, the hydrogen atoms of such coordinated water molecules can participate in hydrogen bonding. This provides considerable flexibility for the stabilization in crystal lattices of hydrated structures of many different types.

2. *Gas hydrates and clathrates.* Gaseous molecules such as Ar, Kr, Cl_2, SO_2, CH_3Cl, and CH_4 can be trapped in the cavities of a water structure or even in a hydrogen-bonded lattice of an organic hydroxy compound such as hydroquinone (*p*-dihydroxybenzene). Substances of the latter type, called *clathrates*, are inclusion compounds in which small guest molecules are trapped in a lattice having holes of a suitable size. Clathrates provide a method for handling the noble gases in a condensed phase.

3. *Salt hydrates.* Salt hydrates are formed when tetraalkylammonium (R_4N^+) or trialkylsulfonium (R_3S^+) salts crystallize from aqueous solution. The water molecules are hydrogen bonded to fluoride anions or to anion oxygen atoms. The cations and parts of anions occupy cavities randomly and incompletely.

4. *Clathrate hydrates of strong acids.* Strong acids can be stabilized as clathrate hydrates; examples of such strong acid clathrate hydrates are $HPF_6 \cdot 7.67H_2O$, $HBF_4 \cdot 5.75H_2O$, and $HClO_4 \cdot 5.5H_2O$.

1.4 Protic Acids and Solvents

Hydrogen compounds generate protons only when dissolved in media that can solvate the protons, since solvation is needed to provide the energy for bond rupture (Section 1.1). Compounds that furnish hydrogen ions in suitable solvents, such as water, are called *protic acids*.

Solvents that provide a source of protons through self-ionization are called *protic solvents*. In such self-ionization processes one solvent molecule solvates a proton originating from another solvent molecule. The most familiar example of a protic solvent is water, which undergoes the familiar self-ionization

$$2H_2O \rightleftarrows H_3O^+ + OH^-$$

$$K_w = [H^+][OH^-] = 1 \times 10^{-14} \, M^2 \tag{1.14}$$

In aqueous solution the hydrogen ion concentration is often given as pH (from the French *puissance d'hydrogène*),[15] where

$$pH = -\log_{10}[H^+] \tag{1.15}$$

In equation 1.15 [H$^+$] is the hydrogen ion activity.[16] Other protic solvents of interest include liquid ammonia, liquid hydrogen fluoride, and pure sulfuric acid, which undergo the following self ionization processes:

$$2NH_3 \rightleftarrows NH_4^+ + NH_2^- \qquad K_{50°C} = [NH_4^+][NH_2^-] \approx 10^{-30} \tag{1.16}$$

$$2HF \rightleftarrows H_2F^+ + F^- \qquad K = [H_2F^+][F^-] \approx 10^{-10} \qquad (1.17)$$

$$2H_2SO_4 \rightleftarrows H_3SO_4^+ + HSO_4^-$$
$$K_{10°C} = [H_3SO_4^+][HSO_4^-] = 1.7 \times 10^{-4} \, M^2 \, kg^{-2} \qquad (1.18)$$

The anions NH_2^-, F^-, and HSO_4^- function as bases in liquid ammonia, hydrogen fluoride, and sulfuric acid, respectively, just as the hydroxide anion, OH^-, functions as a base in water.

The standard hydrogen electrode involving the H^+/H_2 redox system is taken as the reference for all other oxidation–reduction processes. The hydrogen electrode consists of the hydrogen half-cell

$$H^+(aq) + e = \tfrac{1}{2}H_2(g) \qquad (1.19)$$

The potential ($E°$) of this system is defined to be zero (0.000 V) at all temperatures when an inert metallic electrode dips into a solution of hydrogen ions of unit activity (i.e., pH = 0 or $[H^+] = 1 \, M$) in equilibrium with H_2 gas at 1 atmosphere pressure. The potentials of all other electrodes are referred to this defined zero.

For pure water (pH = 7), the H^+ concentration is only $10^{-7} \, M$ rather than $1 \, M$. Thus the corresponding redox system referred to the standard hydrogen electrode is

$$H^+(aq: 10^{-7} \, M) + e^- = \tfrac{1}{2}H_2(g) \qquad E_{1/2} = 0.414 \, V \qquad (1.20)$$

Thus in the absence of overvoltage (i.e., lack of reversibility at certain metal surfaces), H_2 is liberated from pure water by reducing agents whose potentials ($E°$) are more negative than -0.414 V. For example, certain metal ions (e.g., U^{3+} for which the U^{4+}/U^{3+} potential is -0.61 V) are oxidized by pure water, thereby liberating hydrogen. Note that H_2 is a better reductant in pure water than in $1 \, M H^+$.

Acid–base reactions in aqueous solution are excellent examples of very rapid *diffusion-controlled* reactions. Thus all the protons in water are undergoing rapid migration from one oxygen atom to another and the lifetime of an individual H_3O^+ ion in water is only approximately 10^{-13} second.

The proton is normally present in acids (e.g., HBF_4 and H_2PtCl_6) as H_3O^+, although protons more heavily solvated through hydrogen bonding such as $H_5O_2^+$ ($=H(H_2O)_2^+$) are rather common and even $H_7O_3^+$ and $H_9O_4^+$ are known. Crystalline $H_3O^+ClO_4^-$ ("hydrated perchloric acid") and $NH_4^+ClO_4^-$ are isomorphous, indicating analogous roles of H_3O^+ and NH_4^+ in crystalline structures. A few anhydrous strong acids are known in which the anion contains light electronegative atoms, which can participate in hydrogen bonding; a good example of this is anhydrous $H_4Fe(CN)_6$, which contains $M—C\equiv N—H\cdots N\equiv C—M$ hydrogen bonds and can be considered to contain coordinated hydrogen isocyanide, HNC, with its hydrogen atom participating in the hydrogen bonding. The proton can also be solvated by alcohols; the ion $(CH_3OH)_2H^+$ is found in $(CH_3OH)_2H^+BF_4^-$. The solvation of protons by certain aprotic aromatic diamines, such as 1,8-bis(dimethylamino)naphthalene

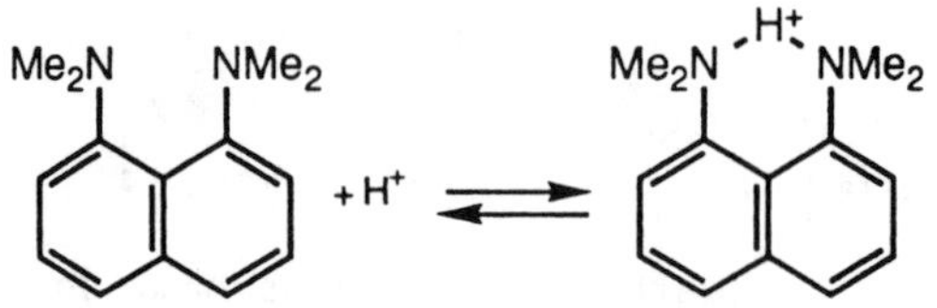

Figure 1.2. Action of the proton sponge 1,8-bis(dimethylamino)naphthalene.

(called *proton sponge*), depends on the formation of a stable six-membered ring from hydrogen bonding (Figure 1.2).[17]

The hydroxide ion and other monovalent oxygen-containing anions can also be solvated through hydrogen bonding. The ion $H_3O_2^- = H(OH)_2^-$ is isoelectronic with HF_2^- and has a very short and thus strong hydrogen bond with an $O\cdots O$ distance of approximately 2.3 Å.

The extent and even the nature of ionization of a potential protic acid, HX, is very solvent dependent, as indicated by the following examples:

1. Perchloric acid, $HClO_4$, is a strong acid in water but a very weak acid in the very acidic solvent H_2SO_4.
2. Trifluoroacetic acid, CF_3CO_2H, is a strong acid in water but nonacidic in H_2SO_4.
3. Phosphoric acid, H_3PO_4, is an acid of medium strength in water but a base in H_2SO_4.

These observations indicate that the more acidic the (protic) solvent, the weaker the acid. Also hydrofluoric acid, HF, is a very weak acid ($pK_a = 3.2$) in dilute aqueous solution because of strong hydrogen bonding of H_3O^+ to F^- so that HF dissociates in water but into tight ion pairs.

The strengths of oxoacids with the general formula H_nXO_m can be estimated by the following rules:

$$m - n = 0 \Rightarrow pK_1 = 8.5 \pm 1.0 \text{ (weak acid)} \tag{1.21a}$$

$$m - n = 1 \Rightarrow pK_1 = 2.8 \pm 0.9 \text{ (medium acid)} \tag{1.21b}$$

$$m - n = 2 \Rightarrow pK_1 \ll 0 \text{ (very strong acid)} \tag{1.21c}$$

The difference between successive pK's for a polybasic acid is 4 to 5 units; this difference arises from the electrostatic effects of the negative charge from the dissociation of one proton on the remaining protons. Apparent exceptions to this oxoacid strength rule (1–21) occur in the following cases.

1. The phosphorus oxoacids H_3PO_2 and H_3PO_3 actually have the structures $H_2P(O)(OH)$ and $HP(O)(OH)_2$, respectively, with P—H bond(s), and are, therefore, monobasic and dibasic acids, respectively, rather than tribasic acids. Their effective $m - n$ values are thus both 1 in accord with the observed pK_a values of 2 for H_3PO_2 and 1.8 for H_3PO_3.

2. Carbonic acid, H_2CO_3, with an apparent $m - n$ value of 1, is a much weaker acid ($pK_a = 6.38$) than predicted from the rule (equation 1.21b) because in solution it is present as loosely hydrated CO_2 to the extent of $\sim 99.8\%$.

Many metal ions in aqueous solution function as acids because of the acidity of coordinated water molecules. Thus the acidities of iron cations in aqueous solution can be attributed to the following equilibria:

$$Fe^{III}(H_2O)_6^{3+} \rightleftharpoons Fe^{III}(H_2O)_5(OH)^{2+} + H^+ \qquad K_{Fe^{3+}} \approx 10^{-3} \qquad (1.22a)$$

$$Fe^{II}(H_2O)_6^{2+} \rightleftharpoons Fe^{II}(H_2O)_5(OH)^+ + H^+ \qquad K_{Fe^{2+}} \ll K_{Fe^{3+}} \qquad (1.22b)$$

The Fe^{2+} ion, with a lower positive charge, is thus less acidic in aqueous solution than the Fe^{3+} ion. Such hydrolysis reactions of Fe^{3+} impart the characteristic yellow to red-brown coloration to aqueous solutions of iron(II) salts; the unhydrolyzed ion $Fe(H_2O)_6^{3+}$ is pale purple, as seen in crystalline hydrated iron(III) salts such as the nitrate or sulfate.

The concepts of hydrogen ion concentration and pH apply to dilute aqueous solutions of acids. These concepts can be generalized to acidities of acids in other media and at high concentrations by means of the *Hammett acidity function*, H_0, which is defined as follows where B refers to an *indicator base*[18]:

$$H_0 = pK_{BH^+} - \log\left(\frac{[BH^+]}{[B]}\right) \qquad (1.23)$$

In very dilute solutions

$$K_{BH^+} = \left(\frac{[B][H^+]}{[BH^+]}\right) \qquad (1.24)$$

$$\Rightarrow H_0 = -\log\left(\frac{[B][H^+]}{[BH^+]}\right) - \log\left(\frac{[BH^+]}{[B]}\right) = -\log[H^+] = pH \qquad (1.25)$$

indicating that H_0 is a generalization of pH. The values of H_0 for strong acids extending from dilute solutions to the pure acid can be determined by using suitable organic weak bases such as *p*-nitroaniline with suitable indicators in various solvents and concentrations or by methods based on NMR. The H_0 values of strong acids in aqueous solution up to about 8 M are very similar owing to protonation of water, so that $(H_2O)_n H^+$ formed by the following equation is the actual protonating agent:

$$H^+ X^- + n H_2O \rightarrow (H_2O)_n H^+ + X^- \qquad (1.26)$$

This effect is called a *solvent leveling effect*.

By using H_0 values (given in parentheses), the following acidity series for the strongest pure liquid acids can be generated: $HSO_3F/(14.1 \text{ mol}\%)SbF_5$ ($H_0 = -26.5$) > $HF/(0.6 \text{ mol}\%)SbF_5$($H_0 = -21.1$) > $HSO_3F \sim H_2S_2O_7$ ($H_0 = -15$) > CF_3SO_3H($H_0 = -14.1$) > H_2SO_4($H_0 = -12.1$) > pure liquid HF($H_0 = -10.2$) > $HF/(1 \text{ } M)NaF$($H_0 = -8.4$) > H_3PO_4($H_0 = -5.0$)

$> HCO_2H(H_0 = -2.2)$. Note that the acidity of pure HF is increased by addition of an acceptor but decreased by addition of NaF owing to the following reactions:

$$2HF + SbF_5 \rightarrow H_2F^+SbF_6^- \tag{1.27a}$$

$$HF + NaF \rightarrow Na^+HF_2^- \tag{1.27b}$$

This relates to the role of SbF_5 as an acid and F^- as a base in liquid HF.

Acids with H_0 values more negative than -6 are called *superacids*, since on the H_0 scale they are more than a million (10^6) times stronger than an aqueous solution of a strong acid.[19] The strongest of the superacids, namely HSO_3F/SbF_5, is a very complicated system that has been investigated by NMR and Raman spectroscopy. The acidity arises from formation of protonated fluorosulfuric acid, namely the $H_2SO_3F^+$ ion. Solutions of HSO_3F/SbF_5 are very viscous; their viscosity can be reduced without destroying their acidity by dilution with liquid sulfur dioxide.

Superacids can protonate a variety of substances not normally considered to be bases, such as formic acid, formaldehyde, and fluorobenzene.[20] Transition metal derivatives are often protonated at the metal (e.g., $Fe(CO)_5$ to $HFe(CO)_5^+$ or $(C_5H_5)_2Fe$ to $(C_5H_5)_2FeH^+$). Carbonic acid is protonated to give the trihydroxycarbonium ion, $C(OH)_3^+$. Superacids can protonate saturated hydrocarbons as indicated by hydride abstraction, H–D exchange, and formation of carbonium ions. The protonation of saturated hydrocarbons can involve such intermediates with three-center H—C—H bonds as

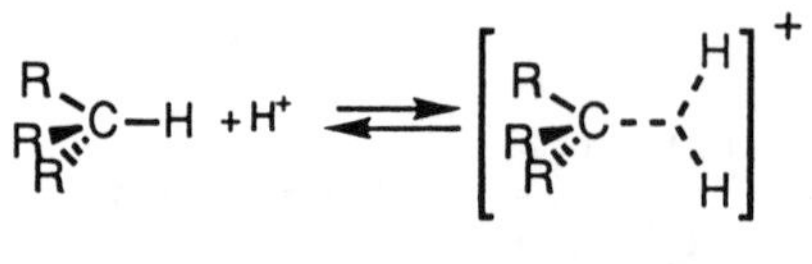

$$\tag{1.27}$$

The observation of H/D exchange in molecular hydrogen upon treatment with superacids suggests that even H_2 is a strong enough base to be protonated by superacids, possibly to give planar H_3^+ with a three-center H—H—H bond.

The properties of some of the most important strong acids are summarized in Sections 1.4.1–1.4.8. In some cases further information on these compounds not related to their acidities is given in the later chapters of the book on the individual elements.

1.4.1 Hydrogen Fluoride (HF)

Hydrogen fluoride (bp $+19.7°C$), obtained from sulfuric acid and calcium fluoride, is a common fluorine source for the synthesis of fluorine compounds. It attacks glass but can be handled in equipment made of suitable metals (copper or Monel alloy) or fluoropolymers such as Teflon or Kel-F. Its self-ionization involves the following equilibria, which account for its extreme acidity:

$$2HF \rightleftharpoons H_2F^+ + F^-; \qquad F^- + HF \rightleftharpoons HF_2^-;$$

$$HF_2^- + HF \rightleftharpoons H_2F_3^-, \text{ etc.} \tag{1.29}$$

1.4.2 Other Hydrogen Halides (HX: X = Cl, Br, I)

The strong acids formed from the nonfluoride halides are all gases with pungent odors under normal conditions. Their self-ionization in the liquid state is small in contrast to hydrogen fluoride, possibly because of the much weaker hydrogen bonding to the heavier halogens relative to fluorine.

1.4.3 Perchloric Acid (HClO$_4$).[21]

Concentrated aqueous perchloric acid is 70–72% by weight $HClO_4$; the water azeotrope contains 72.5% $HClO_4$ and boils at 203°C. Anhydrous $HClO_4$ (bp 43°C/7 torr) is obtained by the distillation of the concentrated aqueous acid over anhydrous $Mg(ClO_4)_2$; it reacts vigorously, often explosively, with organic matter. Pure (100%) $HClO_4$ decomposes upon standing for several days at room temperature to give $HClO_4 \cdot H_2O$ ($= H_3O^+ ClO_4^-$) and Cl_2O_7. Perchlorate salts of organic and organometallic cations are often dangerously explosive; substitution of the explosive perchlorate anion with the nonexplosive trifluoromethanesulfonate anion, $CF_3SO_3^-$ with a similar shape is strongly recommended whenever possible.

1.4.4 Nitric Acid (HNO$_3$)

The concentrated aqueous acid (70% HNO_3) is colorless but becomes yellow from NO_2 generated by photochemical decomposition according to the following equation:

$$4HNO_3 \xrightarrow{h\nu} 4NO_2\uparrow + 2H_2O + O_2 \tag{1.30}$$

"Fuming" nitric acid contains excess NO_2 and "red fuming" nitric acid contains N_2O_4. Pure (100%) nitric acid (bp 82.6°C) is obtained from solid KNO_3 and 100% H_2SO_4 at 0°C and isolated by vacuum distillation; it must be stored below 0°C to avoid decomposition according to equation 1.30. Nitric acid has the highest self-ionization constant of the pure liquid acids; its self-ionization can be described by the following equations:

$$2HNO_3 \rightleftharpoons H_2NO_3^+ + NO_3^- \tag{1.31a}$$

$$H_2NO_3^+ \rightleftharpoons H_2O + NO_2^+ \tag{1.31b}$$

$$2HNO_3 \rightleftharpoons NO_2^+ + H_2O + NO_3^- \tag{1.31a+b}$$

Pure nitric acid is a good ionizing solvent, but only NO_2^+ (nitronium) and NO_3^- salts are soluble. Dilute ($< 2\ M$) aqueous nitric acid has little oxidizing power, but concentrated nitric acid is a powerful oxidizing agent attacking all metals except gold, platinum, iridium, and rhenium. A few metals, however, (e.g., Al, Fe, Cu) become passive, possibly from a protective oxide film. *Aqua regia* is a 3:1 mixture by volume of concentrated HCl and HNO_3; it contains Cl_2 and ClNO and dissolves gold and platinum metals through complexation with chloride, generating $AuCl_4^-$, $PtCl_6^{2-}$, and so on. Similarly, HNO_3/HF dissolves tantalum through complexing action of fluoride to give TaF_6^-, whereas tantalum is resistant to pure nitric acid. An HNO_3/H_2SO_4 mixture is a good nitrating agent for organic compounds (e.g., for conversion of benzene to nitrobenzene) because of the formation of nitronium ion, NO_2^+ (see equations 1.31—the added H_2SO_4 enhances protonation of HNO_3 to $H_2NO_3^+$, the source of NO_2^+).

1.4.5 Sulfuric Acid (H_2SO_4)

Sulfuric acid (bp $\sim 270°C$, dec), is prepared from elemental sulfur on a gigantic scale by the oxidation of sulfur to SO_2 followed by further oxidation to SO_3 catalyzed by $HOSO_2ONO$ or platinum metal; the SO_3 is then hydrated to give H_2SO_4. Sulfuric acid undergoes extensive self-ionization (equation 1.18). Sulfuric acid is not a strong oxidizing agent but dehydrates organic compounds. Addition of SO_3 to sulfuric acid gives "fuming" sulfuric acid $(SO_3)_n \cdot H_2O$, also called *oleum*; depending on the amounts of SO_3, oleum can contain varying amounts of $H_2S_2O_7$, $H_2S_3O_{10}$, and $H_2S_4O_{13}$ as indicated by Raman spectra. Dissolution in concentrated sulfuric acid of the weak acid $B(OH)_3$ gives the strong acid $B(HSO_4)_3$.

1.4.6 Fluorosulfuric Acid (HSO_3F)

Fluorosulfuric acid, HSO_3F (bp 162.7°C) is one of the strongest pure liquid acids and is obtained either by reaction of SO_3 with hydrogen fluoride or by treatment of KHF_2 or CaF_2 with oleum. It can be distilled in glass if freed from hydrogen fluoride. Fluorosulfuric acid is hydrolyzed relatively slowly in water in contrast to chlorosulfuric acid, HSO_3Cl, which is hydrolyzed violently. Fluorosulfuric acid undergoes self-ionization in the standard manner:

$$2HSO_3F \rightleftarrows H_2SO_3F^+ + SO_3F^- \qquad (1.32)$$

but its self-ionization is much lower than H_2SO_4. In addition to its solvent and acid properties fluorosulfuric acid is a convenient laboratory fluorinating agent, converting $KClO_4$ into ClO_3F and K_2CrO_4 into CrO_2F_2.

1.4.7 Trifluoromethanesulfonic or "Triflic" Acid (CF_3SO_3H)

Triflic acid (bp 162°C) is a very strong acid, which is recommended as a nonexplosive substitute for perchloric acid. It is very hygroscopic and forms a monohydrate, presumably $H_3O^+CF_3SO_3^-$.

1.4.8 Tetrafluoroboric Acid (HBF_4) and Hexafluorophosphoric Acid (HPF_6)

The acids HBF_4 and HPF_6 do not exist in unsolvated forms because the central boron and phosphorus atoms would have the impossible coordination numbers of 5 and 7, respectively. However, both these acids are readily available in concentrated aqueous solutions and can be isolated as oxonium salts (e.g., $H_5O_2^+BF_4^-$). These acids are available in an anhydrous form as diethyl etherates, actually diethyloxonium salts such as $Et_2OH^+BF_4^-$.

1.5 Binary and Ternary Metallic Hydrides

Binary compounds of a metal with a nonmetal such as hydrogen are of one of the following fundamental types:

Saline, with an ionic structure containing metal cations and nonmental anions. Saline binary compounds are formed by the most electropositive metals.

Covalent, in which the metal is chemically bonded to the nonmetal by means of covalent bonds, generally but not necessarily two-electron, two-center bonds.

Interstitial, in which atoms of the nonmetal are distributed in a metallic lattice. Interstitial binary compounds are often formed by transition metals.

Key features of metal hydrides[22-24] of these three types are summarized in Sections 1.5.1–1.5.3. Metal hydrides are of general interest in energy storage applications based on the reversible chemical storage of hydrogen.[25]

1.5.1 Saline (Ionic) Binary Metal Hydrides

The position of the hydrogen atom relative to the other atoms of the periodic table (see Table I.1) is ambiguous. Thus the ionization of hydrogen to form the proton H^+ resembles the ionization of alkali metals to form their M^+ monocations, suggesting an analogy between hydrogen and the alkali metals. However, the chemistry of the H^+ ion is very different from that of the alkali metal cations M^+, largely because of the small size of a bare proton. Also, since hydrogen is one electron lighter than a noble gas, namely helium, the hydride ion H^- has the favored noble gas configuration of helium. This suggests an analogy between hydrogen and the halogens, which also very readily form the monoanions X^-.

Despite this analogy between hydrogen and the halogens, the energetics of forming the monoanions are very different for hydrogen and for the halogens. Thus the formation of the hydride ion from elemental hydrogen is endothermic, in contrast to the exothermic formation of halide ions from the corresponding elemental halogens (e.g., bromine) as indicated by the following equations:

$$\tfrac{1}{2}H_2(g) \to e^- \to H^-(g) \qquad \Delta H = +151 \text{ kJ/mol} \qquad (1.33a)$$

$$\tfrac{1}{2}Br_2(g) + e^- \to Br^-(g) \qquad \Delta H = -232 \text{ kJ/mol} \qquad (1.33b)$$

The endothermic conversion of H_2 to hydride ion requires the highly exothermic ionization potentials of the most electropositive metals, such as the alkali and the alkaline earth metals, for the overall formation of ionic metal hydrides to be exothermic. For this reason only the most electropositive metals—namely the alkali metals and the heavier alkaline earth metals (Ca, Sr, Ba)—form stable saline hydrides.

The ionic nature of saline hydrides such as NaH and CaH_2 is indicated by their high conductivities at or just below their melting points, as well as the liberation of H_2 at the *anode* rather than the cathode upon electrolysis of solutions of alkali metal hydrides (e.g., LiH) in molten alkali metal halides. X-ray and neutron diffraction studies on the saline hydrides indicate that H^- has a crystallographic radius between that of F^- and Cl^-. However, the apparent crystallographic radius of H^- in saline hydrides depends upon the electropositivity of the metal countercation, ranging from 1.30 Å in MgH_2 with the rutile (TiO_2) structure to 1.52 Å in CsH with the NaCl structure. This indicates that the rather diffuse H^- is relatively easily compressible.

The saline halides when pure are white crystalline solids. However, in practice they are often gray owing to contamination with traces of pure metal. Saline hydrides are obtained by heating the metal in a hydrogen atmosphere at temperatures ranging from 300 to 700°C. The reactivity of the alkali metals toward hydrogen to form the corresponding saline hydrides, M^+H^-, decreases in the sequence Li > Cs > K > Na. The thermal stability of the alkali metal hydrides decreases in the sequence LiH (dec 550°C) > NaH ∼ KH (dec ∼210°C) > RbH ∼ CsH (dec ∼170°C). Reactive forms of the saline hydrides can be obtained by hydrogenolysis of organometallic compounds (M = Li, etc.) in solution, for example,

$$H_2 + MR \to MH + HR \qquad (1.34)$$

The saline hydrides liberate hydrogen upon treatment with water or other weak protonic acids, such as:

$$NaH + H_2O \to NaOH + H_2\uparrow \qquad (1.35)$$

The reactions of saline hydrides with water is often so violent that some saline hydrides (e.g., RbH, CsH, BaH_2) can ignite spontaneously in moist air. Some saline hydrides, particularly NaH and CaH_2, are useful as drying agents (e.g., for the preparation of anhydrous solvents).

The hydride ion is a strong base corresponding to the very low acidity of molecular hydrogen, H_2. Thus NaH and KH are often useful as a strong bases in organic chemistry. The hydride ion is among the strongest reducing agents, as indicated by thermodynamics:

$$H_2 + 2e^- = 2H^- \qquad E^\circ = \sim -2.25 \text{ V} \qquad (1.36)$$

However, sodium hydride is slow and inefficient as a reducing agent. The reducing power of NaH in tetrahydrofuran or 1,2-dimethoxyethane can be enhanced by addition of a sodium alkoxide (e.g., t-amyl-ONa) and a metal salt (e.g., nickel acetate).[26]

1.5.2 Covalent Binary Metal Hydrides

Less electropositive main group metals form metal hydrides with structures constructed from M—H bonds. Examples of metal hydrides of this type include polymeric $[BeH_2]_n$ with four-coordinate Be and Be—H—Be three-center, two-electron bridge bonds (Section 10.4) and polymeric $[AlH_3]_n$ with six-coordinate Al and Al—H—Al three-center, two-electron bridge bonds (Section 9.6).

1.5.3 Interstitial Binary Metal Hydrides

Interstitial metal hydrides are formed by both the f-block metals (lanthanides and actinides) and the d-block transition metals. These are obtained by direct combination of the metals (or alloys in the case of ternary, quaternary, etc., metal hydrides) with hydrogen. In general, interstitial metal hydrides are less dense than the parent metal; that is, the volume of the metal increases upon combination with hydrogen. The following interstitial metal hydrides are particularly important.

1.5.3.1 Lanthanide Hydrides

Nonstoichiometric phases of nominal compositions LnH_2 and LnH_3 (Ln = lanthanide) are known. These phases are pyrophoric and react vigorously with water. All these phases appear to contain Ln^{3+} ions except for EuH_2 and YbH_2, which contain the corresponding Ln^{2+} ions. The other lanthanide dihydrides LnH_2 (Ln $\neq$ Eu, Yb) appear to contain *trivalent* lanthanide cations Ln^{3+} with the "extra" electron per lanthanide atom in the conduction band, thereby giving these dihydrides metallic properties.

1.5.3.2 Actinide Hydrides

The most important of the actinide hydrides is uranium trihydride (UH_3), which is obtained as a pyrophoric black powder in the rapid and exothermic reaction of uranium metal with hydrogen at 250–300°C. Uranium hydride decomposes at a somewhat higher temperature to regenerate a reactive, finely divided form of uranium metal; it is often more suitable than uranium metal for the synthesis of uranium compounds. The formation of uranium hydride or deuteride is also useful for purifying and regenerating hydrogen or deuterium gas, respectively.

1.5.3.3 Hydrides of d-Block Transition Metals

Titanium, zirconium, and hafnium absorb H_2 exothermically to give nonstoichiometric products such as $TiH_{1.7}$ and $ZrH_{1.9}$, which are useful reducing agents in metallurgical processes. Palladium and its alloys with gold and silver have high affinities for hydrogen.[27] Thus H_2 diffuses through a membrane of Pd at a high rate relative to other gases allowing it to be separated in the pure state from gaseous mixtures. Copper hydride, CuH, can be obtained as an insoluble precipitate with the wurtzite structure from the treatment of a Cu^{2+} solution with hypophosphorus acid or as a soluble amorphous material from the reaction of CuI with $LiAlH_4$ in pyridine.

1.5.4 Ternary Metal Hydrides and Polyhydrometallate Anions

Some ternary hydrides containing an electropositve metal such as Li or Mg as well as a transition metal such as Mg_2NiH_4, Li_4RhH_5, or Mg_2FeH_6 can be made by direct combination of the elements.[28] If the electropositive metal such as Li or Mg is assumed to ionize completely to Li^+ and Mg^{2+}, respectively, then the transition metal in the polyhydrometallate "anion" generally has the favored noble gas electronic configuration with 18 valence electrons. Thus the 18 valence electrons of the iron atom in the FeH_6^{4-} "anion" in Mg_2FeH_6 arises from 8 (the neutral Fe atom) + 6 (the 6 hydrogen atoms) + 4 (the -4 charge) = 18 electrons.

The enneahydrorhenate(VII) anion, ReH_9^{2-}, is obtained upon reduction of perrhenate and can be isolated as a potassium or tetraalkylammonium salt.[29] It is a rare example of a central metal atom forming nine two-electron, two-center bonds, with the nine bonds being directed toward the vertices of a tricapped trigonal prism. Note that the central rhenium atom in this structure has the favored 18 valence electron noble gas electronic configuration arising from 7 (the neutral Re atom) + 9 (the nine hydrogen atoms) + 2 (the -2 charge) = 18.

Some ternary hydrides, such as the hydrides obtained from H_2 and the alloys FeTi, Mg_2Ni, and $LaNi_5$, are of interest as low volume hydrogen storage materials in both mobile and stationary applications.[30]

References

1. Vasaru, G.; Ursu, D.; Mihǎilǎ, A.; Szentgyörgyi, A., *Deuterium and Heavy Water*, Elsevier, Amsterdam, 1975.

2. Ray, H. K., Ed., *Separation of Hydrogen Isotopes*, American Chemical Society, Washington, DC, 1978.

3. Evans, E. A., *Tritium and Its Compounds*, Butterworths, London, 1974.

4. Ache, H. J., Chemical Aspects of Fusion Technology, *Angew. Chem. Int. Ed. Engl.*, 1989, **29**, 1–20.

5. Eisenberg, R., Parahydrogen-Induced Polarization: A New Spin on Reactions with H_2, *Acc. Chem. Res.*, 1991, **24**, 110–116.

6. Kubas, G. J., Molecular Hydrogen Complexes: Coordination of a σ-Bond to Transition Metals, *Acct. Chem. Res.*, 1988, **21**, 120–128.

7. Pimentel, G. C.; McClellan, A. L., *The Hydrogen Bond*, Freeman, San Francisco, 1960.

8. Hamilton, W. C.; Ibers, J. A., *Hydrogen Bonding in Solids*, Benjamin, New York, 1968.

9. Joesten, M. D.; Schaad, L. J., *Hydrogen Bonding*, Dekker, New York, 1974.

10. Pople, J. A.; Schneider, W. G.; Bernstein, H. J., *High Resolution Nuclear Magnetic Resonance*, McGraw-Hill, New York, 1959, Chapter 15.

11. Legon, A. C.; Millen, D. J., Directional Character, Strength, and Nature of the Hydrogen Bond in Gas-Phase Dimers, *Acc. Chem. Res.*, 1987, **20**, 39–46.

12. Taylor, R.; Kennard, O.; Versichel, W., *J. Am. Chem. Soc.*, 1984, **106**, 244–248.

13. Taylor, R.; Kennard, O., Hydrogen Bond Geometry in Organic Crystals, *Acct. Chem. Res.*, 1984, **17**, 320–326.

14. Eisenberg, D. S.; Kauzmann, W., *The Structure and Properties of Water*, Oxford University Press, New York, 1969.

15. Sörenson, S. P. L., Über die Messung und die Bedeutung der Wasserstoffionenkonzentration bei Enzymatischen Prozessen, *Biochem. Z.*, 1909, **21**, 131.

16. Westcott, C. C., *pH Measurements*, Academic Press, New York, 1978.

17. Staab, H. A.; Saupe, T., Proton Sponges and the Geometry of Hydrogen Bonds: Aromatic Nitrogen Bases with Exceptional Basicities, *Angew. Chem. Int. Ed. Engl.*, 1988, **27**, 865–879.

18. Rochester, C. H., *Acidity Functions*, Academic Press, London, 1970.

19. Gillespie, R. J., Fluorosulfuric Acid and Related Superacid Media, *Acct. Chem. Res.*, 1968, **1**, 202–209.

20. Olah, G. A.; White, A. M.; O'Brien, D. H., Protonated Hereroaliphatic Compounds, *Chem. Rev.*, 1970, **70**, 561–591.

21. Pearson, G. S., Perchloric Acid, *Adv. Inorg. Chem. Radiochem.*, 1966, **8**, 177–224.

22. MacKay, K. M. *Hydrogen Compounds of the Metallic Elements*, E. and F. Spon, London, 1966.

23. Mueller, W. M.; Blackledge, J. P.; Libowitz, G. G., Eds., *Metal Hydrides*, Academic Press, New York, 1968.

24. Wiberg, E.; Amberger, E. *Hydrides of the Elements of Main Groups I–IV*, Elsevier, Amsterdam, 1971.

25. Bogdanović, B.; Ritter, A.; Spliethoff, B., Active MgH_2–Mg System for Reversible Chemical Energy Storage, *Angew. Chem. Ind. Ed. Engl.*, 1990, **29**, 223–234.

26. Caubière, P., Complex Reducing Agents (CRAs)—Versatile Novel Ways of Using Sodium Hydride in Organic Synthesis, *Angew. Chem. Int. Ed. Engl.*, 1983, **22**, 599–613.

27. Lewis, F.A., *The Palladium–Hydrogen System*, Academic Press, London, 1967.

28. Bronger, W., Complex Transition Metal Hydrides, *Angew. Chem. Int. Ed. Engl.*, 1991, **30**, 759–768.

29. Abrahams, S. C.; Ginsberg, A. P., Knox, K., *Inorg. Chem.*, 1964, **3**, 558–567.

30. Hoffman, K. C.; Reilly, J. J.; Salzano, F. J.; Waide, C. H.; Wiswall, R. H.; Winsche, W. E., Metal Hydride Storage for Mobile and Stationary Applications, *Int. J. Hydrogen Energy*, 1976, **1**, 133–151.

2

Carbon

2.1 General Aspects of Carbon Chemistry

Carbon is the basis of organic chemistry; there are more compounds of carbon than of any other element except hydrogen and possibly fluorine. However, most of the chemistry of carbon is the province of organic chemistry and thus is not covered in this book. The "inorganic" chemistry of carbon discussed in this book includes the allotropic forms of elemental carbon, metal carbides, simple molecular carbon halides and oxides, simple cyano derivatives and carbon–sulfur derivatives. Carbon chemistry has the following features.

1. Carbon normally forms a total of four bonds including both σ and π bonds. The coordination number of a carbon forming four bonds can be 2 ($\equiv$C—such as in alkynes or $=$C$=$ such as in allene), 3 ($=$C$\diagup$ such as in alkenes), or 4 (such as in alkanes) with linear, planar triangular, and tetrahedral geometries, respectively. The carbon atom in carbon monoxide ($:$C$^-\equiv$O$^+:$) and the isocyanide carbon atom in isocyanides ($:$C$^-\equiv$N$^+$—R) have coordination numbers of only one. Stable trivalent carbon "free radicals," $R_3C\cdot$, can be isolated if R is an aryl group. Tris(pentachlorophenyl)methyl, $(C_6Cl_5)_3C\cdot$, is particularly stable.[1]

2. Exceptional examples of carbon atoms with apparent coordination numbers greater than 4 include five-coordinate carbon atoms in bridging alkyl groups in metal alkyls (e.g., $Al_2(CH_3)_6$), five- and six-coordinate carbon atoms in deltahedral carboranes of the general formula $C_2B_{n-2}H_n (6 \leq n \leq 12)$ and

in transition metal cluster compounds (e.g., five-coordinate carbon in the center of an Fe_5 square pyramid in $Fe_5(CO)_{15}C$, six-coordinate carbon in the center of an Ru_6 octahedron in $Ru_6(CO)_{17}C$, and eight-coordinate carbon in the center of a Co_8 square antiprism in the $Co_8C(CO)_{18}^{2-}$ anion).

3. The high carbon–carbon bond strength leads to the unusual stability of catenated carbon compounds, which are the basis of organic chemistry. Thus the C—C bond strength of ~ 356 kJ/mol is comparable to the C—O bond strength of ~ 336 kJ/mol. This contrasts with carbon's heavier congener silicon as well as with sulfur, where Si—O bonds (~ 368 kJ/mol) and S—O bonds (~ 330 kJ/mol) are much stronger than Si—Si bonds (~ 226 kJ/mol) and S—S bonds (~ 226 kJ/mol).

2.2 Carbon Isotopes and Elemental Carbon

2.2.1 Isotopes of Carbon

Table 2.1 summarizes some of the important properties of the carbon isotopes. Note that only the rare ($\sim 1\%$) naturally occurring stable carbon isotope, namely ^{13}C, has a nuclear spin and is observable by NMR spectroscopy.[2] The organic chemist is fortunate that $\sim 99\%$ of natural carbon is the isotope ^{12}C, with no nuclear spin, so that proton and carbon-13 NMR spectra of organic compounds are not complicated by spin-spin splitting arising from adjacent carbon atoms. The radioisotope[3] ^{14}C is made by thermal neutron irradiation of lithium or aluminum nitride using the following reaction:

$$^{14}_{7}N + ^{1}_{0}n \rightarrow {^{14}_{6}C} + {^{1}_{1}H} \tag{2.1}$$

It decays back to stable $^{14}_{7}N$ by β-emission according to the following equation with a half-life of 5570 years:

$$^{14}_{6}C \rightarrow {^{14}_{7}N} + \beta^{-} \tag{2.2}$$

Cosmic rays generate thermal neutrons which lead to formation of $^{14}CO_2$ in the atmosphere by equation 2.1; metabolism of this $^{14}CO_2$ by organisms while alive but not after death is the basis of radiocarbon dating.[4]

Table 2.1 Properties of the Carbon Isotopes

Isotope	Natural Abundance (%)	Half-Life	Spin	NMR Sensitivity*
^{12}C	98.89	Stable	0	0
^{13}C	1.11	Stable	$\frac{1}{2}$	0.0159
^{14}C	1.2×10^{-10}	5570 years, β^-	0	0

* Relative to the proton for equal numbers of nuclei.

Table 2.2 Properties of the Carbon Allotropes

Allotrope	C Hybridization	Structure
Diamond	sp^3	Infinite three-dimensional lattice
Graphite	sp^2	Infinite two-dimensional planar layers of hexagons
Fullerenes	sp^2	Finite C_n cages with 12 pentagons and $n - 12$ hexagons

2.2.2 Allotropes of Carbon

Carbon forms allotropes of three types: diamond, graphite, and the recently discovered fullerenes. The properties of these allotropes are compared in Table 2.2.

3.2.2.1 Diamond[5]

Diamonds are found naturally in Kimberly, South Africa, in ancient volcanic pipes embedded in a relatively soft, dark-colored basic rock in low concentrations (~ 1 part per 1.5×10^7). The structure of diamond consists of an infinite three-dimensional lattice of four-coordinate sp^3 carbon atoms in which there are interlocking six-membered C_6 rings similar to those in cyclohexane or adamantane and C—C distances of 1.514 Å. Diamond is normally found in the cubic form, but a hexagonal form (lonsdaleite) is found in certain meteorites and is available synthetically. Diamond is the hardest natural solid known and has the highest melting point ($\sim 4000°C$) and thermal conductivity as well as the lowest molar entropy (2.4 J mol^{-1} K^{-1}) of any element. Diamond is an insulator. Since the density of diamond (3.51 g/cm^3) far exceeds that of graphite (2.22 g/cm^3), high pressures (e.g., > 125 kbar) can be used to convert graphite to diamond even though graphite is thermodynamically more stable than diamond by 2.9 kJ/mol. To attain useful rates for the high pressure conversion of graphite to diamond, a transition metal catalyst such as chromium, iron, or platinum is used. Diamond burns in air at elevated temperatures (600–800°C).

2.2.2.2 Graphite[6,7]

Graphite consists of planar layers of hexagons of sp^2-hybridized carbon atoms with a separation of 3.35 Å between layers. The relatively slight forces between graphite layers are consistent with its softness and lubricity. The infinite delocalization of π electrons in the graphite structure leads to electrical conductivity. The average C—C bond order in graphite is 1.33, corresponding to a C—C distance of 1.415 Å. By contrast, the average C—C bond order in benzene is 1.5, corresponding to the expected shorter C—C distance of 1.39 Å. The most stable and common form of graphite is hexagonal α-graphite with an ABABAB... stacking of the graphite layers. However, rhombohedral

β-graphite with an ABCABC... stacking of the layers is also known. Strong oriented graphite fibers can be obtained by pyrolysis of fibers of organic polymers such as poly(acrylonitrile). Amorphous carbons (carbon black, soot, charcoal, etc.) are microcrystalline forms of graphite; their physical properties depend upon their surface area and the nature of their surface (e.g., presence and type of oxygenated groups).

2.2.2.3 Fullerenes

Fullerenes[8,9] are large molecular carbon cages, which are isolated by extraction of specially prepared soot with organic solvents such as benzene. A rich source of fullerene is soot made by arcing graphite rods in a helium atmosphere at 200 torr pressure.[10] The most common fullerene is C_{60}, which has a truncated icosahedral "soccer ball" structure with icosahedral (I_h) symmetry (Figure 2.1). Less symmetrical fullerenes with larger number of carbon atoms are known such as C_{70}, which is next most abundant to C_{60}; C_{76} and C_{84} are also known.

The fullerene cage structures are based on polyhedra in which three edges meet at each vertex (i.e., polyhedra of *degree* 3). Fullerene polyhedra always have 12 pentagonal faces that never share edges; the remaining faces are hexagons. Fullerene carbon atoms are sp^2 hybridized like olefinic carbon atoms. The C_{60} fullerene surface can formally be considered to contain 30 carbon–carbon double bonds. However, conjugation between these double bonds is reduced by the curvature of the fullerene surface. Fullerenes undergo addition and metal complexation reactions similar to polyolefins with independent carbon–carbon double bonds and electronegative substituents. Reduction of the fullerene C_{60} with the heavier alkali metals leads to superconductors of the stoichiometry M_3C_{60} (M = K, Rb) with superconducting critical temperatures (T_c's) up to 33 K.[11]

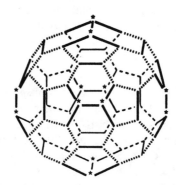

Figure 2.1. A schematic diagram of the truncated icosahedron "soccer ball" found in C_{60}. The hexagonal faces are shown by dotted lines and the six unique edges shared by two hexagonal faces rather than one hexagonal and one pentagonal face are starred.

2.2.3 Graphite Intercalation Compounds[12,13]

The large distance between the parallel planes of carbon atoms in graphite (3.35 Å) implies relatively weak interlayer bonding in accord with the softness of graphite crystals. In addition, molecules and ions can penetrate between the layers of graphite to form interstitial or lamellar compounds of variable stoichiometry. These so-called *intercalation* compounds are of two types: nonconducting compounds, in which the planarity of the graphite layers is destroyed, and conducting compounds, in which the planarity of the graphite layers is preserved.

Two examples of nonconducting graphite intercalation compounds are the so-called *graphite oxide* and *graphite fluoride*; formation of such intercalation compounds requires very vigorous treatment of graphite to destroy the planarity of the graphite layers. Thus pale lemon-colored graphite oxide is obtained by treatment of graphite with fuming nitric acid or a mixture of nitric acid and potassium chlorate. Graphite oxide has the approximate stoichiometry $[C_8O_2OH]_n$, containing C=O and C—OH groups; it is fairly acidic. Graphite fluoride[14,15] is obtained by treatment of graphite with elemental fluorine under various conditions.[16] The initial product from treatment of graphite with F_2 at 2 atm pressure in liquid HF has the stoichiometry $C_xF(5 > x > 2)$ but transforms to an ionic material of stoichiometry $C_{12}^+HF_2^-$ at 20°C. Treatment of graphite with fluorine at elevated temperatures (400–600°C) gives a product of stoichiometry $(CF)_n$ sometimes called "carbon monofluoride." The white color and lack of conductivity of $(CF)_n$ is consistent with the addition of a single fluorine atom to each carbon atom of the graphite layer, thereby converting each carbon atom from planar sp^2 to tetrahedral sp^3 hybridization with concurrent conversion of the mobile π electrons required for conductivity to immobile σ electrons in the newly formed C—F bonds.

The electrically conducting graphite intercalation or "lamellar" compounds are obtained by insertion of various *guest* atoms, molecules, or ions between the layers of the graphite *host* without disturbing the planarity of the layers and thus maintaining intact the infinite network of π electrons in each layer responsible for the electrical conductivity of graphite. Such graphite intercalation compounds can be made by the following methods:

1. Electrical intercalation either by cathodic reduction or anodic oxidation.
2. Reduction of graphite with alkali metals in the vapor phase or in liquid ammonia; products of stoichiometries such as C_8M, $C_{24}M$, $C_{48}M$, or $C_{60}M$ (M = K, Rb, Cs) are obtained depending on the amount of alkali metal used.
3. Oxidation of graphite with reagents such as $FeCl_3$ or ReF_6.

The *stage* of a graphite intercalation compound is the number of carbon layers per layer of guest species; thus the highest concentration of guest occurs in a stage 1 compound. For example, KC_{24} with potassium atoms between alternate graphite layers is a stage 2 compound, whereas KC_8 with potassium atoms between every graphite layer is a stage 1 compound.

2.3 Carbides

Carbides are compounds in which carbon is combined with elements of similar or lower electronegativity.[17] Carbides may be prepared by the following general methods.

1. Direct union of the elements at high temperatures, generally above 2200°C.
2. High temperature reaction of a metal compound, generally the oxide, with elemental carbon.
3. Heating a metal in a suitable hydrocarbon vapor.
4. Reactions of metal compounds with acetylene in solution. Reactions of acetylene with solutions of compounds of the coinage metals precipitate the insoluble explosive acetylides M_2C_2 (M = Cu, Ag, Au). Similarly, reactions of zinc and cadmium alkyls with acetylene gives the corresponding metal acetylides, MC_2 (M = Zn, Cd), according to the following equation:

$$R_2M + 2HC\equiv CH \rightarrow MC_2 + 2RH \tag{2.3}$$

Metal carbides can be classified into the same three general types as metal hydrides, as discussed in Chapter 1 (Section 1.5), namely saline, covalent, and interstitial carbides. The important features of each of these types of metal carbides are outlined in sections 1.3.1–1.3.4.

2.3.1 Saline Carbides

As is the case with saline hydrides, saline carbides are formed by the most electropositive metals such as the alkali and alkaline earth metals. Saline carbides are readily hydrolyzed by water or dilute acids and can be classified further as follows by the carbon anion implied by the hydrocarbon obtained upon hydrolysis.

2.3.1.1 Methane Corresponding to C^{4-}

Methane is produced from the hydrolyses of Be_2C and Al_4C_3.

2.3.1.2 Acetylene Corresponding to C_2^{2-}

Many carbides produce acetylene upon hydrolysis and are thus formulated with acetylide, C_2^{2-}, ions. The most important of these carbides is calcium carbide, CaC_2, which is made on a large scale by the higher temperature endothermic reaction of lime, CaO, and coke, C:

$$CaO + 3C \xrightarrow{\text{2200--2250°C}} CaC_2 + CO \qquad \Delta H = 465.7 \text{ kJ/mol} \tag{2.4}$$

The exothermic reaction of calcium carbide with water ($\Delta H = -120$ kJ/mol) is used to generate acetylene for use in acetylene lamps for underground illumination where suitable electric power sources are not available. The $M^{II}C_2$ derivatives derived from electropositive metals such as CaC_2 have structures derived from the NaCl structure containing discrete C_2^{2-} anions with C—C distances in the range 1.19–1.24 Å.

2.3.1.3 Propyne Corresponding to C_3^{4-}

Mg_2C_3 produces propyne, $CH_3C{\equiv}CH$, upon hydrolysis and can thus be formulated with C_3^{4-} anions. A related species is red Li_4C_3, which can be obtained in hexane solution by metallation of propyne with excess n-butyllithium.[18]

2.3.2 Covalent Carbides

Silicon and boron give completely covalent carbides, since they approach carbon closely in size and electronegativity. Important covalent carbides include carborundum and two boron carbides.

2.3.2.1 SiC (carborundum)

Carborundum, SiC, is an extremely hard, infusible, and chemically stable compound used as a grinding agent. In this form SiC is obtained by heating silicon dioxide and coke in an electric furnace. Silicon carbide ceramic fibers with useful mechanical properties are obtained by the pyrolysis of poly(dimethylsilylene), $[—Si(CH_3)_2—]_n$, through a polycarbosilane intermediate $[—CH_2Si(CH_3)—]_n$.[19,20] The structure of silicon carbide is related to that of diamond, featuring infinite three-dimensional arrays of tetrahedral silicon and carbon atoms with only Si—C bonds (i.e., no Si—Si or C—C bonds).

2.3.2.2 $B_4C = B_{12}C_3$

B_4C is also an extremely hard, infusible, and chemical inert substance, which is made by heating B_2O_3 and coke in an electric furnace. Its structure has B_{12} icosahedra and C_3 chains in an NaCl structure; some of the carbon in the C_3 chains can be replaced by boron, leading to stoichiometries closer to $B_{13}C_2$.[21]

2.3.2.3 BC_3

Lustrous metallic-appearing BC_3 is obtained by reaction of BCl_3 with benzene at 800°C. The graphitelike layer structure of BC_3 allows it to intercalate strongly oxidizing and strongly reducing species.

2.3.3 Interstitial Carbides

Interstitial carbides are infusible, extremely hard, refractory materials that retain the luster and conductivity of free metals.[22] The carbon atoms in interstitial carbides typically occupy octahedral interstices in a close-packed lattice of metal atoms although the arrangement of metal atoms is not always the same as that of the free metal. Metals with radii greater than 3.15 Å (e.g., Ti, Zr, Hf; V, Nb, Ta; Mo, W) form interstitial carbides of the stoichiometry MC. Metals with smaller radii (e.g., Cr, Mn, Fe, Co, Ni) do not form carbides of the stoichiometry MC but instead form metal-richer carbides with more complicated structures. Interstitial carbides, unlike ionic carbides, are very inert chemically. The extreme hardness and inertness of WC and TaC lead to their extensive use in high speed cutting tools.

2.3.4 Carbide Ligands

Isolated carbon atoms appear as ligands in transition metal chemistry in complexes such as those given in Sections 2.3.4.1–2.3.4.3.

2.3.4.1 Metal Carbonyl Carbide Clusters[23]

Isolated carbon atoms, generally with apparent coordination numbers larger than 4, hence participating in multicenter bonding, can occur in the center of metal carbonyl cluster polyhedra such as the Fe_5 square pyramid in $Fe_5C(CO)_{15}$, the Ru_6 octahedron in $Ru_6C(CO)_{17}$, the Co_6 trigonal prism in $Co_6C(CO)_{15}^{2-}$, and the Co_8 tetragonal antiprism in $Co_8C(CO)_{18}^{2-}$.

2.3.4.2 Early Transition Metal Halide and Lanthanide Carbide Clusters

Clusters such as $Gd_{10}Cl_{18}C_4$ (Ref. 24) and $KZr_6Cl_{15}C$ (Ref. 25) have carbon atoms in the center of metal octahedra.

2.3.4.3 Metal Porphyrin Carbides

An isolated carbon atom appears between two metal porphyrin units in $TppFe=C=FeTpp$ (Tpp = tetraphenylporphyrin).[26]

2.4 Carbon Halides

Carbon tetrafluoride, CF_4 (mp $-184°C$, bp $-128°C$), a very stable gas, is the end product of the fluorination of carbon compounds. Carbon tetrafluoride can also be made by the fluorination of silicon carbide, which proceeds according to the following equation:

$$SiC + 4F_2 \rightarrow SiF_4 + CF_4 \tag{2.5}$$

The SiF_4 is removed from the CF_4 by passing the product gases through 20%

aqueous NaOH, which reacts with the SiF_4 to give a water-soluble sodium silicate but leaves the CF_4 unaffected. This major difference in hydrolytic reactivity of CF_4 and SiF_4 is a consequence of accessible d orbitals on silicon but not on carbon.

Organic compounds containing only carbon and fluorine are called *fluorocarbons*; the chemistry of fluorocarbons is a vast area, far beyond the scope of this book.[27-30] General routes to saturated fluorocarbons include the fluorination of hydrocarbons with higher valent fluorides such as CoF_3 and the fluorination of chlorocarbons with SbF_3, often in the presence of catalytic amounts of Sb(V). Since such fluorination procedures usually add fluorine to all carbon–carbon double and triple bonds, giving products with exclusively carbon–carbon single bonds, unsaturated fluorocarbons normally require indirect methods of synthesis, generally using dehalogenation reactions by metals such as zinc or dehydrohalogenation reactions with bases. Fluorocarbons have diverse chemical and physiological properties. Thus poly(tetrafluoroethylene), $(-CF_2CF_2-)_n$, commonly known as Teflon, is very chemically inert and has a low coefficient of friction. Octafluoroisobutylene, $(CF_3)_2C=CF_2$, is highly toxic, but other fluorocarbons, such as perfluorodecahydronaphthalene, $C_{10}F_{18}$, are so physiologically inactive that they are the active components of blood substitutes because of the relatively high solubility of oxygen in fluorocarbons.[31]

Carbon tetrachloride, a liquid (mp $-23°C$, bp $76°C$), is a common solvent and is useful for converting metal oxides to metal chlorides at elevated temperatures ($300-500°C$). It is thermodynamically unstable with respect to hydrolysis but is kinetically very inert to hydrolysis because of the absence of empty acceptor orbitals (e.g., d orbitals) for nucleophilic attack. However, heating carbon tetrachloride with free alkali metals or other strong reducing agents can lead to dangerous explosions.

Organic compounds containing only carbon and chlorine are called *chlorocarbons*. The presence of relatively large chlorine atoms on a carbon network makes the properties of chlorocarbons very different than those of corresponding hydrocarbons. In general, chlorocarbons with trigonal sp^2 carbon atoms are more stable than chlorocarbons with tetrahedral sp^3 carbon atoms. As a result, certain chlorocarbons having only sp^2 carbon atoms (e.g., perchlorofulvene and perchlorofulvalene) are very stable under normal conditions, whereas their hydrocarbon analogues polymerize rapidly. Chlorocarbon free radicals[1] with trivalent carbon atoms are among the most stable carbon free radicals. For example, the trivalent carbon atom in $(C_6Cl_5)_3C\cdot$ is stable not only to air but even to permanganate oxidation.

Carbon tetrabromide is a pale yellow, water-insoluble solid (mp $90.1°C$, bp $189.5°C$), which can be obtained by bromination of CH_4 with HBr or Br_2 or by reaction of CCl_4 with $AlBr_3$ at $100°C$. *Carbon tetraiodide* is a bright red solid, which decomposes readily to I_2 and $I_2C=CI_2$, again indicating the stability of sp^2 carbon atoms relative to sp^3 carbon atoms in halocarbons other than fluorocarbons. Carbon tetraiodide is made by an electrophilic halogen exchange reaction:

$$CCl_4 + 4C_2H_5I \xrightarrow{AlCl_3} CI_4 + 4C_2H_5Cl \tag{2.6}$$

The increasing instability of the carbon tetrahalides with increasing atomic weight of the halogen relates to the steady decrease in C—X bond energies from 485 kJ/mol for the C—F bond to 213 kJ/mol for the C—I bond.

Carbonyl halides, $O{=}CX_2$ (X = F, Cl, Br, I), are all known. The fluoride and chloride are made by reaction of carbon monoxide with the corrresponding halogen; the bromide is made by the partial hydrolysis of CBr_4 with concentrated sulfuric acid. Carbonyl chloride, $O{=}CCl_2$ (bp + 7.6°C), are known as *phosgene* (from $\phi\omega s = ph\bar{o}s$ for *light* and $-\gamma\epsilon\nu\eta s = -gen\bar{e}s$ for *form*, relating to its formation from $CO + Cl_2$ in sunlight), is a toxic gas smelling like "new-mown hay." Phosgene was used rather ineffectively as a chemical warfare gas in World War I. Thiocarbonyl chloride or "thiophosgene," $S{=}CCl_2$, is an evil-smelling orange toxic liquid (bp 73°C).

2.5 Carbon Oxides, Carbonic Acid, and Oxocarbon Anions

2.5.1 Carbon Oxides

The oxides of carbon include the very stable and abundant CO and CO_2, the unstable but isolable C_3O_2, and more complicated carbon oxides derived from organic backbones by removal of all hydrogen—for example, $C_{12}O_9$ from the complete dehydration of mellitic acid (benzene hexacarboxylic acid). Oxocarbon anions, $C_nO_n^{2-}$, with apparent "aromatic" properties are also known.[32] The properties of these carbon–oxygen compounds are summarized in Sections 2.5.1.1–2.5.1.4.

2.5.1.1 Carbon Monoxide

Carbon monoxide (mp -205°C, bp -191°C, $:C^-{\equiv}O^+$, is a colorless, odorless, but very toxic gas, which is sparingly soluble in water. Carbon monoxide is formed by burning carbon is a deficiency of oxygen. It, therefore, is a dangerous product of many combustion processes. Carbon monoxide is also made by the "water gas reaction" from steam and coal as "synthesis gas," a 1:1 CO/H_2 mixture arising from the following reaction at elevated temperatures:

$$C + H_2O \xrightarrow{\Delta} CO + H_2 \tag{2.7}$$

In the laboratory carbon monoxide can be generated from the dehydration of formic acid, HCO_2H, using concentrated sulfuric acid at 140°C. Transition metals combine with carbon monoxide to form *metal carbonyls* (Figure 2.2), which play a key role in organometallic and coordination chemistry. The

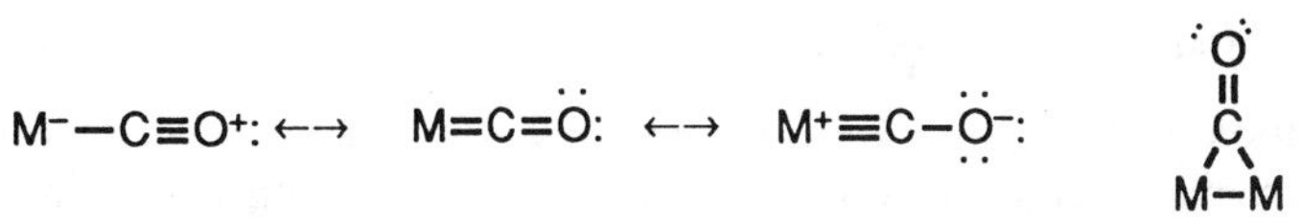

$$M^- —C\equiv O^+: \leftrightarrow \quad M=C=\overset{..}{\underset{..}{O}}: \quad \leftrightarrow \quad M^+\equiv C-\overset{..}{\underset{..}{O}}{}^-:$$

Figure 2.2. The bonding of CO to transition metals in metal carbonyls as terminal and bridging ligands.

toxicity of carbon monoxide is a consequence of its binding to the iron in blood hemoglobin, thereby inhibiting the essential function of hemoglobin as an oxygen carrier. Carbon monoxide can bond to transition metals as a terminal or a bridging ligand. Simple neutral binary metal carbonyls are volatile covalent compounds, which readily decompose upon heating to carbon monoxide and the free metal. Metal carbonyls thus are often useful sources of finely divided and/or pure metals such as chromium, molybdenum, iron, and nickel. The central transition metal in the simple binary metal carbonyls normally has the favored 18 valence electron configuration of the next noble gas leading to $M(CO)_6$ (M = Cr, Mo, W), $Fe(CO)_5$, and $Ni(CO)_4$ as stable compounds. Since, however, the central vanadium atom in the binary vanadium carbonyl $V(CO)_6$, has only 17 valence electrons, $V(CO)_6$ is paramagnetic with one unpaired electron. However, $V(CO)_6$ is much less stable than the isostructural $Cr(CO)_6$. Carbon monoxide reacts with alkali metals to give the so-called *alkali metal carbonyls*, which contain salts of the acetylenediolate ion, $O^- —C\equiv C—O^-$, which upon hydrolysis can trimerize to give hexahydroxybenzene derivatives.

2.5.1.2 Carbon Dioxide

Carbon dioxide, $:\overset{..}{O}=C=\overset{..}{O}:$ (mp $-57°C$ at 5.2 atm, bp $-79°C$; subl), is obtained from the combustion of carbon and hydrocarbons in excess air or oxygen or by the pyrolysis ("calcination") of $CaCO_3$ (limestone). The photosynthesis in plants reduces CO_2 to organic matter but the similar reduction of CO_2 in a nonliving system ("in vitro") appears to be very difficult. However, CO_2 can be reduced electrochemically to methanol, formate, oxalate, methane, and/or CO depending on the conditions. Numerous transition metal complexes of CO_2 are known[33–39] which exhibit the modes of metal–CO_2 bonding depicted in Figure 2.3.

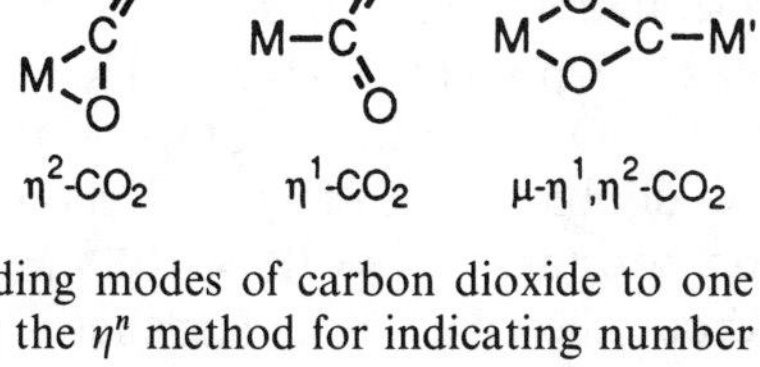

Figure 2.3. Some bonding modes of carbon dioxide to one or two transition metals (M and M') illustrating the η^n method for indicating number of attachment points.

3.5.1.3 "Carbon Suboxide"

Carbon suboxide, $C_3O_2(= :\ddot{O}{=}C{=}C{=}C{=}\ddot{O}:)$, also called 1,2-propadiene-1,3-dione, is an evil-smelling unstable gas (bp $+6.8°C$), which is obtained by the dehydration of malonic acid with P_4O_{10} in vacuum at 140–150°C according to the following equation:

$$HO_2CCH_2CO_2H \xrightarrow{P_4O_{10}} :\ddot{O}{=}C{=}C{=}C{=}\ddot{O}: + 2H_2O \qquad (2.8)$$

Carbon suboxide polymerizes readily at room temperature to a yellow solid and above 100°C to a ruby-red, water-soluble solid. Reaction of C_3O_2 with water readily regenerates malonic acid. Photolysis of C_3O_2 gives $C_2O(= :C{=}C{=}\ddot{O}:)$ as a reactive intermediate, which reacts with olefins by carbon atom insertion:

$$H_2C{=}CH_2 + :C{=}C{=}\ddot{O}: \xrightarrow{h\nu} H_2C{=}C{=}CH_2 + CO \qquad (2.9)$$

2.5.1.4 More Complicated Carbon Oxides

Some more complicated carbon oxides are known with carbon backbones familiar in organic chemistry. An example is benzene hexacarboxylic acid (mellitic acid) trianhydride, $C_{12}O_9 = C_6[(CO)_2O]_3$ (2.1), a white sublimable solid obtained by dehydration of the corresponding acid with acetyl chloride in a sealed tube at 160°C.

2.1

2.5.2 Carbonic Acid

Carbon dioxide is the anhydride of carbonic acid but hydrolyzes only slowly at pH 7 to H_2CO_3. In equeous solution CO_2 is physically dissolved and loosely solvated so that only $\sim 0.2\%$ is present as $H_2CO_3/HCO_3^-/CO_3^{2-}$. This relates to the stability of the $:\ddot{O}{=}C{=}\ddot{O}:$ structure for CO_2 and the reluctance of water to add to one of the C=O double bonds of CO_2. Carbonic acid is a weak acid with the true dissociation constants $K_1 = 1.6 \times 10^{-4}$ and $K_2 = 4.7 \times 10^{-11}$. Determination of K_1 without allowing for the fact that $\sim 99.8\%$ of an aqueous solution of CO_2 is undissociated CO_2 gives the fictitious dissociation constant of 4.16×10^{-7} for "K_1" of H_2CO_3. Pure carbonic acid, H_2CO_3, cannot be isolated because of its ready dehydration to CO_2. However, colorless crystals of a white crystalline etherate $H_2CO_3 \cdot (CH_3)_2O$ (mp $-47°C$), can be isolated by treatment of Na_2CO_3 at $-35°C$ with hydrogen chloride in dimethyl ether. This etherate decomposes above $-26°C$ into CO_2, H_2O, and $(CH_3)_2O$.

2.5.3 Oxocarbon Anions[25,40]

Oxocarbon anions, $C_nO_n^{2-}$ ($3 \leq n \leq 6$) are derived from the cyclic dihydroxy compounds depicted in Figure 2.4. Deltic and squaric acids are made by hydrolysis of suitable cyclopropene and cyclobutene derivatives (Figure 2.5). To avoid cleavage of the three-membered ring in deltic acid, much milder conditions (e.g., n-butanol in diethyl ether) are used for the preparation of deltic acid than for squaric acid. Squaric acid can also be made by the electrolytic reduction of carbon monoxide in dimethylformamide at $18°C$ and 230 atm with tetrabutylammonium bromide as the supporting electrolyte, using the stainless steel pressure vessel as the cathode and an aluminum anode. Rhodizonic acid

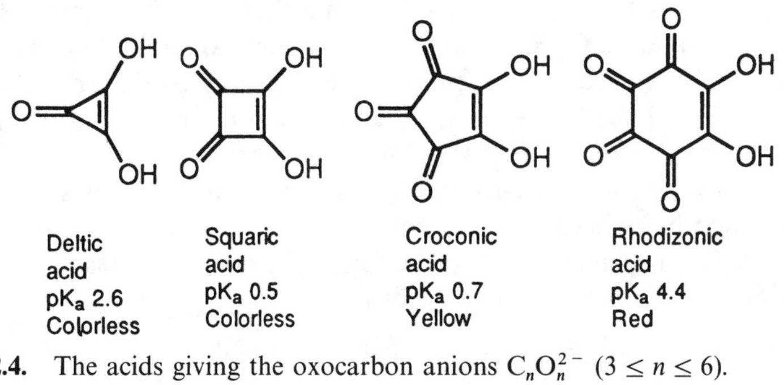

Deltic acid	Squaric acid	Croconic acid	Rhodizonic acid
pK_a 2.6	pK_a 0.5	pK_a 0.7	pK_a 4.4
Colorless	Colorless	Yellow	Red

Figure 2.4. The acids giving the oxocarbon anions $C_nO_n^{2-}$ ($3 \leq n \leq 6$).

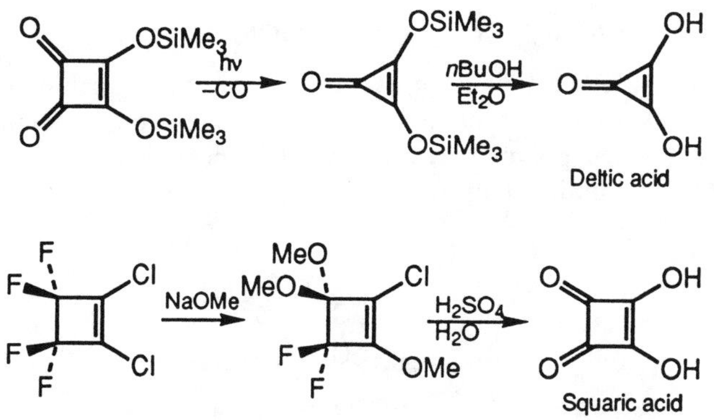

Figure 2.5. Syntheses of squaric acid and deltic acid.

is obtained from the microbiological oxidation of myoinositol or by the chemical oxidation (HNO_3 followed by O_2/CH_3CO_2K) of hexahydroxycyclohexane. Alternatively, carbon monoxide can be reduced with potassium metal under mild conditions to give potassium ethynediolate, $K_2C_2O_2$, which upon heating undergoes cyclotrimerization to the potassium derivative of hexahydroxybenzene, $K_6C_6O_6$, which in turn can be oxidized to potassium rhodizonate, $K_2C_6O_6$. Further oxidation of rhodizonate in basic solution results in ring contraction to give potassium croconate. The preparation of rhodizonic and croconic acids from elemental carbon was known in the early nineteenth century, long before their structures had been elucidated; therefore, the trivial names of these acids are based on the characteristic colors of their salts: ρωδιζειν (rhōdiżein) = rose red and κρωκς (krōkos) = yellow. Rhodizonates are used as spot test reagents for alkaline earth and other divalent ions such as Pb^{2+}.

2.6 Cyanides and Related Compounds

2.6.1 Cyanogen

Cyanogen, $:N{\equiv}C—C{\equiv}N:$, is a flammable and toxic gas (mp $-27.9°C$, bp $-21.2°C$), which is kinetically stable but endothermic (297 kJ/mol). It has a linear structure, and its C—C bond is weak enough to allow dissociation into $\cdot CN$ radicals. Cyanogen can be prepared by various oxidation reactions of cyano derivatives. Thus cyanogen is obtained from the oxidation of hydrogen cyanide (see next section) with oxygen over a silver catalyst, with chlorine over activated carbon or silica, or with nitrogen dioxide over CaO/glass. Cyanogen is also obtained by oxidation of aqueous cyanide ion with copper(II) or with persulfate ($S_2O_8^{2-}$) in acid solution; the following solid state reaction between $Hg(CN)_2$ and $HgCl_2$ is yet another way:

$$Hg(CN)_2 + HgCl_2 \rightarrow Hg_2Cl_2 + :N{\equiv}C—C{\equiv}N: \tag{2.10}$$

Cyanogen disproportionates in basic solution according to the following equation:

$$:N{\equiv}C—C{\equiv}N: + 2OH^- \rightarrow CN^- + OCN^- + H_2O \tag{2.11}$$

Because of its endothermic nature, cyanogen burns in a stoichiometric amount of pure oxygen with one of the hottest known flames from a chemical reaction (5050 K). Cyanogen polymerizes readily when heated to give paracyanogen (Figure 2.6) as a infinite chain polymer.

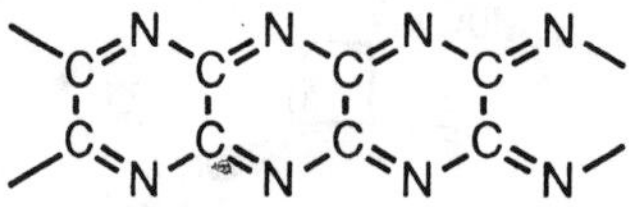

paracyanogen

Figure 2.6. Structure of paracyanogen.

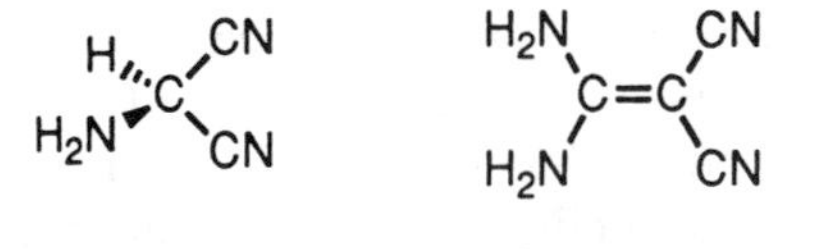

Figure 2.7. Some oligomers of hydrogen cyanide.

2.6.2 Hydrogen Cyanide

Hydrogen cyanide, $H-C\equiv N$:, is a colorless, extremely toxic, volatile liquid (mp $-13.4°C$, bp $+25°C$), with a high dielectric constant (107 at $25°C$). It functions as a weak acid ($K_a = 4.9 \times 10^{-10}$). Hydrogen cyanide is made by acidifying aqueous solutions of cyanides or, industrially, by the exothermic reaction of methane with ammonia in a fast flow/rapid quench system according to the following equation:

$$CH_4 + NH_3 \xrightarrow[\text{1200°C}]{\text{Pt catalyst}} HCN + 3H_2 \qquad \Delta H = -247 \text{ kJ/mol} \qquad (2.12)$$

Hydrogen cyanide polymerizes readily under a variety of conditions. Hydrogen cyanide oligomers include the trimer aminomalononitrile and the tetramer diaminomaleonitrile (Figure 2.7).

2.6.3 Cyanide Ion

Cyanide ion, :$C^-\equiv N$:, is made by treatment of hydrogen cyanide with base (e.g., NaOH or Na_2CO_3 for the preparation of sodium cyanide) or from sodium, ammonia, and carbon according to the following sequence of reactions:

$$2Na + 2NH_3 \rightarrow H_2 + 2NaNH_2 \text{ (sodium amide or "sodamide")} \qquad (2.13a)$$

$$2NaNH_2 + C \rightarrow 2H_2 + Na_2NCN \text{ (disodium cyanamide)} \qquad (2.13b)$$

$$Na_2NCN + C \rightarrow 2NaCN \qquad (2.13c)$$

The cyanide ion in crystalline alkali metal cyanides is rotationally disordered and thus is effectively spherical, with a radius of 1.92 Å intermediate between Cl^- and Br^-. Cyanide ion forms strong complexes in aqueous solution with transition metals such as $Fe(CN)_6^{4-}$ (hexacyanoferrate(II) or "ferrocyanide"); the transition metal in these complexes always bonds to the carbon atom of the cyano group.[41] The cyanide ion occupies a high position in the spectrochemical series, giving rise to a large nephelauxetic effect and producing a large trans effect. Unidentate cyanide ion always bonds to the metal through its carbon atom, but $M-C\equiv N-M$ links are present in many solid cyanides such as AuCN, $Zn(CN)_2$, and the Prussian blues, which are highly colored, mixed oxidation state Fe(II/III) derivatives. Anhydrous acids corresponding to the more stable complex transition metal cyanides can be isolated (e.g., $H_3Rh(CN)_6$ and $H_4Fe(CN)_6$; such acids have hydrogen-bonded $M-C\equiv N-H \cdots N\equiv C-M$ units (Section 1.4). Related species are the hydrogen isocyanide complexes such as $HNCW(CO)_5$, which is obtained by protonation of the coordinated cyano group in $W(CO)_5CN^-$. Such protonation reactions indicate the basicity of the nitrogen atoms in coordinated cyano groups. Sodium cyanide is used in the extraction of gold and silver from their ores, since it forms very stable linear cyano complexes $[:N\equiv C-M-C\equiv N:]^-$ (M=Ag, Au) with these metals. Organic isocyanides, $R-N^+\equiv C^-:$ (R=alkyl or aryl group), also form extensive series of transition metal complexes including zerovalent derivatives such as $(PhNC)_6Cr$. The transition metal always bonds to the carbon rather than the nitrogen atom of isocyanides.[42]

2.6.4 Cyanamide

Cyanamide, $H_2\ddot{N}-C\equiv N:$, is a colorless crystalline solid (mp 46°C), prepared by hydrolysis of its calcium salt, CaNCN, under mild conditions (e.g., aqueous CO_2), to avoid further hydrolysis of cyanamide. Calcium cyanamide is made on a large industrial scale for a direct application fertilizer, weed killer, and cotton defoliant by a high temperature reaction of calcium carbide with molecular nitrogen:

$$CaC_2 + N_2 \xrightarrow{\sim 1000°C} CaNCN + C \qquad \Delta H = -297 \text{ kJ/mol} \quad (2.14)$$

The cyanamide dianion in CaNCN, namely $CN_2^{2-} = :\ddot{N}^-=C=\ddot{N}^-:$, is isoelectronic and isostructural with carbon dioxide ($CO_2 = :\ddot{O}=C=\ddot{O}:$). Cyanamide dimerizes in aqueous alkaline solution at 80°C to form dicyandiamide, $(H_2N)_2C=NCN$ (mp 209°C, dec), which undergoes further conversion to a trimer, 1,3,5-triamino-1,3,5-triazine or "melamine," by heating in ammonia (Figure 2.8). Melamine, which is a useful ingredient in polymers and plastics, can also be made by heating urea under pressure:

$$6H_2NC(=O)NH_2 \xrightarrow{300°C/100 \text{ atm}} 6NH_3 + 3CO_2 + C_3N_3(NH_2)_3 \quad (2.15)$$

The NH_3 and CO_2 produced in this reaction can be recycled to make more urea starting material.

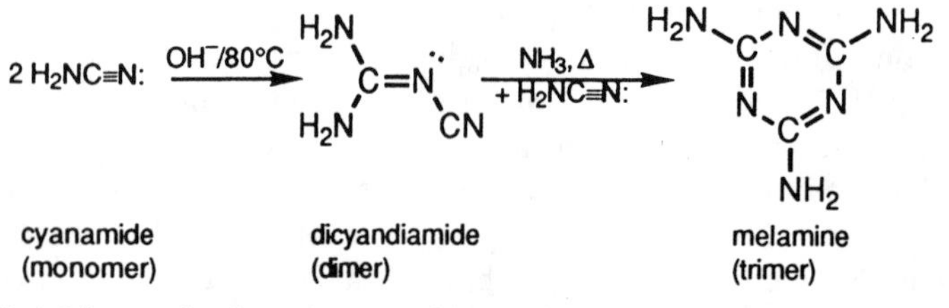

Figure 2.8. Oligomerization of cyanamide to give dicyandiamide and melamine.

2.6.5 Cyanogen Halides

Cyanogen chloride, Cl—$C\equiv N$:, a toxic gas (mp $-69°C$, bp $+13°C$), is obtained by reaction of HCN with Cl_2 or by electrolysis of an aqueous solution containing HCN and NH_4Cl. It readily undergoes trimerization to give cyanuric chloride, $C_3N_3Cl_3$, which has a 1,3,5-triazine structure similar to that of melamine (Figure 2.8). The chlorine atoms in cyanuric chloride are reactive toward nucleophilic substitution. Fluorination of cyanuric chloride with NaF in tetramethylenesulfone gives cyanuric fluoride, $C_3N_3F_3$, which can be cracked thermally to give cyanogen fluoride, F—$C\equiv N$: (bp $-46°C$), which polymerizes at room temperature. Cyanogen bromide (mp $52°C$, bp $61.4°C$) and cyanogen iodide (mp $147°C$) are volatile solids, which can be made by treatment of cyanide ion with the free halogen in aqueous solution. Cyanogen iodide can also be obtained from $Hg(CN)_2$ and I_2.

2.6.6 Cyanic Acid and Cyanates

The cyanate ion, $:\ddot{O}{=}C{=}\ddot{N}^-:$, like the cyanamide dianion, is isoelectronic and isostructural with carbon dioxide; it is made by the mild oxidation of aqueous cyanide ion, for example,

$$KCN(aq) + PbO(s) \rightarrow Pb(s) + KOCN(aq) \qquad (2.16)$$

Sodium cyanate can also be made by heating sodium carbonate with urea in the absence of a solvent:

$$Na_2CO_3 + 2O{=}C(NH_2)_2 \xrightarrow[dry]{heat} 2NaOCN + CO_2 + 2NH_3 + H_2O \quad (2.17)$$

Unstable free cyanic acid, HOCN (bp $23.5°C$, $K_a = 1.2 \times 10^{-4}$) can be obtained by hydrolysis of cyanuric chloride (see preceding section) to "cyanuric acid," 1,3,5-trihydroxy-1,3,5-triazine, $C_3N_3(OH)_3$, followed by pyrolysis. Cyanic acid undergoes facile hydrolysis in aqueous solution to give carbon dioxide and ammonia:

$$HOCN + H_2O \rightarrow NH_3 + CO_2 \qquad (2.18)$$

In covalent derivatives or metal complexes, cyanate can bond either through oxygen (cyanates) or nitrogen (isocyanates), thereby providing examples of linkage isomerism.[43] The facile isomerization of ammonium cyanate ($NH_4^+OCN^-$) to urea ($O{=}C(NH_2)_2$), which was first found in the nearly nineteenth century, is regarded as the first link between inorganic and organic chemistry.

2.6.7 Thiocyanates and Selenocyanates

The thiocyanate ion,[44] $:\ddot{S}{=}C{=}\ddot{N}^-:$, also called the "rhodanid" ion in the early German literature because of the intense red color of its Fe(III) derivative, is obtained by the fusion of alkali metal cyanides with sulfur. It is also the product of the detoxification of cyanide ion in living systems. Most thiocyanate salts are very soluble in both water and liquid ammonia and are useful soluble sources of metal ions in liquid ammonia. Thiocyanogen, $:N{\equiv}C{-}S{-}S{-}C{\equiv}N:$, is the prototypical example of a *pseudohalogen*, that is, a group of atoms that behaves like a halogen, X_2 (X = Cl, Br, I). Thiocyanogen, $:N{\equiv}C{-}S{-}S{-}C{\equiv}N:$ (mp $\sim -7°C$), is made by bromination of metal thiocyanate in an inert (nonprotonic) solvent, for example,

$$2AgSCN + Br_2 \rightarrow 2AgBr + :N{\equiv}C{-}S{-}S{-}C{\equiv}N: \qquad (2.19)$$

Thiocyanogen is rapidly hydrolyzed by water and rapidly and irreversibly polymerized in the pure state to the brick red solid "parathiocyanogen" $(SCN)_x$. However, thiocyanogen can be handled in solutions in inert solvents including glacial acetic acid as well as carbon tetrachloride or carbon disulfide.

Selenocyanates are also known; they are analogous to thiocyanates but less stable.

2.6.8 Fulminic Acid

Fulminic acid, $:C^-{\equiv}N^+{-}OH \leftrightarrow :C{=}\ddot{N}{-}OH$, is a "bivalent" carbon derivative like carbon monoxide; it is formally an "oxime" of carbon monoxide. Free fulminic acid is unstable, but its explosive mercury salt, which is used as a detonator, is obtained by treatment of elemental mercury with a mixture of ethanol and nitric acid. The fulminate ion, $:C^-{\equiv}N^+{-}O^-$, like carbon monoxide and cyanide ion, functions as a strong field ligand toward transition metals to form complexes such as $Fe(CNO)_6^{4-}$ analogous to the ferrocyanide ion, $Fe(CN)_6^{4-}$. However, fulminato metal complexes, like mercury fulminate, are frequently explosive. The likewise explosive silver fulminate (AgCNO) and silver cyanate (AgOCN) are of historical interest: they constitute the first pair of isomers discovered and led to the first proposal of the concept of isomerism; by Liebig in 1823.

2.7 Carbon–Sulfur Derivaties

The raw material for much of carbon–sulfur chemistry is *carbon disulfide*, $CS_2 = :\ddot{S}{=}C{=}\ddot{S}:$, a very flammable and reactive toxic liquid (mp $-112°C$, bp $+46°C$, flash point $-30°C$), which can be synthesized from elemental carbon or methane and sulfur at high temperatures.[45] Chlorination of carbon disulfide according to the following equation provides a source of carbon tetrachloride on an industrial scale:

$$CS_2 + 3Cl_2 \rightarrow CCl_4 + S_2Cl_2 \qquad (2.20)$$

The carbon atom in carbon disulfide is susceptible to nucleophilic attack by HS^-, RO^-, RNH_2, R_2NH, and so on. Nucleophilic attack of carbon disulfide with SH^- in basic solution gives the yellow *trithiocarbonate ion*, CS_3^{2-}. Further reaction of the trithiocarbonate ion with elemental sulfur gives the ion CS_4^{2-}, shown to have the disulfide structure $[S_2C\text{—}S\text{—}S]^{2-}$. The free acids H_2CS_3 and H_2CS_4 can be obtained from these anions as red liquids, stable only at low temperatures. Reaction of carbon disulfide with secondary amines in basic solution leads to *dithiocarbamates* according to the following equation (R = methyl, ethyl, etc.)[46]:

$$CS_2 + R_2NH + NaOH \rightarrow R_2NCS_2^- Na^+ + H_2O \qquad (2.21)$$

Zinc, manganese, and iron dithiocarbamates are extensively used as agricultural fungicides and zinc dithiocarbamates are used as accelerators in rubber vulcanization. Oxidation of aqueous dimethyldithiocarbamate with Cl_2, H_2O_2, or $S_2O_8^{2-}$ gives *tetramethylthiuram disulfide*, $Me_2NC(=S)\text{—}S\text{—}S\text{—}C(=S)NMe_2$. The lability of its sulfur–sulfur bond makes tetramethylthiuram disulfide a useful source of free radical for polymerization initiators and vulcanization accelerators. Tetraethylthiuram disulfide, $Et_2NC(=S)\text{—}S\text{—}S\text{—}C(=S)NEt_2$, is the drug Antabuse, which is used to treat alcoholism by making the body allergic to ethanol. Carbon disulfide is a very versatile ligand toward transition metals, forming a variety of types of metal complex including complex types analogous to those found for carbon dioxide (Figure 2.3).

Carbon monosulfide, CS, is obtained by action of a high frequency discharge on CS_2 vapor. Unlike carbon monoxide, carbon monosulfide is not stable (dec $> -160°C$). However, numerous transition metal complexes of carbon monosulfide are known [47–49]; such complexes are generally called *metal thiocarbonyls*, by analogy to metal carbonyls. Selenocarbonyls containing the CSe group are also known.[47] The inaccessibility of free CS restricts the variety of metal thiocarbonyls that can be prepared, and relatively few complexes with more than one CS ligand are known. For example, $M(CS)_6$ (M = Cr, Mo, W) and $Fe(CS)_5$ analogous to $M(CO)_6$ and $Fe(CO)_5$, respectively, remain unknown. Sources of CS ligands in metal complexes include CS_2 and $CSCl_2$, for example:

$$(Ph_3P)_3RhCl + CS_2 \rightarrow (Ph_3P)_2Rh(CS)Cl + Ph_3PS \qquad (2.22)$$

$$Na_2W(CO)_5 + S{=}CCl_2 \rightarrow W(CO)_5CS + 2NaCl \qquad (2.23)$$

A problem with the latter reaction is the separation of $W(CO)_5CS$ from $W(CO)_6$ formed from side reactions, since these compounds have very similar properties. Coordinated CS is reactive toward electrophilic attack at sulfur and nucleophilic attack at carbon, for example:

$$L_nM(CS) + RI \rightarrow [L_nM(CSR)]^+ I^- \quad (R = Me, etc.) \qquad (2.24)$$

$$L_nM(CS) + R_2N^- \rightarrow [L_nM\text{—}C(=S)MR_2]^- \qquad (2.25)$$

The greater reactivity of metal thiocarbonyl derivates relative to corresponding metal carbonyl derivates provides the best method for separating $W(CO)_5CS$ from $W(CO)_6$ (e.g., equation 2.23). Thus $W(CO)_5CS$ reacts with iodide ion to give ionic *trans*-$W(CO)_4(CS)I^-$ under conditions that do not affect

$W(CO)_6$; this reaction, coupled with regeneration of $W(CO)_5CS$ from *trans*-$W(CO)_4(CS)I^-$ by treatment with $AgSO_3CF_3$ in a CO atmosphere, provides a useful method for separating $W(CO)_6/W(CO)_5CS$ mixtures.

References

1. Ballester, M., Inert Free Radicals: A Unique Trivalent Carbon Species, *Acc. Chem. Res.*, 1985, **18**, 380–387.

2. Breitmaier, E.; Voelter, W., *^{13}C NMR Spectroscopy*, Verlag Chemie, Weinheim, 1978.

3. Raaen, V. F.; Ropp, G. A.; Raaen, H., *Carbon-14*, McGraw-Hill, New York, 1968.

4. Michels, J. M., *Dating Methods in Archeology*, Seminar Press, New York, 1973.

5. Bruton, E., *Diamonds*, NAG Press, London, 1970.

6. Mantell, C. L., *Carbon and Graphite Handbook*, Wiley-Interscience, New York, 1968.

7. Reynolds, W. N., *Physical Properties of Graphite*, Elsevier, Amsterdam, 1968.

8. Kroto, H. W.; Allaf, A. W.; Balm, S. P., C_{60}: Buckminsterfullerene, *Chem. Rev.*, 1991, **91**, 1213–1225.

9. Hammond, G. S.; Kuck, V., Eds., *Fullerenes*, ACS Symposium Series No. 481, American Chemical Society, Washington, DC, 1992.

10. Parker, D. H.; Wurz, P., Chatterjee, K.; Lykke, K. R.; Hunt, J. E.; Pellin, M. J.; Hemminger, J. C.; Gruen, D. M.; Stock, L. M., *J. Am. Chem. Soc.* 1991, **113**, 7499–7503.

11. Haddon, R. C., Electronic Structure, Conductivity, and Superconductivity of Alkali Metal Doped C_{60}, *Acc. Chem. Res.*, 1992, **25**, 127–133.

12. Ebert, L. B., Intercalation Compounds of Graphite, *Ann. Rev. Mater. Sci.*, 1976, **6**, 181–211.

13. Selig, H.; Ebert, L. B., Graphite Intercalation Compounds, *Adv. Inorg. Chem. Radiochem.*, 1980, **23**, 281–327.

14. Watanabe, N.; Nakajima, T.; Touhara, H., *Graphite Fluorides*, Elsevier, Amsterdam, 1988.

15. Nakajima, T.; Watanabe, N. *Graphite Fluorides and Carbon–Fluorine Compounds*, CRC Press, Boca Raton, FL, 1990.

16. Mallouk, T.; Bartlett, N., *J. Chem. Soc. Chem. Commun.*, 1983, 103–105.

17. Frad, W. A., Metal Carbides, *Adv. Inorg. Chem. Radiochem.*, 1968, **11**, 153–247.

18. West, R.; Jones, P. C., *J. Am. Chem. Soc.*, 1969, **91**, 6156–6161.

19. Yajima, S., Hayashi, J., Omori, M.; Okamura, K., *Nature*, 1976, **261**, 683–685.

20. Toreki, W., Polymeric Precursors to Ceramics—A Review, *Polym. News*, 1991, **16**, 6–14.

21. Will, G.; Kossobutzki, K. H., *J. Less Common Met.*, 1976, **47**, 43–48.

22. Johansen, H. A., Recent Developments in the Chemistry of Transition Metal Carbides and Nitrides, *Survey Prog. Chem.*, 1977, **8**, 57–81.

23. Tachikawa, M.; Muetterties, E. L., Metal Carbide Clusters, *Prog. Inorg. Chem.*, 1981, **28**, 203–238.

24. Satpathy, S.; Anderson, O. K., *Inorg. Chem.*, 1985, **24**, 2604–2608.

25. Ziebarth, R. P.; Corbett, J. D., *J. Am. Chem. Soc.*, 1987, **109**, 4844–4850.

26. Lançon, D.; Kadish, K. M., *Inorg. Chem.*, 1984, **23**, 3942–3947.

27. Sheppard, W. A.; Sharts, C. M., *Organic Fluorine Chemistry*, Benjamin, New York, 1969.

28. Hudlický, M., *Organic Fluorine Chemistry*, Plenum Press, New York, 1971.

29. Hudlický, M., *Chemistry of Organic Fluorine Compounds*, Ellis Horwood, Chichester, 1976.

30. Banks, R. E., *Preparation, Properties, and Industrial Applications of Organofluorine Compounds*, Ellis Horwood, Chichester, 1982.

31. Riess, J. G.; Le Blanc, M., Solubility and Transport Phenomena in Perfluorochemicals Relevant to Blood Substituion and Other Biomedical Applications, *Pure Appl. Chem.*, 1982, **54**, 2383–2406.

32. West, R., Ed., *Oxocarbons*, Academic Press, New York, 1980.

33. Vol'pin, M. E.; Kolomnikov, I. S., The Reactions of Organometallic Compounds with Carbon Dioxide, *Organometallic Reactions*, 1975, **5**, 313–386.

34. Eisenberg, R.; Hendriksen, D. E., The Binding and Activation of Carbon Monoxide, Carbon Dioxide, and Nitric Oxide, and Their Homogeneously Catalyzed Reactions, *Adv. Catal.*, 1979, **28**, 79–172.

35. Ibers, J. A., Reactivities of Carbon Disulfide, Carbon Dioxide, and Carbonyl Sulfide Towards Some Transition Metal Systems, *Chem. Soc. Rev.*, 1982, **11**, 57–73.

36. Darensbourg, D. J.; Kudaroski, R. A., The Activation of Carbon Dioxide by Metal Complexes, *Adv. Organomet. Chem.*, 1983, **22**, 129–168.

37. Palmer, D. A.; van Eldik, R., The Chemistry of Metal Carbonato and Carbon Dioxide Complexes, *Chem. Rev.*, 1983, **83**, 651–731.

38. Ayers, W. M., Ed., *Catalytic Activation of Carbon Dioxide*, ACS Symposium Series 363, American Chemical Society, Washington, DC, 1988.

39. Behr, A., *Carbon Dioxide Activation by Metal Complexes*, VCH, Weinheim, Germany, 1988.

40. West, R., Niu, J., Oxocarbons and Their Reactions, in *The Chemistry of the Carbonyl Groups*, Vol. 2, J. Zabicky, Ed., Wiley-Interscience, London, 1970, pp. 241–275.

41. Sharpe, A. G., *The Chemistry of Cyano Complexes of the Transition Metals*, Academic Press, London, 1976.

42. Malatesta, L.; Bonati, F., *Isocyanide Complexes of Metals*, Wiley, London, 1969.

43. Norbury, A. H., Coordination Chemistry of the Cyanate, Thiocyanate, and Selenocyanate Ions, *Adv. Inorg. Chem. Radiochem.*, 1975, **17**, 231–402.

44. Newman, A. A., Ed., *Chemistry and Biochemistry of Thiocyanic Acid and Its Derivatives*, Academic Press, London, 1975.

45. Gattow, G.; Behrendt, W., *Carbon Sulfides and Their Inorganic and Complex Chemistry*, Thieme, Stuttgart, 1977.

46. Thorne, G. D.; Ludwig, L. A., *The Dithiocarbamates and Related Compounds*, Elsevier, Amsterdam, 1962.

47. Butler, I. S., Transition Metal Thiocarbonyls and Selanocarbonyls, *Acc. Chem. Res.*, 1977, **10**, 359–365

48. Yaneff, P. V., Thiocarbonyl and Related Compounds of the Transition Metals, *Coord. Chem. Rev.*, 1977, **23**, 183–220.

49. Broadhurst, P. V., Transition Metal Thiocarbonyl Complexes, *Polyhedron*, 1985, **4**, 1801–1846.

CHAPTER

3

Silicon, Germanium, Tin, and Lead

3.1 General Aspects of the Chemistry of Silicon, Germanium, Tin, and Lead

The chemistry of the heavier congeners of carbon, namely silicon, germanium, tin, and lead, is quite different from that of carbon.[1-3] Whereas the properties of carbon are strictly nonmetallic, the metallic character of its heavier congeners increases upon descending this column of the periodic table with the result that tin and particularly lead are clearly metallic. This column of the periodic table containing elements with four valence electrons forms the so-called covalent divide (see Introduction),[4] separating the hypoelectronic elements (i.e., those with fewer than four valence electrons) from the hyperelectronic elements (i.e., those with more than four valence electrons). The following properties of these elements are particularly significant.

1. *Electronegativity.* An alternation of electronegativity has been suggested for this column of the periodic table, with germanium considered to be more electronegative than either silicon or tin based on ambiguous interpretation of physical data.[5] However, the major aspects of the chemistry of the elements in this column of the periodic table can be rationalized on the basis of decreasing electronegativity in the series Si → Pb.

2. *Catenation.* The tendency for catenation (i.e., formation of networks E—E bonds) for these elements decreases in the sequence C ≫ Si > Ge ≈ Sn ≫ Pb relating to the strengths of the corresponding E—E bonds.

3. *Multiple bonding.* The accessibility of d orbitals for the elements Si → Pb makes $d\pi$–$p\pi$ multiple bonding feasible as well as the $p\pi$–$p\pi$ multiple bonding

found in carbon chemistry (e.g., alkenes, alkynes, arenes). Multiply bonded derivatives with $p\pi$–$p\pi$ bonds involving silicon corresponding to alkenes such as $R_2Si{=}SiR_2$ are isolable only when the pendant R groups are bulky groups such as 2,4,6-trimethylphenyl (mesityl) and tris(trimethylsilyl)methyl; similar less stable compounds with multiple $p\pi$–$p\pi$ bonding are known for germanium and tin.[6,7] Multiple $d\pi$–$p\pi$ bonding is found in derivatives, with Si—O and Si—N leading to major differences in the structures and chemical properties of Si—O and Si—N derivatives compared with their carbon analogues. Thus $(H_3Si)_3N$ is nonbasic and planar, in contrast to the basic pyramidal carbon analogue $(CH_3)_3N$.

4. *Valence shell expansion.* The availability of d orbitals in the valence shell of silicon and its heavier congeners, in contrast to carbon, allows expansion of the coordination sphere to give five-coordinate trigonal bipyramidal derivatives with dsp^3 hybridization and six-coordinate octahedral derivatives with d^2sp^3 hybridization.[8] Examples of five-coordinate derivatives of these elements (E) include the anions EX_5^- and $R_nEX_{5-n}^-$ stabilized in lattices by large cations, adducts of halides or substituted halides with donor ligands (L) of the type EX_4L, and, in the case of tin, polymeric compounds $[R_3SnX]_n$ with bridging halogen atoms. Octahedral coordination is found in the SiF_6^{2-} anion and related hexahalometallates(IV), EX_6^{2-}.

5. *The divalent state.* The divalent state becomes increasingly stable in the series Si → Pb, so that it is the dominant oxidation state of lead.

6. *Stereochemistry.* All elements form tetrahedral compounds in the +4 oxidation state. Tetrahedral compounds of germanium and tin with four different substituents (e.g., (α-naphthyl)Ge(H)(Me)(Ph) and MeSn(p-anisyl)-(α-naphthyl)($CH_2CH_2C(OH)Me_2$)) may be chiral and resolvable into optically active enantiomers, as is the case with their carbon analogues.[9] These elements in their +2 oxidation states have a lone pair of electrons which can be stereochemically active (see Introduction). Thus tin(II) derivatives are frequently pseudotetrahedral SnX_3^- derivatives with the lone pair in the fourth coordination position.

3.2 Elemental Silicon, Germanium, Tin, and Lead

Silicon is second only to oxygen in weight percentage in the earth's crust ($\sim 28\%$) and is found in diverse silicate minerals.[10,11] Germanium, tin, and lead are relatively rare elements (Ge, 1.5 ppm; Sn, 2.1 ppm; Pb, 13 ppm) but are rather well known because of the technical importance. Tin and lead have been known since antiquity because of the ease of obtaining them from natural sources. Lead, the most abundant of all heavy elements, is the stable end product of all natural radioactive decay series. Lead is a heavy metal poison, which complexes with the oxo groups in enzymes.[12]

Elemental silicon in 96–99% purity is made on a large scale by the reduction of SiO_2 (quartzite or sand) with high purity coke in an electric arc furnace.

Excess SiO_2 is used to prevent accumulation of SiC. Addition of scrap iron to the SiO_2/C reaction mixture gives ferrosilicon alloys, which are used in metallurgy for the deoxidation of steel. Very pure elemental silicon for use as semiconductors in transistors and integrated circuits is obtained from $SiCl_4$ or $HSiCl_3$, which is purified by exhaustive fractional distillation and then reduced by very pure metallic zinc or magnesium. Alternative routes to semiconductor grade silicon include the thermal decomposition of an SiI_4/H_2 mixture on a hot tungsten filament and the epitaxial growth of a single crystal layer by pyrolysis of SiH_4 (Section 3.4.1). The final purification of semiconductor grade silicon can be effected by zone-refining of cylindrical single crystals obtained from the melt.

The graphite structure is unique to carbon, reflecting the greater stability of $p\pi-p\pi$ bonds in carbon relative to its heavier congeners (see Section 3.1). Thus the common forms of elemental silicon and germanium are infinite three-dimensional covalently bonded networks isostructural with diamond. Tin has two crystalline modifications with the equilibria:

$$\alpha\text{-Sn("gray")} \underset{}{\overset{13°C}{\rightleftharpoons}} \beta\text{-Sn ("white")} \underset{}{\overset{232°C}{\rightleftharpoons}} \text{Sn (liquid)} \tag{3.1}$$

Covalent gray or α-tin has the cubic diamond structure, whereas metallic white or β-tin has a distorted close-packed tetragonal lattice. For this reason β-tin is denser ($d_{20°C} = 7.31$ g/cm^3) than α-tin ($d_{20°C} = 5.75$ g/cm^3). Lead exists only as a cubic close-packed metallic form. This property is related to the preference of lead for the divalent state and the low stability of the Pb–Pb covalent bond.

These elements have several spin $\frac{1}{2}$ isotopes available for NMR work such as ^{29}Si, ^{115}Sn, ^{117}Sn, ^{119}Sn, and ^{207}Pb of relative abundances 4.7, 0.35, 7.67, 8.68, and 21.11%, respectively, and sensitivities for equal numbers of nuclei of 0.00785, 0.0350, 0.0453, 0.0518, and 0.00913, respectively, relative to the proton.[13] Tin is the only element with three stable spin $\frac{1}{2}$ isotopes, and germanium is the only element of this group that has no spin $\frac{1}{2}$ isotopes, although the quadrupolar nucleus ^{73}Ge (I = $\frac{9}{2}$, relative abundance 7.61%, relative sensitivity to the proton 0.0014) has been used for NMR work. The tin isotope ^{119}Sn is also an excellent isotope for Mössbauer spectroscopy.[14]

3.3 Anions and Related Binary Compounds of Silicon, Germanium, Tin, and Lead

Reduction of germanium, tin, and lead compounds with alkali metals in liquid ammonia gives anionic metal clusters with the element in a negative formal oxidation state. These anionic clusters are commonly called *Zintl anions*, because they were discovered by Zintl in the 1930s from potentiometric titration studies. Such Zintl ions include trigonal bipyramidal Sn_5^{2-} and Pb_5^{2-}; capped square antiprismatic Ge_9^{4-}, Sn_9^{4-}, and Pb_9^{4-}; and tricapped trigonal prismatic Sn_9^{3-} (Figure 3.1); these highly oxygen-sensitive anions can be isolated as

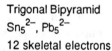

Trigonal Bipyramid
$Sn_5{}^{2-}$, $Pb_5{}^{2-}$
12 skeletal electrons

Tricapped Trigonal Prism
Paramagnetic $Sn_9{}^{3-}$
21 skeletal electrons

Capped Square Antiprism
$Ge_9{}^{4-}$, $Sn_9{}^{4-}$, $Pb_9{}^{4-}$
22 skeletal electrons

Figure 3.1. Polyhedra found in Zintl anions containing germanium, tin, and lead.

crystalline salts by addition of cryptand ligands to complex the alkali metal counterions.[15] These ions are all diamagnetic except for the paramagnetic $Sn_9{}^{3-}$, which has an unpaired electron. The germanium, tin, or lead vertices of each of these anions contribute two skeletal electrons apiece, since two of the four electrons of their sp^3 manifolds are required for an external lone pair. This gives skeletal electron counts of $2n + 2$ for closed deltahedra (polyhedra in which all faces are triangles) and $2n + 4$ for so-called *nido* polyhedra in which only one face is not a triangle. However, the paramagnetic deltahedral $Sn_9{}^{3-}$ has one extra electron. These electron-counting rules are analogous to those for the polyhedral boranes and carboranes (Section 8.3.4).

Thermal reactions of alkali or alkaline earth metals with silicon or germanium or their oxides give various silicides or germanides, but most of these products do not contain true anions and are semiconductors.[16,17] Compounds of this type include ME (M = alkali metal; E = Si, Ge, Sn, Pb) = $(M^+)_4(E_4^{4-})$ with isolated E_4^{4-} tetrahedra isoelectronic with neutral P_4 (Section 5.2), $Li_{12}Si_7$ with Si_5 rings and trigonal planar centered Si_4 units in the lattice, and Ba_2Si with an *anti*-$PbCl_2$ structure with Si^{4-} ions octahedrally coordinated by Ba^{2+} ions.

Silicon carbide, SiC (Section 2.3), and silicon nitride,[18] Si_3N_4, are of interest as refractory ceramics. Silicon nitride is obtained by direct combination of silicon with nitrogen at 1100–1400°C or by heating a mixture of silicon dioxide and coke in a N_2/H_2 atmosphere to 1450–1600°C. Silicon nitride is very hard (9 Mohs), almost inert chemically, and retains its strength, shape, and corrosion resistance at about 1000°C. Its structure consists of an infinite network of $SiN_{4/3}$ building blocks, where the subscript "4/3" means that each silicon atom is bonded to four nitrogen atoms (as expected for tetrahedral sp^3 coordination of silicon) and each nitrogen atom is bonded to three silicon atoms (as expected for tetrahedral sp^3 coordination of nitrogen with a stereochemically active lone pair). This subscripting convention is used throughout this book to indicate the stoichiometries and structures of infinite solids.

Silicon, germanium, and tin form sulfides of stoichiometry ES_2. The sulfides SiS_2 and GeS_2 have infinite chain structures of tetrahedral $ES_{4/2}$ units with

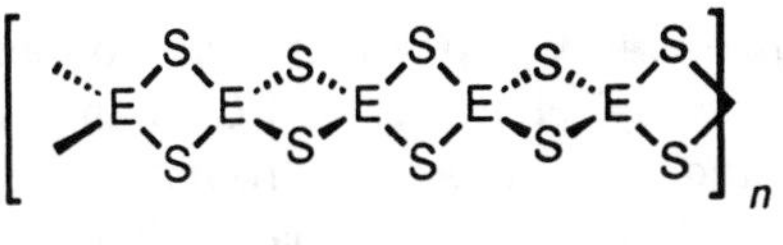

Figure 3.2. Infinite chain structure of SiS_2 and GeS_2.

tetracoordinates Si or Ge atoms (Figure 3.2). However, SnS_2 has a CaI_2 structure with $SnS_{6/3}$ structural units in which each tin atom is octahedrally coordinates to six sulfur neighbors.

3.4 Hydrides and Halides of Silicon, Germanium, Tin, and Lead

3.4.1 Hydrides

Silanes of the general formula Si_nH_{2n+2} are formally analogous to alkanes but are much less stable and more reactive.[19,20] Thus, only SiH_4 and Si_2H_6 are indefinitely stable at room temperature, and the thermal stability of silanes decreases with increasing chain length. Silanes are spontaneously flammable or even *explosive* in air, for example:

$$SiH_4 + 2O_2 \rightarrow SiO_2 + 2H_2O \tag{3.2}$$

Silanes are stable to water and dilute mineral acids but are hydrolyzed rapidly by aqueous bases:

$$Si_2H_6 + (4 - n)H_2O \rightarrow 2SiO_2 \cdot nH_2O + 7H_2\uparrow \tag{3.3}$$

The most important silane is *monosilane*, SiH_4 (m.p. $-185°C$, bp $-112°C$), particularly because of its use as a source of pure silicon for semiconductors. On the laboratory scale, SiH_4 can be prepared by the reaction of $LiAlH_4$ with $SiCl_4$ under mild conditions or with SiO_2 at 150–170°C. On a large scale, SiH_4 can be prepared by reduction of alkali metal silicates or SiO_2 with hydrogen at 175°C and 400 atm or with aluminum metal in an $NaCl/AlCl_3$ eutectic (mp 120°C).

Silanes are strong reducing agents and react explosively with halogens at 25°C. However, controlled replacement of hydrogen by halogen in SiH_4 can be effected in the presence of AlX_3 to give halosilanes (silyl halides) such as H_3SiCl. Potassium metal reacts with SiH_4 to give colorless crystalline $KSiH_3$, which has an NaCl-type structure containing K^+ cations and SiH_3^- anions.

Germane, GeH_4 (mp $-165°C$, bp $-88°C$), can be synthesized by reaction of GeO_2 with $LiAlH_4$ in the solid state or with $NaBH_4$ in acid solution. Germanes are less flammable than silanes but still are rapidly oxidized in air. Germanes are more resistant to hydrolysis than silanes, and GeH_4 is unaffected by 30% aqueous sodium hydroxide.

Stannane, SnH_4 (mp $-146°C$, bp $-52.5°C$), is obtained by interaction of $SnCl_4$ with $LiAlH_4$ in diethyl ether at $-30°C$. Stannane is much less stable than SiH_4 or GeH_4 and decomposes at room temperature to give β-tin. Stannane is stable to dilute acids and bases, but 2.5 M aqueous sodium hydroxide causes some decomposition to metallic tin and $Sn(OH)_6^{2-}$. Stannane is easily oxidized and thus a good reducing agent; it converts benzaldehyde to benzyl alcohol and nitrobenzene to aniline.

The existence of *plumbane*, PbH_4, is doubtful. Studies with radioactive lead tracers suggest that trace quantities of plumbane are formed by acid hydrolysis of magnesium–lead alloys or cathodic reduction of lead salts.

3.4.2 Halides

The fluorides EF_4 (E = Si, Ge, Sn) are made by fluorination of other halides; PbF_4 is obtained by the reaction of PbF_2 with excess fluorine and is itself a strong fluorinating agent. Hydrolysis of SiF_4 and GeF_4 with excess water gives the hydrous oxides. Reaction of the fluorides EF_4 with excess aqueous HF gives the octahedral anions EF_6^{2-}. Attack of glass by hydrofluoric acid gives SiF_4 and subsequently SiF_6^{2-}. The fluorides SiF_4 (bp $-86°C$) and GeF_4 (bp $-43.5°C$, subl.) are gases under ambient conditions but $SnF_4 = SnF_{4/2}F_2$ is a solid polymer in which each tin atom is coordinated to four bridging ($F_{4/2}$) and two terminal (F_2) fluorine atoms.

Silicon tetrachloride, $SiCl_4$ (bp 58°C), is obtained by chlorination of elemental silicon or ferrosilicon at a red heat. *Hexachlorodisilane*, Si_2Cl_6 (bp 150°C), is obtained by reaction of $SiCl_4$ with elemental silicon at high temperatures or by the chlorination of certain silicides such as $CaSi_2$. Higher silicon chlorides,[21] which have branched structures, are obtained from the photolysis of $HSiCl_3$ or from amine-catalyzed redistribution reactions such as

$$5Si_2Cl_6 \xrightarrow{\text{0.1\% Me}_3\text{N}} Si_6Cl_{14} + 4SiCl_4 \tag{3.4a}$$

$$3Si_3Cl_8 \xrightarrow{\text{Me}_3\text{N}} Si_5Cl_{12} + 2Si_2Cl_6 \tag{3.4b}$$

Mixtures of higher silicon chlorides can be separated by fractional distillation.

All silicon chlorides are readily and completely hydrolyzed by water, although by using limited amounts of water the hydrolysis of $SiCl_4$ can be controlled to give $Cl_3SiOSiCl_3$ and/or $(Cl_3SiO)_2SiCl_2$. Reactions of Si_2Cl_6 with compounds having oxygen bonded to nitrogen, phosphorus, or sulfur (e.g., amine oxides, phosphine oxides, sulfoxides) lead to their deoxygenation. Reactions of this type are useful for converting optically active tertiary phosphine oxides $R''R'RP{=}O$ to the corresponding tertiary phosphines $R''R'RP:$ and occur with inversion of configuration at the phosphorus tetrahedron. This suggests a mechanism involving nucleophilic attack by the $SiCl_3^-$ anion, which is isoelectronic and presumably isostructural with neutral PCl_3.

Silicon tetrachloride undergoes hydrogenolysis upon treatment with hydrogen to form *trichlorosilane*, $HSiCl_3$ (bp $31.8°C$), by the equation

$$SiCl_4 + H_2 \rightarrow HSiCl_3 + HCl \tag{3.5}$$

Alternatively, $HSiCl_3$ can be made by treatment of HCl with elemental silicon at elevated temperatures:

$$Si + 3HCl \xrightarrow{350°C} HSiCl_3 + H_2 \tag{3.6}$$

The Si—H bond in $HSiCl_3$ adds across the C=C double bonds of olefins in the presence of small amounts of a platinum catalyst such as H_2PtCl_6 to form alkyltrichlorisilanes in the so-called *Speier reaction*, for example:

$$HSiCl_3 + H_2C\text{=}CH_2 \xrightarrow{\text{Pt catalyst}} CH_3CH_2SiCl_3 \tag{3.7}$$

In addition, $HSiCl_3$ in the presence of tertiary amines can reduce tertiary phosphine oxides stereospecifically to the corresponding tertiary phosphines similar to Si_2Cl_6 (see equations 3.4).

Germanium tetrachloride, $GeCl_4$ (mp $-49.5°C$, bp $83.1°C$), is made by chlorination of elemental germanium at elevated temperatures or by treatment of GeO_2 with concentrated hydrochloric acid; it can be separated by distillation from the latter reaction. Germanium tetrachloride is less sensitive toward hydrolysis than $SiCl_4$; thus it is only partially hydrolyzed in aqueous $6-9\ M$ hydrochloric acid. *Tin tetrachloride*, $SnCl_4$ (mp $-33°C$, bp $114.1°C$), and lead tetrachloride, $PbCl_4$ (mp $-15°C$), are hydrolyzed completely only in pure water; they form chloroanions of the type ECl_6^{2-} in excess aqueous hydrochloric acid. *Lead tetrachloride*, $PbCl_4$, is a strong chlorinating agent and readily decomposes to $PbCl_2 + Cl_2$ above $50°C$; this indicates the stability of Pb(II) relative to Pb(IV).

3.4.3 Halide Complexes

The octahedral *hexafluorosilicate(IV) anion*, SiF_6^{2-}, is the only silicon fluoroanion obtained in aqueous solution. It is obtained by treatment of hydrous silica or glass with hydrofluoric acid or by the hydrolysis of SiF_4 according to the following equation:

$$2SiF_4 + 2H_2O \rightarrow SiO_2 + SiF_6^{2-} + 2H^+ + 2HF \tag{3.8}$$

The trigonal bipyramidal *pentafluorosilicate(IV) anion*, SiF_5^-, is not stable in aqueous solution but can be obtained as salts with large monovalent cations of the type R_4E^+ (E = N, P, As). The ^{19}F NMR spectra for the SiF_5^- anion as well as for organic derivatives of the types $RSiF_4^-$ and $R_2SiF_3^-$ indicate trigonal bipyramidal structures with exchange processes occurring around $-60°C$ similar to those in PF_5 and other five-coordinate phosphorus derivatives (Section 5.1). Octahedral anions of the type EF_6^{2-} are also formed by germanium, tin, and lead.

The *hexachlorosilicate(IV) anion*, $SiCl_6^{2-}$, is not known, at least in aqueous solution, because of the sensitivity of Si—Cl bonds to hydrolysis. However, its germanium and tin analogues, ECl_6^{2-} (E = Ge, Sn), can be obtained by reactions of the corresponding tetrachlorides, ECl_4, with hydrochloric acid or chloride ion. The *hexachloroplumbate(IV) anion*, $PbCl_6^{2-}$, can be obtained by treatment of a solution of $PbCl_2$ in concentrated aqueous hydrochloric acid with chlorine at low temperatures and isolated as M_2PbCl_6 salts (M = Na, K, Rb, Cs, NH_4). These salts are much more stable than $PbCl_4$ and serve as useful stable sources of Pb(IV).

Tin halides form a varity of six-coordinate complexes with Lewis bases and related ligands such as β-diketonates of the type *trans*-$SnCl_2(\beta$-diketonate$)_2$, dithiocarbamates of the type $SnCl_2(S_2CNR_2)_2$, and L_2SnX_4 complexes with neutral Lewis bases.[22] Related six-coordinate silicon tetrahalide complexes include *trans*-$SiF_4(py)_2$, *cis*-$SiCl_4(2,2$ bpy$)$, and L_2SiCl_4 (py = pyridine; bpy = 2,2'-bipyridyl, L = Me_3P or py). Reactions of organosilicon halides of the type R_3SiX with 2,2'-bipyridyl result in halogen abstraction to give five-coordinate cationic complexes of the type $[R_3Si(bpy)]^+$.

3.5 Silicates and Related Oxygen Compounds of Silicon

Oxygen compounds of silicon[23] are unusually important because they are ubiquitous in nature as are the vast variety of silicate minerals. In fact, silicate mineral chemistry may be viewed as the pure inorganic analogue of organic natural products chemistry. The fundamental structural unit in silicate structures is the $SiO_{4/2}$ tetrahedron ($= SiO_2$ if no elements other than silicon and oxygen are present). Silicate structures have been determined mainly by X-ray diffraction methods. In recent years such structural information has been supplemented by ^{29}Si and ^{27}Al magic angle spinning solid state NMR work on microcrystalline and amorphous solids and glasses.

3.5.1 Silicon Dioxide (Silica)[24, 25]

Silicon dioxide (silica) has been studied more than any other chemical compound except for water, and at least a dozen polymorphs of pure SiO_2 are known. Pure silicon dioxide may be described as $SiO_{4/2}$ in which all silicon atoms are bonded for four oxygen atoms and each oxygen atom bridges two silicon atoms. However, there is a form of silica called *stishovite*, obtained at high temperatures and pressures (e.g., 250–1300°C at 35,000–120,000 atm), which has a rutile (TiO_2) structure constructed from $SiO_{6/3}$ rather than $SiO_{4/2}$ units in which each silicon atom is bonded to six oxygen atoms and each oxygen atom bridges three silicon atoms. Stishovite is denser and chemically more inert than normal silica but reverts to normal silica upon heating. Stishovite is found in nature at Meteor Crater (in Arizona) presumably having formed under transient shock pressures following meteorite impact.

The thermodynamically favored form of SiO_2 under ambient conditions is *quartz*, which has a helical structure with two slightly different Si—O distances (1.597 and 1.617 Å) and a 144° Si—O—Si angle. Enantiomeric crystals of quartz can be obtained and separated mechanically. Pure enantiomeric crystals of quartz are optically active. Organic racemic mixtures can sometimes be separated into their enantiomers by chromatography on a column of a pure enantiomer of quartz. At 573°C, α-quartz is transformed into β-quartz, which has the same general structure but with less distortion (Si—O—Si = 155°). The thermal transformation of α-quartz to β-quartz preserves its optical activity. Further heating of quartz to 867°C results in the transformation of β-quartz into β-tridymite, which involves breaking Si—O bonds to enable the $SiO_{4/2}$ tetrahedra to rearrange to a simpler more open hexagonal structure of lower density. The quartz–tridymite transformation is a high activation energy process that results in loss of the optical activity of the quartz. Further heating of β-tridymite to 1470°C gives β-*cristobalite*, which has a structure related to that of diamond (Section 2.2) with silicon atoms in the diamond carbon positions and an oxygen atom midway between each pair of silicon atoms.

Silica also occurs in amorphous forms. Thus slow cooling of molten silica gives an amorphous material that is glassy in appearance; it has no long-range order but shows a disordered array of polymeric chains, sheets, and/or three-dimensional units. *Silica gel* is a form of amorphous silica obtained by the hydrolysis of alkoxides such as $Si(OEt)_4$; it often contains ~4% water. Magic angle solid state ^{29}Si NMR results suggest the presence of $Si(OSi\equiv)_4$, $Si(OSi\equiv)_3OH$, and $Si(OSi\equiv)_2(OH)_2$ groups in silica gel. Silica gel is useful as a drying agent (when dehyrated) and as supports for chromatography and catalysis.

3.5.2 Alkali Silicates and Silicates in Aqueous Solution

Sodium silicates are made on a large scale by fusion of SiO_2 with Na_2CO_3 at ~1500°C for various applications including detergents.[26, 27]

Raman and ^{29}Si NMR spectroscopy indicate that in aqueous solution the free SiO_4^{4-} has a limited lifetime in accord with its high charge density; stable silicate anions, even in solution, are polymeric species. The components in aqueous silicate solutions can be identified by trimethylsilylation of their hydroxy groups to give volatile derivatives, which can then be separated and analyzed by gas–liquid chromatography. Such studies indicate that aqueous silicate solutions contain linear, cyclic, and cage polymeric ions in proportions depending on the pH, concentration, and temperature. Examples of silicate structures inferred from such studies include the trigonal prismatic "double-three rings" structure and the cubic "double-four rings" structure (Fig. 3.3). Actual species isolated from such studies include trigonal prismatic Si_6O_9-$(OSiMe_3)_6$, cyclic $Si_6O_6(OSiMe_3)_{12}$, and cubic $Si_8O_{12}(OMe)_8$.

3.5.3 Chemistry of Silicate Minerals and Related Structures

A major portion (~85%) of the earth's crustal rocks and their breakdown products, including soils, clays and sands, is composed almost entirely of silicate

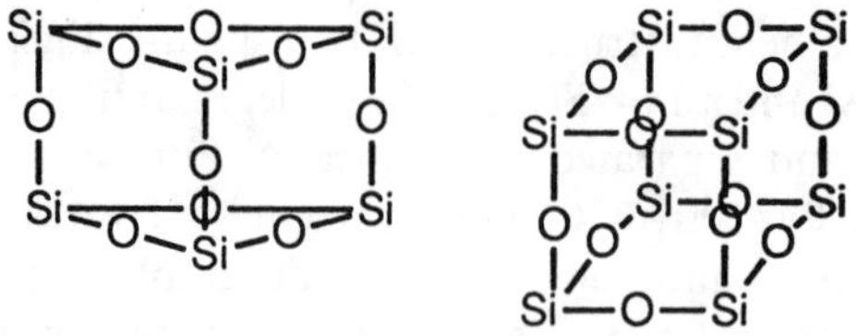

"Double-three rings"
Trigonal Prism

"Double-four rings"
Cube

Figure 3.3. Some silicate structures.

minerals and silica. The predominance of silicates and aluminosilicates is reflected in the fact that oxygen, silicon, and aluminum are the most abundant elements in the earth's crust. In almost all the silicate mineral structures, each silicon atom is bounded to four oxygen atoms in SiO_4 tetrahedron as building blocks.

Silicate minerals can be classified on the basis of the number of oxygen atoms of the SiO_4 tetrahedron which are shared with adjacent silicons in Si—O—Si structural units (Table 3.1).

3.5.3.1 Simple Orthosilicates and Other Acyclic Discrete Silicates (neso-Silicates and soro-Silicates)

A number of crystalline orthosilicates with discrete SiO_4^{4-} ions in their crystal structures are known. Examples include derivatives of the general type $M_2^{II}SiO_4$ such as Be_2SiO_4 (*phenacite*) and Zn_2SiO_4 (*willemite*), in which Be or Zn is four-coordinate; $ZrSiO_4$ (*zircon*), in which Zr is eight-coordinate; and more complicated structures such as *olivine* ($9Mg_2SiO_4 \cdot Fe_2SiO_4$) and the *garnets*, $M_3^{II}M_2^{III}(SiO_4)_3$ (M^{II} = Ca, Mg, Fe; M^{III} = Al, Cr, Fe), in which M^{II} is eight-coordinate and M^{III} is six-coordinate. Discrete binuclear $Si_2O_7^{6-}$ pyrosilicate or disilicate anions are found in *thortvetite* ($Sc_2Si_2O_7$) and similar lanthanide (Ln) disilicates $Ln_2Si_2O_7$, *barysilite* ($Pb_3Si_2O_7$), and *hemimorphite* ($Zn_4(OH)_2$-Si_2O_7; the Si—O—Si angle can vary from 131° to 180° in these structures.

Table 3.1 Classification of Silicate Minerals

Shared Oxygen Atoms of the SiO_4 Units	Structure	Name
0	Discrete SiO_4 units	*neso*-Silicates
1	Discrete Si_2O_7 units	*soro*-Silicates
2	Closed $(SiO)_n$ ring structures	*cyclo*-Silicates
2	Infinite chains and ribbons	*ino*-Silicates
3	Infinite sheets	*phyllo*-Silicates
4	Infinite three-dimensional networks	*tecto*-Silicates

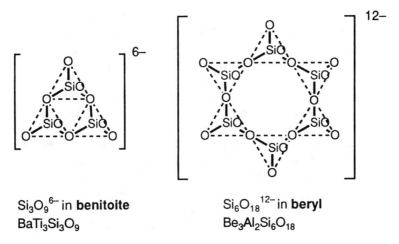

Si$_3$O$_9^{6-}$ in **benitoite**
BaTi$_3$Si$_3$O$_9$

Si$_6$O$_{18}^{12-}$ in **beryl**
Be$_3$Al$_2$Si$_6$O$_{18}$

Figure 3.4. Schematic representations of cyclic silicate anions found in benitoite and beryl.

3.5.3.2 Cyclic Silicates (cyclo-Silicates)

Cyclic silicate anions of the general formula $(SiO_3)_n^{2n-} = (SiO_2O_{2/2})_n^{2n-}$ are found in minerals such as *benitoite* (BaTiSi$_3$O$_9$: $n = 3$) and α-*wollastonite* (Ca$_3$Si$_3$O$_9$: $n = 3$) with six-membered Si$_3$O$_3$ rings and *beryl* (Be$_3$Al$_2$Si$_6$O$_{18}$: $n = 6$) with twelve-membered Si$_6$O$_6$ rings; these anions are depicted schematically in Figure 3.4.

3.5.3.3 Infinite Chain and Ribbon Silicates (ino-Silicates): Asbestos

The two main infinite chain silicates are the *pyroxenes*, which contain infinite single-strand chains $(SiO_3)_n^{2n-} = (SiO_2O_{2/2})_n^{2n-}$ (Figure 3.5), and the *amphiboles*, which contain infinite double-strand chains $(Si_4O_{11})_n^{6n-} = [(SiO_2O_{2/2}) - (SiOO_{3/2})]_n^{3n-}$. The general formula $(SiO_3)_n^{2n-}$ of the infinite chain pyroxenes is the same as that of the cyclic silicates already discussed, just as the general formula $(CH_2)_n$ for infinite chain polyethylene is the same as that of the cycloalkanes in organic chemistry. Silicates constructed from $(SiO_3)_n^{2n-} = (SiO_2O_{2/2})_n^{2n-}$

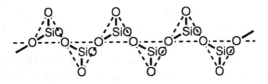

Figure 3.5. Schematic representation of the infinite $(SiO_3)_n^{2n-} = (SiO_2O_{2/2})_n^{2n-}$ chains found in pyroxenes.

units, whether rings or chains, are often called "metasilicates." Examples of pyroxenes include *β-wollastonite*, $CaSiO_3$, *enstatite*, $MgSiO_3$, *diopside*, $CaMg$-$(SiO_3)_2$, *spodumene*, $LiAl(SiO_3)_2$, (an important lithium ore), and *jadeite*, $NaAl(SiO_3)_2$. The amphiboles always have some OH groups attached to the cations; an example of an amphibole is *tremolite*, $Ca_2Mg_3(Si_4O_{11})_2(OH)_2$. These chain silicate structures have strong forces within the chains but weak interchain forces. As a result, they tend to form fibers by cleaving readily in the directions parallel to the chains. A generic but imprecise term for fibrous silicates of this type is *asbestos* (from Greek $\alpha\sigma\beta\epsilon\sigma\tau\sigma$, unquenchable, from the nonflammability of asbestos), a term used to describe fibrous silicates such as *chrysotile* (white serpentine), $Mg_6Si_4O_{10}(OH)_2$, and *crocidolite* (blue amphibole), $Na_2Fe_3^{II}Fe_2^{III}Si_8O_{22}(OH)_2$.[28] Such materials were frequently used in the past for fire-resistant fibrous construction materials, but are now considered to be dangerous because exposure to the airborne fibers has been shown to produce cancer.

3.5.3.4 Infinite Sheet Silicates (*phyllo-Silicates*): Clays and Micas

Infinite sheet silicates are constructed from the anions $(Si_2O_5)_n^{2n-} = (SiOO_{3/2})_n^{2n-}$ with cations between the infinite sheets. These silicates cleave readily into thin plates and include the *clays* and *micas*. Infinite sheet clay minerals include *montmorillonite*, *kaolin*, and *talc*, all of which are very abundant in nature. The layers in clay structures can intercalate molecules and ions and thus frequently have ion exchange properties, often with high catalytic applications. *Pillared clays* are clays intercalated with large cations such as $Nb_6Cl_{12}^{2+}$ or even $[Al_{13}O_4(OH)_{24}(H_2O)_{12}]^{7+}$ (Section 9.3), which form the pillars, which increase the interlayer distance. The expanded layers in pillared clays lead to improved diffusion, sorption and catalytic properties.

3.5.3.5 Three-Dimensional Framework Silicates (*tecto-Silicates*): Aluminosilicates Including Zeolites

Infinite condensation of SiO_4^{4-} tetrahedra in one dimension gives infinite $(SiO_3)_n^{2n-} = (SiO_2O_{2/2})_n^{2n-}$ chains with two terminal and 2/2 bridging oxygens per silicon atom, and in two dimensions it gives infinite $(Si_2O_5)_n^{2n-} = (SiOO_{3/2})_n^{2n-}$ sheets with only one terminal and 3/2 bridging oxygens per silicon atom, as indicated earlier. Continuing this infinite condensation in all three dimensions so that all oxygen atoms are bridging oxygen atoms leads to $(SiO_2)_n = (SiO_{4/2})_n$; this, of course, is silica. However, substitution of some of the $Si_{4/2}$ units in this three-dimensional infinite framework by isoelectronic $AlO_{4/2}^-$ units leads to an anionic framework with positive counterions distributed throughout; such materials, called *aluminosilicates*, are exemplified by the *feldspars*, *ultramarines*, and *zeolites*. Feldspars, which are major components of igneous rocks, include *orthoclase*, $KAlSi_3O_8 = K(AlO_2)(SiO_2)_3$, *albite*, $NaAlSi_3O_8 = Na(AlO_2)(SiO_2)_3$, *anorthite*, $CaAl_2Si_2O_8 = Ca(AlO_2)_2(SiO_2)_2$, and *celsian*, $BaAl_2Si_2O_8 = Ba(AlO_2)_2(SiO_2)_2$. *Ultramarines*, such as the semiprecious deep-blue gem *lapis lazuli*, are aluminosilicates incorporating sulfur as the colored radical anions S_3^- and S_2^-; the color of ultramarine shifts from green to blue as the ratio of S_3^- to S_2^- increases.[29]

The most important class of three-dimensional framework silicates includes the *zeolites* (from Greek $\zeta\epsilon\iota\nu = zein$, to boil, and $\lambda\iota\theta o s = lithos$, stone, because zeolites appeared to boil in a blowpipe flame); zeolites are also called *porotectosilicates*.[30, 31] Zeolites are aluminosilicates with framework structures enclosing cavities occupied by large ions and water molecules, both of which have considerable freedom of movement, thereby allowing ion exchange and reversible dehydration. The framework of the zeolites is an open arrangement of corner-sharing tetrahedra in which some $SiO_{4/2}$ tetrahedra are replaced by $AlO_{4/2}^-$ tetrahedra with sufficient cations in the cavities to achieve electroneutrality. The idealized formula of a zeolite can thus be represented as $M_{x/n}^{n+}[(AlO_2)_x(SiO_2)_y] \cdot nH_2O$. Magic angle ^{29}Si NMR studies on zeolites indicate five distinct $Si(OAl\equiv)_n(OSi\equiv)_{4-n}$ $(0 \leq n \leq 4)$ structures corresponding to the tetrahedral $SiO_{4/2}$ units. Aluminum and silicon alternate on the tetrahedral sites such that there are only Si—O—Si and Si—O—Al links but no Al—O—Al links (Loewenstein's rule). Zeolites are of considerable interest, since structures having cavities, tunnels, and pores of precisely defined atomic scale dimensions can be synthesized reproducibly. Such zeolites can be used as molecular sieves to remove water or other small molecules selectively, to separate normal from branched-chain paraffins, to generate highly dispersed metal catalysts, and to promote specific size-dependent chemical reactions.

There are a vast number of known zeolites, including approximately 40 natural and more than 100 synthetic zeolites. *Faujasite*, $Na_{13}Ca_{11}Mg_9KAl_{55}$-$Si_{137}O_{384} \cdot 235H_2O$, is an example of a naturally occurring zeolite. *Molecular sieves* are zeolites that have been dehydrated by heating in vacuum at $\sim350°C$. Zeolites are made commercially by crystallizing aqueous gels of mixed alkaline silicates and aluminates at 60–100°C.[32, 33] *High silica zeolites* include ZSM-5 (Mobil Oil) and Silicalite (Union Carbide); they are made hydrothermally using large tetraalkylammonium ions as templates for crystal growth. The general formula of high silica zeolites is $H_x[(AlO_2)_x(SiO_2)_{96-x}] \cdot 16H_2O$ $(x \approx 3)$ and they have pore diameters around 5.5 Å.

The catalytic, sorption, and ion exchange properties of zeolites are strongly dependent on the Al/Si ratio. Treatment of such zeolites with $SiCl_4$ or $AlCl_3$ can be used to modify the Al/Si ratio by removing aluminum or silicon, respectively, as their volatile chlorides. Silica can be selectively leached from zeolites by aqueous potassium hydroxide. Silicon can also be replaced with gallium (as $GaO_{4/2}^-$ units) by treatment with sodium gallate (Section 9.3).

3.6 Other Oxygen Compounds of Silicon, Germanium, Tin, and Lead

3.6.1 Oxides and Hydroxides of Germanium, Tin, and Lead

Structural modifications of GeO_2 are known corresponding to the quartz, cristobalite, and stishovite structures of silica; GeO_2 is acidic but less so than SiO_2. There are three different structural modifications of SnO_2, of which the rutile form with six-coordinate tin (*cassiterite*) is the most common; SnO_2 is amphoteric. Only the rutile structure is known for PbO_2, which is relatively inert to chemical attack other than by reducing agents to give Pb(II). There is little evidence for true hydroxides $E(OH)_4$ (E = Ge, Sn, Pb); hydrolysis products of these elements in their +4 oxidation states are best regarded as hydrous oxides.

3.6.2 Oxoanions and Hydroxoanions of Germanium, Tin, and Lead

The chemistry of germanates is much less extensive than that of silicates. Germanium exhibits coordination numbers greater than 4 more readily than silicon. For example, in the potassium germanate of stoichiometry $K_2Ge_8O_{17}$, two of the eight germanium atoms are five-coordinate; the remaining germanium atoms are four-coordinate. Hydroxogermanates containing the anion $Ge(OH)_6^{2-}$ are known. Major germanate ions found in dilute aqueous solution include $GeO(OH)_3^-$, $GeO_2(OH)_2^{2-}$, and the polynuclear ion $\{[Ge(OH)_4]_8(OH)_3\}^{3-}$.

Stannates and plumbates of stoichiometry K_2MO_3 are obtained by fusion of the corresponding oxides MO_2 (E = Sn, Pb) with K_2O. The structures of these stannates and plumbates have chains of edge-shared $MOO_{4/2}$ ($=$"MO_3") square pyramids in which two-thirds of the oxygen atoms bridge two metal atoms. Hydrated stannates and plumbates of the stoichiometry $K_2MO_3 \cdot 3H_2O$ (M = Sn, Pb) contain octahedral hexahydroxometallate(IV) anions $M(OH)_6^{2-}$.

3.6.3 Other Oxygen Derivatives: Alkoxides, Carboxylates, and so on

Alkoxides of the stoichiometry $E(OR)_4$ can be obtained from the corresponding tetrachlorides by standard methods of alkoxide synthesis, for example:

$$ECl_4 + 4ROH + 4R_3N \rightarrow E(OR)_4 + 4R_3NH^+Cl^- \tag{3.9}$$

The silicon compounds $Si(OR)_4$ are also called *silicate esters*; the methyl ester, $Si(OMe)_4$, is highly toxic and can destroy the optic nerve, leading to blindness. The surface hydroxyl groups of silica or glass can be removed using analogous alkylation reactions. In general, alkoxides of these elements are readily hydrolyzed by water; such hydrolysis reactions are the basis of *sol–gel methods* for the preparation of oxide ceramics.[34]

Carboxylates of the general stoichiometry $E(CO_2R)_4$ (E = Si, Ge, Sn, Pb) are known. The lead tetracarboxylates are important as oxidants in organic chemistry in view of the facile reduction of Pb(IV) to Pb(II). Thus colorless, moisture-sensitive lead tetraacetate, $Pb(O_2CMe)_4$, is a strong but selective oxidizing agent,[35, 36] which is obtained by treatment of Pb_3O_4 (Section 3.8.4) with glacial acetic acid or by electrolytic oxidation of Pb(II) in acetic acid solution. Lead tetrakis(trifluoroacetate), $Pb(O_2CCF_3)_4$, is an even stronger oxidizing agent than lead tetraacetate and can oxidize saturated hydrocarbons (e.g., n-heptane, to give n-$C_7H_{15}CO_2CF_3$). The mechanisms of lead tetracarboxylate oxidations can either involve free radicals or cationic species such as $Pb(CO_2CH_3)_3^+$.

The remaining elements (Si, Ge, Sn) also form tetracarboxylates. In the case of the tetraacetates of the type $E(CO_2CH_3)_4$, the silicon and germanium compounds are four-coordinate with monodentate acetate groups, the lead compound is eight-coordinate with bidentate acetate groups, and the tin compound has an intermediate structure with highly distorted eight-coordinate tin.

Tin(IV) is basic enough to form derivatives of a number of the strong inorganic oxoacids. Tin(IV) sulfate, $Sn(SO_4)_2 \cdot 2H_2O$, is obtained by oxidizing tin(II) sulfate; it is extensively hydrolyzed by water. Anhydrous tin(IV) nitrate, $Sn(NO_3)_4$, is a colorless volatile solid obtained by reaction of N_2O_5 with $SnCl_4$; it reacts rapidly with organic matter because of the strong oxidizing properties of the covalently bonded nitrate groups. The structure of tin(IV) nitrate contains discrete $Sn(NO_3)_4$ molecules with four bidentate nitrate groups, leading to eight-coordinate tin.

3.7 Organometallic Derivatives of Silicon, Germanium, Tin, and Lead

Organometallic derivatives of silicon, germanium, tin, and lead in their usual +4 oxidation states have the general formula $R_{4-n}EX_n$ ($0 \leq n \leq 3$) in which R is an alkyl or aryl group and X is another monovalent group such as hydrogen, halogen, alkoxy, dialkylamino, alkylthio, or even transition metal groups such as $M(CO)_5$ (M = Mn, Re) or $M'(CO)_2Cp$ (M' = Fe, Ru; Cp = cyclopentadienyl or substituted cyclopentadienyl).

3.7.1 Organosilicon and Organogermanium Compounds: Silicones

The Si—C bond dissociation energies are less than those of C—C bonds but are still quite high (250–335 kJ/mol). The chemical reactivity of Si—C bonds is greater than that of C—C bonds for the following reasons.

1. The polarity of the $Si^{\delta+}$—$C^{\delta-}$ bond allows easier nucleophilic attack on silicon and electrophilic attack on carbon than is the case with C—C bonds.

2. Displacement reactions on silicon are facilitated by the ability of this element to form five-coordinate transition states using d orbitals. However, reactions of organosilicon compounds have no mechanistic pathway analogous to first-order S_N1 reactions of carbon compounds because of the difficulty in forming siliconium ion intermediates of the type R_3Si^+.

Other important features of organosilicon (and organogermanium) chemistry include the following.

1. Groups of the type R_3Si and R_3Ge migrate much more readily than alkyl or aryl groups. In fact, the migratory ability of R_3Si groups has been compared to that of protons.

2. Free radicals are less important in organosilicon chemistry than in carbon chemistry.

Alkyl- and arylsilicon halides (alkyl- and arylhalosilanes),[37] R_nSiX_{4-n}, are obtained by reactions of $SiCl_4$ with a deficiency of the corresponding Grignard reagent, RMgX, and are generally liquids. The methyl derivatives, $(CH_3)_nSiCl_{4-n}$, can be prepared on an industrial scale by the following exothermic reaction, known as the *Rochow reaction*:

$$CH_3Cl + Si(Cu) \xrightarrow{300°C} (CH_3)_nSiCl_{4-n} \qquad (3.10)$$

A typical $(CH_3)_nSiCl_{4-n}$ product from this reaction contains $\sim 70\%$ $(CH_3)_2SiCl_2$ (bp 70.3°C), $\sim 12\%$ CH_3SiCl_3 (bp 66.4°C), and $\sim 5\%$ $(CH_3)_3SiCl$ (bp 57.9°C) and can be separated by fractional distillation. Relative product yields can be altered by modifying the reaction conditions or adding HCl to favor formation of $MeSiHCl_2$.

Alkyl- and arylsilicon halides are readily hydrolyzed by water to give silanols $R_nSi(OH)_{4-n}$, which readily condense under the hydrolysis conditions to form siloxanes with Si—O—Si bonds,[38] for example:

$$(CH_3)_3SiCl + H_2O \rightarrow (CH_3)_3SiOH \text{ (bp 99°C)} + HCl \qquad (3.11a)$$

$$2(CH_3)_3SiOH \rightarrow (CH_3)_3Si—O—Si(CH_3)_3 \text{ (bp 100.8°C)} + H_2O \qquad (3.11b)$$

Such hydrolysis reactions can lead to linear, cyclic, and cross-linked polymers of varying molecular weights and properties; these polymers are generically called *silicones* and have high thermal stabilities, high dielectric strengths, and high resistance to oxidation and chemical attack. Thus hydrolysis of $(CH_3)_2SiCl_2$ usually gives high polymers of the stoichiometry $[(CH_3)_2SiO]_n$. Addition of $(CH_3)_3SiCl$ to $(CH_3)_2SiCl_2$ before hydrolysis gives linear siloxanes with $(CH_3)_3Si—O—$ as a chain-terminating group, for example, $(CH_3)_3SiO-[(CH_3)_2SiO]_nSi(CH_3)_3$ $(0 \leq x \leq 4)$. Conversely, cross-linking is achieved by addition of CH_3SiCl_3 to $(CH_3)_2SiCl_2$ before hydrolysis; such cross-linking occurs through $CH_3SiO_{3/2}$ structural units (Figure 3.6). In addition, hydrolysis

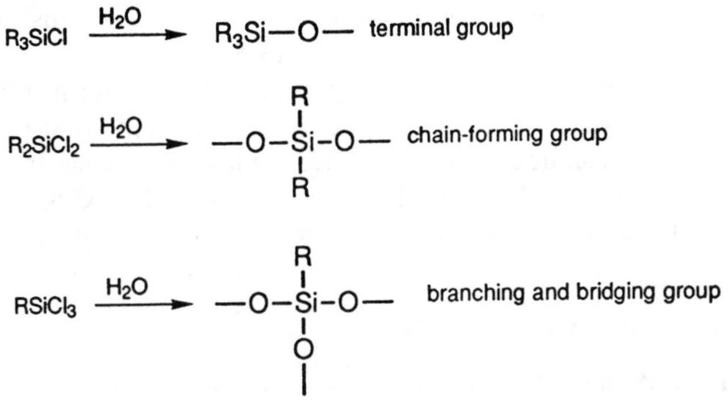

$$R_3SiCl \xrightarrow{\text{H}_2\text{O}} R_3Si\text{—O—} \quad \text{terminal group}$$

$$R_2SiCl_2 \xrightarrow{\text{H}_2\text{O}} \begin{matrix} & R & \\ & | & \\ \text{—O—} & Si & \text{—O—} \\ & | & \\ & R & \end{matrix} \quad \text{chain-forming group}$$

$$RSiCl_3 \xrightarrow{\text{H}_2\text{O}} \begin{matrix} & R & \\ & | & \\ \text{—O—} & Si & \text{—O—} \\ & | & \\ & O & \\ & | & \end{matrix} \quad \text{branching and bridging group}$$

Figure 3.6. Generation of structural units in silicone polymers by hydrolysis of R_nSiCl_{4-n} derivatives.

of $RSiCl_3$ derivatives can lead to organosilsesquioxanes with cage structures based on polyhedra, such as the cube and various prisms, in which three Si—O—Si edges meet at each silicon vertex.[39]

Numerous organosilicon derivatives with Si—Si bonds are known. The simplest such compound is *hexamethyldisilane*, $(CH_3)_3Si\text{—}Si(CH_3)_3$ (bp 114°C), which is made by the following reaction, analogous to the Wurtz coupling in organic chemistry;

$$2(CH_3)_3SiCl + 2Li \xrightarrow{\text{tetrahydrofuran}} (CH_3)_3Si\text{—}Si(CH_3)_3 + 2LiCl$$

$$(3.12)$$

Reactions of R_2SiCl_2 derivatives with alkali metals under similar conditions can give either cyclic polysilanes or polyalkylsilylene polymers.[40] Thus reaction of $(CH_3)_2SiCl_2$ with lithium metal in tetrahydrofuran at 0°C gives dodecamethylcyclohexasilane, $[(CH_3)_2Si]_6$ which melts at 250°C.[41] Reaction of $(CH_3)_2$-$SiCl_2$ with sodium metal gives poly(dimethylsilylene), $[\text{—}Si(CH_3)_2\text{—}]_n$, which undergoes pyrolysis to silicon carbide above 800°C through a poly(carbosilane) intermediate $[\text{—}CH_2\text{—}Si(H)(CH_3)\text{—}]_n$; it thus is a useful precursor for the fabrication of silicon carbide fibers.[42]

Organosilicon compounds with Si=Si double bonds can be isolated if the very reactive Si=Si double bond is protected by very bulky substituents.[7,43] An example of such a compound is yellow $Mes_2Si=SiMes_2$ (Mes = mesityl = 2,4,6-trimethylphenyl), which can be prepared by the following photolytic reaction[44]:

$$2Mes_2Si(SiMe_3)_2 \xrightarrow[254 \text{ nm}]{h\nu} 2Me_3Si\text{—}SiMe_3 + Mes_2Si=SiMes_2 \quad (3.13)$$

The Si=Si double bond in $Mes_2Si=SiMes_2$ undergoes addition reactions upon treatment with hydrogen halides, halogens, alcohols, alkynes, and CH_2 (from diazomethane). In addition, $Mes_2Si{=}SiMes_2$ is very oxygen sensitive owing to the reactivity to atmospheric oxygen of its $Si{=}Si$ bond.

Organogermanium chemistry closely resembles organosilicon chemistry except that organogermanium compounds tend to be somewhat less thermally stable and more chemically reactive. Germanium analogues of the silicon polymers appear to be unknown. Thus hydrolysis of $(CH_3)_2GeCl_2$ is reversible and incomplete leading to the cyclic tetramer, which melts at 92°C.

3.7.2 Organotin Compounds[45,46]

Organotin compounds differ from organosilicon and organogermanium compounds because of the greater tendency of tin to exhibit coordination numbers greater than 4 and to ionize to form cationic species. Thus trialkyltin halides, R_3SnX, are associated in the solid state by anion bridging, leading to trigonal bipyramidal five-coordinate tin. In addition trialkyltin halides form 1:1 and 1:2 adducts with Lewis bases and are toxic in biological systems. Dimethyltin difluoride, $(CH_3)_2SnF_2$, has a polymeric structure with six-coordinate octahedral tin, bridging fluorine atoms (i.e., $(CH_3)_2SnF_{4/2} = (CH_3)_2SnF_2$), and a linear $CH_3{-}Sn{-}CH_3$ unit. However, $(CH_3)_2SnX_2$ (X = Cl, Br) have monomeric structures with four-coordinate tetrahedral tin. Nevertheless, dialkyltin dihalides gives conducting solutions in water which contain octahedral tin cation of the general type $trans\text{-}R_2Sn(H_2O)_4^{2+}$. Organotin carboxylates and dialkylphosphinates exhibit a number of complicated oligomeric structures with octahedrally coordinated tin atoms including $[PhSn(O)O_2CC_6H_{11}]_6$ containing a hexagonal prismatic Sn_6O_6 core, $\{[Bu^nSn(O)O_2CC_6H_{11}]_2Bu^nSn(O_2CC_6H_{11})_3\}_2$ containing an Sn_4O_4 ladder structure, $[Bu^nSn(O)O_2P(C_6H_{11})_2]_4$ containing a cubic Sn_4O_4 core, and $\{Bu^nSn(S)(O_2PPh_2)]_3O\}_2Sn$ containing an $Sn(Sn_3S_3O)_2$ double cube (Figure 3.7).[47] In these structures tin atoms are bonded directly to oxygen and/or sulfur atoms but not to other tin atoms.

A number of organotin compounds containing $Sn{-}Sn$ bonds are known. Treatment of $(CH_3)_2SnCl_2$ with sodium metal in liquid ammonia gives a material of stoichiometry $[(CH_3)_2Sn]_n$ that consists mainly of linear molecules with $12 \leq n \leq 20$ as well as some cyclic $[(CH_3)_2Sn]_6$. Related species with other alkyl and aryl groups include $[Et_2Sn]_n$ ($n = 6, 9$), $[Ph_2Sn]_n$ ($n = 5, 6$), and $[Bu^t_2Sn]_4$. There is no evidence for structures with branched tin chains.

Trialkyltin hydrides, R_3SnH (e.g., R = n-butyl), are obtained by $LiAlH_4$

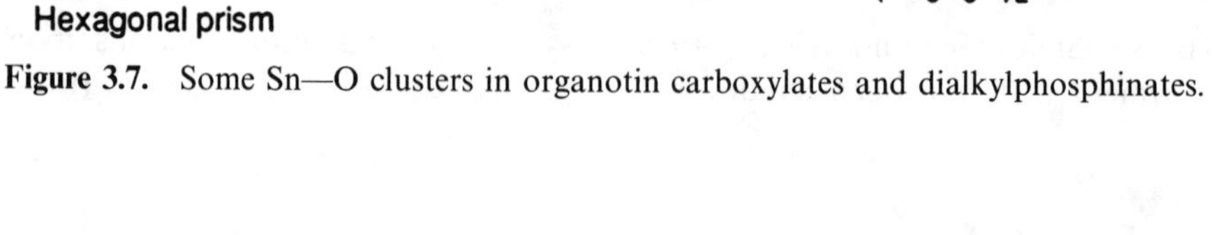

Sn₆O₆
Hexagonal prism   Sn₄O₄ Ladder   Sn₄O₄ Cube   Sn(Sn₃S₃O)₂ Double Cube

Figure 3.7. Some Sn—O clusters in organotin carboxylates and dialkylphosphinates.

reduction of the corresponding trialkyltin halides. They are useful reducing agents in organic chemistry; some of their reductions involve free radical mechanisms. In addition, the Sn—H bond in R_3SnH derivatives can add to the C=C bond in olefins:

$$R_3SnH + R_2'C=CR_2'' \rightarrow R_3SnCR_2'CR_2''H \tag{3.14}$$

This so-called *hydrostannation* reaction follows a radical chain mechanism propagated by $R_3Sn\cdot$ radicals.

3.7.3 Organolead Compounds[48]

Organolead chemistry is much less extensive than organotin chemistry largely because of the weakness of the Pb—C bond. *Tetramethyllead*, Me_4Pb (mp $-30°C$, bp $110°C$), and *tetraethyllead* (Et_4Pb, bp $82°C/13$ torr), have been made in large quantities for use as antiknock compounds in gasolines, but this application has been phased out almost completely in many countries because of the toxicity of lead in the environment.[12] These tetraalkyllead compounds are nonpolar highly toxic liquids, which can be obtained on a large scale either by treatment in an autoclave at $80-100°C$ of a sodium–lead alloy with the corresponding alkyl chloride, RCl (R = Me, Et), or by electrolysis of $NaAlR_4$ with a lead anode and a mercury cathode. The thermal decomposition of tetraalkyllead compounds (e.g., Me_4Pb above $200°C$ or Et_4Pb above $110°C$) provides a source of alkyl radicals.

3.8 Divalent Compounds of Silicon, Germanium, Tin, and Lead

3.8.1 Divalent Silicon

Divalent silicon derivatives are generally thermodynamically unstable under normal conditions. However, halides of divalent silicon, SiX_2 (X = F, Cl) can be obtained by high temperature reactions of the corresponding tetrahalide with elemental silicon followed by quenching the unstable SiX_2 halide at low temperatures. For example, passing SiF_4 vapor over elemental silicon at $\sim 1150°C$ and low pressures gives SiF_2, which is stable for a few minutes at $\sim 10^{-6}$ atm pressure; SiF_2 is diamagnetic with an F–Si–F angle of $101°$.

Organometallic derivatives of divalent silicon include at least formally the binuclear derivatives $R_2Si=SiR_2$ with an Si=Si double bond as discussed in the preceding section. Examples of mononuclear organometallic derivatives of divalent silicon include the very air-sensitive bis(cyclopentadienyl)silicon derivative $(\eta^5\text{-}Me_5C_5)_2Si$, which exists in two forms, one with bent C_5 rings and the other with parallel C_5 rings.[49]

3.8.2 Divalent Germanium

Divalent germanium is considerably more stable than divalent silicon. Thus the halides GeX_2 (X = F, Cl, Br, I) are all stable at room temperature. The fluoride, GeF_2, is a white crystalline solid (mp 111°C), which can be obtained by reaction of hydrogen fluoride or GeF_4 with elemental germanium at elevated temperature; GeF_2 is a fluorine-bridged polymer in which distorted trigonal pyramidal $GeFF_{2/2}$ structural units form spiral chains. Reaction of GeF_2 with fluoride ion gives the hydrolytically stable $Ge^{II}F_3^-$, which is oxidized by air in fluoride solution to give $Ge^{IV}F_6^{2-}$. A related chloride ion $GeCl_3^-$ is also known. Yellow flaky solid GeI_2 is obtained by reaction of GeO_2 with a mixture of hydriodic and hypophosphorus acids; it is stable in dry air and forms yellow phosphine adducts $R_3P{\rightarrow}GeI_2$.[50] Reaction of GeO_2 with elemental germanium at 1000°C gives germanium(II) oxide, GeO. Other insoluble divalent germanium derivatives include the sulfide GeS and Ge(β-diketonate)$_2$ chelates.

3.8.3 Divalent Tin

Divalent tin is stable in aqueous solution, although it undergoes the following hydrolytic reactions not involving a change in the tin oxidation state:

$$Sn^{2+} + H_2O \rightleftarrows Sn(OH)^+ + H^+ \qquad \log K = -3.7 \qquad (3.15a)$$

$$3Sn^{2+} + 4H_2O \rightleftarrows Sn_3(OH)_4^{2+} + 4H^+ \qquad \log K = -6.8 \qquad (3.15b)$$

All tin(II) aqueous solutions are readily oxidized by oxygen to tin(IV) and thus normally contain some tin(IV) unless protected from the air. Hydrous tin(II) oxide can be precipitated from Sn(II) aqueous solutions with aqueous ammonia. This precipitate dissolves in strong aqueous alkalies to form *stannites*, which are strong reducing agents believed to contain $Sn(OH)_6^{4-}$ (E° = −0.93 V for the $Sn(OH)_6^{2-}/Sn(OH)_6^{4-}$ couple in basic solution). Tin(II) hydroxide, $Sn(OH)_2$, can be obtained as a white amorphous solid by the nonaqueous reaction of Me_3SnOH with $SnCl_2$ in tetrahydrofuran:

$$2Me_3SnOH + SnCl_2 \xrightarrow{\text{tetrahydrofuran}} Sn(OH)_2 + 2Me_3SnCl \qquad (3.16)$$

All the halides SnX_2 are known isolable compounds and are obtained by treatment of tin metal with the corresponding hydrogen halide. Many salts of the trigonal pyramidal $SnCl_3^-$ ion are known. The $SnCl_3^-$ ion has a lone pair on the tin atom. For this reason, it functions as a strong field ligand in transition metal chemistry, forming complexes such as $Pt^{II}(SnCl_3)_5^{3-}$ and $Os^{II}(SnCl_3)_6^{4-}$. The structure of tin(II) chloride dihydrate also contains trigonal pyramidal tin, that is, $SnCl_2 \cdot 2H_2O = [SnCl_2(OH_2)] \cdot H_2O$.

 The most important organotin(II) derivatives are the bis(cyclopentadienyl)tin derivatives, $(\eta^5\text{-}R_5C_5)_2Sn$ (R = H, Me, Ph, etc.), which are obtained from $SnCl_2$ and the corresponding cyclopentadienide anion.[51] They all have angular structures except for yellow $(\eta^5\text{-}Ph_5C_5)_2Sn$, in which the large size of the multiple phenyl substituents makes parallel rings more sterically favorable. The air-sensitive red tin(II) derivative, $Sn[CH(SiMe_3)_2]_2$ (mp 136°C), is also known; it forms transition metal complexes such as $[(Me_3Si)_2CH]_2Sn \rightarrow M(CO)_5$ (M = Cr, Mo. W).[52]

3.8.4 Divalent Lead

The most stable oxidation state of lead in aqueous solution is the divalent state; lead has a well-defined chemistry of Pb^{2+} in aqueous solution. In water Pb^{2+} undergoes partial hydrolysis:

$$Pb^{2+} + H_2O \rightleftarrows PbOH^+ + H^+ \qquad \log K = -7.9 \qquad (3.17)$$

However, the initial $PbOH^+$ hydrolysis product aggregates to form $Pb_4(\mu^3\text{-}OH)_4$, which contains a Pb_4O_4 cube. Lead(II) salts insoluble in water include the 1:1 salts $PbSO_4$ and $PbCrO_4$, similar to the corresponding Ba^{2+} salts; lead(II) salts sparingly soluble in water include the halides PbF_2 and $PbCl_2$; lead(II) salts readily soluble in water include the perchlorate $Pb(ClO_4)_2 \cdot 3H_2O$, the nitrate $Pb(NO_3)_2$, and the acetate $Pb(O_2CMe)_2 \cdot 3H_2O$. Lead(II) oxide, PbO, has two modifications; red tetragonal *litharge* (density 9.355 g/cm³), with a layer structure, and yellow orthorhombic *massicott* (density 9.642 g/cm³), with a chain structure. The mixed valence lead(II, IV) oxide, Pb_3O_4 (density 8.924 g/cm³), known as "red lead," is used as a pigment and primer where the risk of lead poisoning is not a concern; it is obtained by heating PbO or PbO_2 in air and behaves chemically as a mixture of PbO and PbO_2. The crystal structure of Pb_3O_4 contains $Pb^{IV}O_2O_{4/2}$ octahedra linked in chains by sharing opposite edges with Pb^{II} atoms. Organometallic derivatives of divalent lead include bis(cyclopentadienyl) derivatives such as $(C_5H_5)_2Pb$, which has a solid state structure containing a zigzag chain, with the lead atoms bridges by half of the cyclopentadienyl rings, and purple monomeric $Pb[CH(SiMe_3)_2]_2$, in which Pb(II) is stabilized by the bulky bis(trimethylsilyl)methyl groups.[50]

References

1. Glockling, F., *The Chemistry of Germanium*, Academic Press, London, 1969.

2. Harrison, P. G., Ed., *Chemistry of Tin*, Blackie, Glasgow, 1989.

3. Kumar Das, V. G.; Ng, S. W.; Gielen, M., *Chemistry and Technology of Silicon and Tin*, Oxford University Press, Oxford, 1992.

4. Stone, H. E. N., Alloy Systematics in Relation to the Long Periodic Table, *Acta Metall.*, 1979, **27**, 259.

5. Allred, A. L., Electronegativity Values from Thermochemical Data, *J. Inorg. Nuclear Chem.*, 1961, **17**, 215–221.

6. Barrau, J.; Escudié, J.; Satgé, J., Multiply Bonded Germanium Species. Recent Developments, *Chem. Rev.*, 1990, **90**, 283–319.

7. Cowley, A. H.; Norman, N. C., The Synthesis, Properties, and Reactivities of Stable

Compounds Featuring Double Bonding Between Heavier Group 14 and 15 Elements, *Prog. Inorg. Chem.*, 1986, **34**, 1–63.

8. Tandura, S. N.; Voronkov, M. G.; Alekseev, N. V., Molecular and Electronic Structure of Penta- and Hexacoordinate Silicon Compounds, *Top. Curr. Chem.*, 1986, **131**, 99–189.

9. Gielen, M., From Kinetics to the Synthesis of Chiral Organotin Compounds, *Acc. Chem. Res.*, 1973, **6**, 198–202.

10. Deere, W. A.; Howie, R. A.; Zussman, J., *An Introduction to the Rock Forming Minerals*, Longmans, London, 1966.

11. Mason, B.; Berry, L. G., *Elements of Mineralogy*, Freeman, San Francisco, 1968.

12. Harrison, R. M.; Laxen, D. P. H., *Lead Pollution*, Chapman and Hall, London, 1981.

13. Harris, R. K.; Kennedy, J. D.; McFarlane, W., in *NMR and the Periodic Table*, R. K. Harris and B. E. Mann, Eds, Academic Press, New York, 1978, Chapter 10.

14. Gibb, T. C., *Principles of Mössbauer Spectroscopy*, Chapman and Hall, London, 1976.

15. Corbett, J. D., Polyatomic Zintl Anions of the Post-Transition Elements, *Chem. Rev.*, 1985, **85**, 383–397.

16. Berezhoi, A. S., *Silicon and Its Binary Systems*, Consultants Bureau, New York, 1960.

17. Evers, J.; Oehlinger, G.; Sextl, G.; Weiss, A., High Pressure Phases of KSi, KGe, RbSi, RbGe, CsSi, and CsGe with the NaPb-type Structure, *Angew. Chem. Int. Ed. Engl*, 1984, **23**, 528–529.

18. Lange, H.; Wötting, G.; Winter, G., Silicon Nitride—From Powder Synthesis to Ceramic Materials, *Angew. Chem. Int. Ed. Engl.*, 1991, **30**, 1579–1597.

19. Aylett, B. J., Silicon Hydrides and Their Derivatives, *Adv. Inorg. Chem. Radiochem.*, 1968, **11**, 249–307.

20. Wiberg, E.; Amberger, E., *Hydrides of the Elements of Main Groups I–V*, Elsevier, Amsterdam, 1971, Chapter 7.

21. Urry, G., Systematic Synthesis in the Polysilane Series, *Acc. Chem. Res.*, 1970, **3**, 306–312.

22. Zubieta, J. A.; Zuckerman, J. J., Structural Tin Chemistry, *Prog. Inorg. Chem.*, 1978, **24**, 251–475.

23. Liebau, F., *Structura. Chemistry of Silicates*, Springer, New York, 1985.

24. Sosman, R. B., *The Phases of Silica*, Rutgers University Press, New Brunswick, NJ, 1965.

25. Iler, R. K., *The Chemistry of Silica*, Wiley, New York, 1979.

26. Falcone, J. S., Ed., *Soluble Silicates*, ACS Symposium Series 194, American Chemical Society, Washington, DC, 1982.

27. Dent Glasser, L. S., Sodium Silicates, *Chem. Brit.*, 1982, **18**, 33–39.

28. Michaelis, L.; Chisick, S. S., et al., Eds., *Asbestos: Properties, Applications, and Hazards*, Wiley, New York, Vol. 1, 1979; Vol. 2, 183.

29. Clark, R. J. H.; Cobbold, D. G., *Inorg. Chem.*, 1978, **17**, 3169–3174.

30. Breck, D. W., *Zeolite Molecular Sieves (Structure, Chemistry, and Uses)*, Wiley, New York, 1974.

31. Barrer, R. M., *Zeolites and Clay Minerals as Sorbents and Molecular Sieves*, Academic Press, London, 1978.

32. Schwachow, F.; Puppe, L., Zeolites—Their Synthesis, Structure, and Applications, *Angew. Chem. Int. Ed. Engl.*, 1975, **14**, 620–628.

33. Davis, M. E.; Lobo, R. F., Zeolite and Molecular Sieve Synthesis, *Chem. Mater.*, 1992, **4**, 756–766.

34. Haas, P. A., Gel Processes for Preoaring Ceramics and Glasses, *Chem. Eng. Prog.*, 1989, **85**, 44–52.

35. Criegee, R., *Oxidations with Lead Tetraacetate*, in *Oxidation in Organic Chemistry*, K. Wiberg, Ed., Academic Press, New York, 1965, pp. 277–366.

36. Rubottom, G. M., Oxidations with Lead Tetraacetate, in *Oxidation in Organic Chemistry*, Part D, W. S. Trahanovsky, Ed., Academic Press, New York, 1982, pp. 1–145.

37. Vorhoeve, R. J. H., *Organohalosilanes: Precursors to Silicones*, Elsevier, Amsterdam, 1967.

38. Voronkov, M. G.; Mileshkevich, V. P.; Yuzhelevskii, *The Siloxane Bond*, Consultants Bureau, New York, 1978.

39. Voronkov, M. G.; Lavrent'ev, Polyhedral Organosilsesquioxanes and Their Homo Derivatives, *Top. Curr. Chem.*, 1982, **102**, 199–236.

40. Miller, R. D.; Michl, J., Polysilane High Polymers, *Chem. Rev.*, 1989, **89**, 1359–1410.

41. Laguerre, M.; Dunogues, J.; Calas, R., One-Step Synthesis of Dodecamethylcyclohexasilane, *J. Chem. Soc., Chem. Commun.*, 1978, 272.

42. Yajima, S.; Hayashi, J.; Omori, M.; Okamura, K., *Nature*, 1976, **261**, 683–685.

43. Raabe, G.; Michl, J., Multiple Bonding to Silicon, *Chem. Rev.*, 1985, **85**, 419–509.

44. West, R.; Fink, M. J.; Michl, J., Tetramesityldisilene, a Stable Compound Containing a Silicon–Silicon Double Bond, *Science*, 1981, **214**, 1343–1344.

45. Poller, R. C., *The Chemistry of Organotin Compounds*, Academic Press, New York, 1970.

46. Davies, A. G.; Smith, P. J., Recent Advances in Organotin Chemistry, *Adv. Inorg. Chem. Radiochem.*, 1980, **23**, 1–77.

47. Holmes, R. R., Organotin Cluster Chemistry, *Acc. Chem. Res.*, 1989, **22**, 190–197.

48. Shapiro, H.; Frey, F. W., *The Organic Compounds of Lead*, Wiley-Interscience, New York, 1968.

49. Jutzi, P.; Holtmann, U.; Kanne, D.; Krüger, C.; Blom, R.; Gleiter, R.; Hyla-Krypsin, I., Decamethylsilicocene—The First Stable Silicon(II) Compound: Synthesis, Structure, and Bonding, *Chem. Ber.*, 1989, **122**, 1629–1639.

50. King, R. B., Secondary and Tertiary Phosphine Adducts of Germanium(II) Iodide, *Inorg. Chem.*, 1963, **2**, 199–200.

51. Connolly, J. W.; Hoff, C., Organic Compounds of Divalent Tin and Lead, *Adv. Organometal. Chem.*, 1981, **19**, 123–153.

52. Davidson, P. J.; Lappert, M. F., Stabilization of Metals in a Low Co-ordinative Environment Using the Bis(trimethylsilyl)methyl Ligand; Coloured Sn^{II} and Pb^{II} Alkyls, $M[CH(SiMe_3)_2]_2$, *J. Chem. Soc., Chem. Commun.*, 1973, 317.

4

Nitrogen

4.1 General Aspects of the Chemistry of Nitrogen

The nitrogen atom has five valence electrons. It can therefore complete its octet in the following ways:

1. Gain of three electrons to form the nitride ion, N^{3-}. This occurs only in the case of the most electropositive elements, particularly the lighter alkali and alkaline earth metals.
2. The formation of sufficient single or multiple electron pair bonds to give the nitrogen its favored octet. Thus nitrogen can complete its octet by forming three single bonds, such as in NH_3 and NF_3, or by forming an appropriate number of multiple bonds such as one double and one single bond in azo compounds $RN{=}NR$ or one triple bond in the case of molecular nitrogen, $:N{\equiv}N:$.
3. A combination of electron gain and bond formation such as in the amide ion, NH_2^-, found in sodium amide, $NaNH_2$.
4. A combination of electron $loss$ and bond formation in the quaternary ammonium ions, R_4N^+, as well as NH_4^+, in which nitrogen completes its electron octet by forming four single bonds and losing one electron to form a monocation (i.e., $5 + 4 - 1 = 8$).

There are a number of stable simple paramagnetic nitrogen compounds in which the nitrogen atom does not complete its octet; such compounds can be formulated with only a septet of nitrogen electrons leading to one unpaired electron. Familiar examples are the nitrogen oxides NO and NO_2; other less familiar examples are dialkyl and diaryl nitroxides, R_2NO (R = CF_3, Me_3C, C_6H_5, etc.)[1] and the related disulfonate anion $(O_3S)_2NO^{2-}$, found in Frémy's salt, $K_2[O_3S)_2NO]$.   Nitrogen is unable to expand its octet to form five- and six-coordinate derivatives because of the lack of accessible d orbitals. Thus whereas NF_4^+ is an isolable species (e.g., as $NF_4^+BF_4^-$), NF_5 is unstable and cannot be isolated or detected.

A common stereochemistry of nitrogen is pyramidal three-coordinate, as found in the amines R_3N, RNH_2, and R_2NH, and ammonia, NH_3. Such compounds undergo rapid inversion with a small potential energy barrier (e.g., 23.5 kJ/mol for NH_3). For this reason, optical isomers of amines with three different substituents, $RR'R''N$, cannot be isolated. Pyramidal three-coordinate nitrogen has a lone pair; interaction of this lone pair with an electron-deficient atom or group leads to four-coordinate tetrahedral derivatives such as the quaternary ammonium cations R_4N^+, amine adducts of boron Lewis acids such as $R_3N \to BX_3$ (R = alkyl; X = F, Cl, Br, H, etc.), amine transition metal complexes auch as $Me_3N \to Cr(CO)_5$ and $[Co(NH_3)_6]^{3+}$, and the amine oxides $R_3N \to O \leftrightarrow R_3N^+ - O^-$ with a very polar nitrogen–oxygen bond. Three-coordinate nitrogen derivatives can become planar when potentially π-bonding groups are bonded to nitrogen. An excellent example of such a three-coordinate planar nitrogen derivative is trisilylamine, $(H_3Si)_3N$, which has strong Si–N $d\pi$–$p\pi$ bonding and no basic properties (compare Section 3.1). Highly electronegative fluorine atoms can reduce the basicity of the nitrogen lone pair so that NF_3 is not basic enough to form adducts with BX_3 derivatives, although $F_3N \to O$ is a stable compound that can be prepared by indirect methods.

Nitrogen is frequently involved in multiple bonding (Figure 4.1). The N≡N distance in N_2 is an extremely short bond (1.094Å) of very high strength. Doubly bonded derivatives of the type X=N—Y are non linear with sp^2-hybridized nitrogen with a stereochemically active lone pair. The nitrite ion and organic nitro compounds are resonance hybrids (Figure 4.1) with an average formal N—O bond order of $\frac{3}{2}$.

The nitrogen–nitrogen single bond is rather weak (e.g., 169 kJ/mol in hydrazine, $H_2N—NH_2$); this weakness has been attributed to repulsion between nonbonding lone electron pairs on adjacent nitrogen atoms. However, organic compounds are known with chains of three to eight nitrogen atoms (e.g., $R_2N—N=N—NR_2$, $RN=N—NR—NR_2$, $RN=N—NR—N=NR$, $RN=N—NR—N=N—NR—N=NR$) and with rings containing up to five consecutive nitrogen atoms with some multiple bonds.

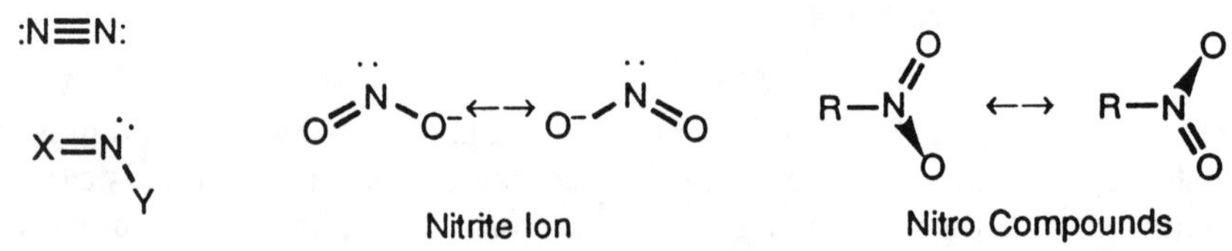

Figure 4.1. Simple nitrogen compounds with multiple bonding.

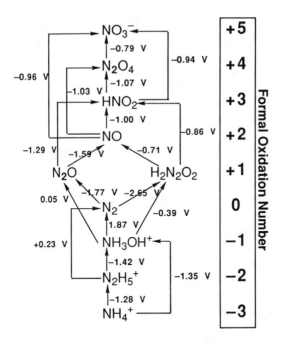

Figure 4.2. Examples of all the integral oxidation states of nitrogen and their redox potentials in *acid* aqueous solution.

Nitrogen is extensively involved in hydrogen bonding, both as a proton donor and as a proton acceptor. For example, hydrogen bonding is very strong in ammonia, NH_3, and hydrazine, $H_2N—NH_2$.

Nitrogen exhibits a variety of formal oxidation states including all possible integral oxidation states as well as some nonintegral oxidation states such as $-\frac{1}{3}$ in the azide ion, N_3^-. Examples of all the integral oxidation states and the redox relationships between them in acid solution are given in Figure 4.2.[2] The stability of nitrogen(0) in elemental nitrogen, N_2, is reflected in the highly positive potentials of redox systems leading to elemental nitrogen. For example, the $+2.85$ V potential in reducing $+1$ nitrogen in hyponitrous acid, $H_2N_2O_2$, to elemental nitrogen is comparable to the $+2.87$ V potential in reducing elemental fluorine to fluoride ion. That is, hyponitrous acid is thermodynamically as strong an oxidizing agent as elemental fluorine.

4.2 Nitrogen Isotopes and Elemental Nitrogen

The earth's atmosphere is 78% N_2 by volume; nitrogen is thus the most abundant element found in an uncombined state. Nitrogen is also found as nitrate in the minerals KNO_3 (*nitre* or *saltpeter*) and $NaNO_3$ (*sodanitre* or *Chilean saltpeter*).

Nitrogen forms two stable isotopes. The most abundant isotope (99.625% natural abundance) is ^{14}N, $I = 1$; the resulting quadrupole moment (2×10^{-2}) makes the NMR spectra[3] of ^{14}N broad except in symmetrical environments such as NF_4^+, NH_4^+, NO_3^-, NO_2^+, and $RN^+\equiv C:^-$. The rare stable isotope (0.375% natural abundance) is ^{15}N, $I = \frac{1}{2}$, so that its NMR spectra, although relatively insensitive, give sharp lines in all cases.[4] Enrichment of ^{15}N is possible by exchange reactions such as

$$^{15}NO(g) + H^{14}NO_3(aq) \rightleftarrows {}^{14}NO(g) + H^{15}NO_3(aq) \qquad K = 1.055$$

$$(4.1)$$

or similar exchange reactions with NH_3/NH_4^+ or NO/NO_2.

Elemental nitrogen or *dinitrogen*, N_2 (mp $-210°C$, bp $-195.8°C$), is a colorless, odorless, tasteless, diamagnetic gas, which is very unreactive as a consequence of its strong $N\equiv N$ triple bond and its low polarity. Ambient temperature reactions of N_2 include the formation of a few nitrides of electropositive metals such as Li_3N and a few N_2 transition metal complexes. High temperature reactions of N_2 include the reaction of N_2 with H_2 to give ammonia ($K_{25°C} = 10^3$ atm^{-2}) in the Haber process at 400–550°C and 100–1000 atm with an iron catalyst, the reaction of N_2 with calcium carbide (CaC_2) at 730°C to give calcium cyanamide ($CaCN_2$), and the direct combination of N_2 with oxygen and with magnesium metal:

$$N_2(g) + O_2(g) \rightarrow 2NO(g) \qquad K_{25°C} = 5 \times 10^{-31} \qquad (4.2)$$

$$N_2(g) + 3\,Mg(s) \rightarrow Mg_3N_2(s) \qquad (4.3)$$

One of the interesting properties of molecular nitrogen ("dinitrogen") is its ability to form transition metal complexes.[5,6] In most such complexes, a lone pair of N_2 is coordinated to the transition metal atom, which can be schematically represented as $:N\equiv N \rightarrow M$. The first example of a metal dinitrogen complex was $[Ru^{II}(NH_3)_5N_2]^{2+}$, which was prepared in 1965 by Allen and Senoff[7] by the reaction of hydrazine hydrate with $RuCl_3$. Dinitrogen complexes of the types $(R_3P)_4M(N_2)_2$ or $(diphos)_2M(N_2)_2$ (M = Mo, W; diphos = chelating ditertiary phosphine such as $Ph_2PCH_2CH_2PPh_2$) are relatively stable and are typically obtained by reduction of the corresponding phosphine metal halide with a strong reducing agent (e.g., zinc or magnesium) in a dinitrogen atmosphere.[8] The electron pair on the remote nitrogen in $:N\equiv N \rightarrow M$ complexes can act as a Lewis base towards electrophiles such as $AlCl_3$ or $AlMe_3$ to give species with structural units $X_3Al \leftarrow N\equiv N \rightarrow M$; the nitrogen–nitrogen bond length in such bridging nitrogen atoms falls in the range of 1.124–1.298 Å.

Although the triple bond in N_2 is very stable, it can be broken by subjecting molecular N_2 to an electrical discharge at 0.1–2 mm pressure to give very reactive "active" nitrogen as ^4S nitrogen atoms.[9] This process is accompanied by a yellow "afterglow" that persists for several seconds, corresponding to preassociation of the nitrogen atoms into excited N_2^* molecules, which emit light upon returning to the N_2 ground state. This afterglow of atomic nitrogen can be quenched with nitric oxide according to the following reaction scheme, which results in light emission as indicated:

$$N(^4S) + NO \rightarrow N_2 + O(^3P) \tag{4.4a}$$

$$N(^4S) + O(^3P) \rightarrow NO^* \rightarrow NO + h\nu \text{ (blue)} \tag{4.4b}$$

$$NO + O(^3P) \rightarrow NO_2^* \rightarrow NO_2 + h\nu \text{ (green-yellow)} \tag{4.4c}$$

4.3 Nitrides

Binary metal nitrides can be classified into the same three general types as binary metal hydrides (Section 1.5) and metal carbides (Section 2.3), namely saline, covalent, and interstitial nitrides. The important features of each of these types of metal nitride are outlined in Sections 4.3.1–4.3.3.

4.3.1 Saline Binary Metal Nitrides

As is the case with saline hydrides and carbides, saline nitrides are formed by the most electropositive metals such as the alkali and alkaline earth metals. Stable saline nitrides include Li_3N and M_3N_2 (M = Mg, Ca, Sr, Ba, Zn). The nitrides of the heavier alkali metals M_3N (M = Na, K, Rb) are poorly characterized, unstable, explosive compounds. Saline nitrides are prepared by direct combination of the metal with nitrogen or by loss of ammonia from amides upon heating. Saline metal nitrides are readily hydrolyzed to the metal hydroxide or oxide and ammonia. Azides of the electropositive metals such as MN_3 (M = Li, Na, K, Rb, Cs) and $M(N_3)_2$ (M = Mg, Ca, Sr, Ba) are also formally saline "nitrides"; they are discussed later under hydrazoic acid (Section 4.4.4). Metal azides are stable in water ionizing to the metal cation and the azide anion, N_3^-, analogous to the corresponding metal halides in contrast to the true saline metal nitrides, which readily hydrolyze to ammonia.

4.3.2 Covalent Nitrides

Covalent nitrides are formed mainly by nonmetals, metalloids, or metals of lower electronegativity. Some examples of binary covalent nitrides include EN (E = B, Al, Ga, In, Tl), E_3N_4(E = Si, Ge), C_2N_2 (cyanogen), P_3N_5, As_4N_4, S_2N_2, and S_4N_4; these binary covalent nitrides are discussed in this book in the chapter on the other element. The EN covalent nitrides (E = B, Al, Ga, In, Tl) are significant because their structures are closely related to the isoelectronic graphite and diamond forms of carbon (Section 2.2.2).

4.3.3 Interstitial Nitrides

Interstitial nitrides are formed by transition metals and have stoichiometries such as MN, M_2N, and M_4N, although they are often nitrogen-deficient, nonstoichiometric compounds. The nitrogen atoms in interstitial metal nitrides often occupy interstrices in the metallic lattice. For this reason the interstitial nitrides are metallic in appearance, hardness, and electrical conductivity. They are chemically inert and extremely hard, and they have very high melting points. Interstitial metal nitrides are prepared by heating the metal in ammonia at elevated temperatures, typically 1100–1200°C. Specific examples of interstitial metal nitrides are VN, Mo_2N, W_2N, and Fe_4N. The interstitial nitrides of stoichiometry MN often have the sodium chloride structure.

4.4 Compounds of Nitrogen with Hydrogen

4.4.1 Ammonia and Ammonium Cations: Liquid Ammonia as a Nonaqueous Protic Ionizing Solvent

The most important binary nitrogen–hydrogen compound is ammonia, NH_3, a colorless alkaline gas, (bp $-33°C$, mp $-78°C$), with a characteristic odor perceptible at concentrations of 20–50 ppm. Its high boiling point relative to that of phosphine, PH_3 (bp $-87.7°C$); indicates strong hydrogen bonding in liquid ammonia (Section 1.3). The strong hydrogen bonding in liquid ammonia also leads to an unusually high heat of evaporation, namely 1.37 kJ/g or 23.35 kJ/mol at the boiling point. As a consequence of its high heat of evaporation, liquid ammonia is readily handled in normal laboratory equipment despite its boiling point far below room temperature.

Liquid ammonia has been the prototypical nonaqueous protic solvent system because of its resemblance to liquid water arising from the polarity of its molecules and strong hydrogen bonding.[10,11] The dielectric constant of liquid ammonia (~ 22 at $-34°C$) is lower than that of water (81 at 25°C) but nevertheless strong enough for ammonia to be a fair ionizing solvent. Liquid ammonia can be used for electrochemical studies even in the liquid phase near its critical pressure (112.5 atm at 133°C).

The lower dielectric constant of liquid ammonia relative to liquid water means that ammonia is a poorer solvent than water for ionic inorganic compounds but a better solvent than water for covalent organic compounds. Examples of salts freely soluble in liquid ammonia include most ammonium salts, nitrates, nitrites, cyanides, and thiocyanates. The solubility of halides in liquid ammonia increases in the sequence F < Cl < Br < I. Silver iodide is of interest because it is very insoluble in water but very soluble in liquid ammonia (2070 g/L) because of the strong complexing tendency of ammonia with the silver ion.

Liquid ammonia undergoes self-ionization like water, although its self-ionization constant is much lower than that of water:

$$2NH_3 \rightleftarrows NH_4^+ + NH_2^- \qquad K_{-50°C} = \sim 10^{-33} \qquad (4.5)$$

Thus ammonium salts are acids and amides (e.g., $NaNH_2$ and KNH_2) are bases in liquid ammonia. The amide bases such as $NaNH_2$ can be obtained by reaction of the corresponding alkali metal with liquid ammonia:

$$2M + 2NH_3 \rightarrow 2MNH_2 + H_2\uparrow \qquad (4.6)$$

A catalyst such as ferric nitrate is necessary for this reaction, since otherwise alkali metals will form stable blue solutions in liquid ammonia (Section 10.2.3). Reactions of alkali metal amides with metal salts in liquid ammonia can result in the formation of insoluble heavy metal amides or the deamination of such heavy metal amides to form the corresponding imides or nitrides, for example:

$$AgNO_3 + KNH_2 \xrightarrow{\text{liq } NH_3} AgNH_2\downarrow + KNO_3 \qquad (4.7a)$$

$$3HgI_2 + 6KNH_2 \xrightarrow{\text{liq } NH_3} Hg_3N_2\downarrow + 6KI + 4NH_3 \qquad (4.7b)$$

Such reactions are analogous to the treatment of metal salts in water with alkali metal hydroxides to give insoluble metal hydroxides, which can be dehydrated into the corresponding insoluble metal oxides. Amphoteric behavior of metal amides in liquid ammonia is observed analogous to the amphoteric behavior of metal hydroxides in water, for example:

$$ZnI_2 + 2KNH_2 \xrightarrow{\text{liq } NH_3} Zn(NH_2)_2\downarrow + 2KI \qquad (4.8a)$$

$$Zn(NH_2)_2 + 2KNH_2 \xrightarrow{\text{liq } NH_3} K_2[Zn(NH_2)_4] \qquad (4.8b)$$

$$K_2[Zn(NH_2)_4] + 2NH_4I \xrightarrow{\text{liq } NH_3} Zn(NH_2)_2\downarrow + 4NH_3 + 2KI \qquad (4.8c)$$

One consequence of the similarity of ammonia and water is the ready miscibility of these compounds. Ammonia forms the hydrates $NH_3 \cdot H_2O$, $NH_3 \cdot 2H_2O$, and $2NH_3 \cdot H_2O$ in which the molecules of NH_3 and H_2O are linked by hydrogen bonds.[12] However, these ammonia hydrates do not contain any discrete NH_4^+ or OH^- ions of NH_4OH. It thus appears that "ammonium hydroxide," NH_4OH, does not really exist. Ammonia is a weak base in aqueous solution:

$$\frac{[NH_4^+][OH^-]}{[NH_3]} = 1.77 \times 10^{-5} \text{ mol/L} \Rightarrow pK_b = 4.75 \qquad (4.9)$$

The proton NMR spectrum of aqueous solutions of ammonia indicates the rapid exchange of hydrogen atoms between ammonia and water by the reaction

$$H_2O + NH_3 \rightleftarrows OH^- + NH_4^+ \qquad (4.10)$$

The reaction of ammonia with oxygen to give elemental nitrogen is thermodynamically favored:

$$4NH_3(g) + 3O_2(g) \rightarrow 2N_2(g) + 6H_2O(g) \qquad K_{25°C} = 10^{228} \qquad (4.11)$$

The flammability limits of ammonia in air are 16–25 vol%. Although the oxidation of ammonia to elemental nitrogen is thermodynamically favored, the oxidation of ammonia in the presence of a platinum or platinum/rhodium catalyst at 500–900°C can be made to give nitric oxide:

$$4NH_3 + 5O_2 \rightarrow 4NO + 6H_2O \qquad K_{25°C} = 10^{168} \qquad (4.12)$$

This reaction is a key step in the manufacture of nitric acid from atmospheric nitrogen via ammonia and nitric oxide by the Ostwald and Haber processes:

$$N_2 \xrightarrow[\text{Haber process}]{H_2} NH_3 \xrightarrow[\text{Ostwald process}]{O_2} NO \xrightarrow{O_2 + H_2O} HNO_3(aq) \qquad (4.13)$$

Many stable crystalline salts of the tetrahedral ammonium ion, NH_4^+, are known; most of these salts are soluble in water. Ammonium salts of strong acids are acidic from hydrolysis according to the following reaction:

$$NH_4^+ + H_2O \rightleftarrows NH_3 + H_3O^+ \qquad K_{25°C} = 5.5 \times 10^{-10} \qquad (4.14)$$

Thus a 1 M aqueous solution of an ammonium salt of a strong acid has a pH of about 4.7.

Ammonium salts resemble those of potassium and rubidium in solubility and structure except when hydrogen bonding is involved. Thus the Pauling radii of NH_4^+ (1.48 Å), K^+ (1.33 Å), and Rb^+ (1.48 Å) are very similar. Ammonium salts of volatile acids, although clearly ionic, are often volatile by dissociation. For example, ammonium chloride can be sublimed upon heating by the following reaction:

$$NH_4Cl(s) \xrightarrow{\Delta} NH_3(g) + HCl(g) \qquad (4.15)$$

Decomposition of ammonium salts of oxidizing acids can often be violent in view of the low formal oxidation state of nitrogen in ammonia (Figure 4.2). For example, heating ammonium dichromate results in a spectacular "volcano" in which an orange mountain of ammonium dichromate emits sparks and green Cr_2O_3 "lava" in the following violent exothermic reaction with reduction of Cr(VI) to Cr(III) by the ammonium ion:

$$(NH_4)_2Cr_2O_7 \xrightarrow{\Delta} N_2 + 4H_2O + Cr_2O_3 \qquad \Delta H = -315 \text{ kJ/mol}$$
$$(4.16)$$

Ammonium permanganate, ammonium nitrate, and ammonium perchlorate can all decompose explosively through oxidation of the ammonium cation by the oxidizing anion. However, cautious heating of ammonium nitrate can provide a source of nitrous oxide:

$$NH_4NO_3 \xrightarrow{\Delta} N_2O + 2H_2O \qquad (4.17)$$

Ammonium perchlorate has been used as an oxidizer in solid propellants.

Substitution of the four hydrogen atoms by alkyl groups in NH_4^+ leads to the tetraalkylammonium cations, R_4N^+, which may be obtained by *quaternization* of a tertiary amine by an alkyl halide:

$$R_3N + RX \rightarrow R_4N^+X^- \qquad (4.18)$$

Such quaternization reactions generally proceed rapidly and exothermically. The corresponding bases, tetraalkylammonium hydroxides, $R_4N^+OH^-$, are strong bases like the alkali metal hydroxides because they cannot readily lose an alkyl group to form the corresponding tertiary amine R_3N in the reverse of equation 4.18. Tetraalkylammonium cations are often used in inorganic chemistry for large, stable, univalent cations, leading to salts with better solubility in organic solvents than corresponding salts of the alkali metal cations. For example, tetra-*n*-butylammonium salts such as $[Bu^n_4N][ClO_4]$ and $[Bu^n_4N][PF_6]$ are useful as supporting electrolytes for electrochemistry in organic solvents. Electrolysis of tetraalkylammonium salts with a mercury cathode leads to crystalline "tetraalkylammonium amalgams," which contain anionic mercury clusters[13] such as $[R_4N]^+[Hg_4]^-$ (Section 11.2).

4.4.2 Hydrazine[14,15]

Hydrazine, N_2H_4 or H_2N-NH_2, is derived from ammonia by replacement of a hydrogen atom by an NH_2 group and may be regarded as the "hydronitrogen" analogue of ethane, H_3C-CH_3. Since hydrazine has two trivalent nitrogen atoms, each with a lone electron pair, it can be protonated twice, giving the $N_2H_5^+$ and $N_2H_6^{2+}$ cations. Salts of the $N_2H_5^+$ cation are stable in water ($K_b = 8.5 \times 10^{-7}$), whereas salts of the $N_2H_6^{2+}$ dication are extensively hydrolyzed in water ($K_{b_2} = 8.9 \times 10^{-16}$).

Anhydrous hydrazine (mp 2°C, bp 113.5°C), is a fuming colorless liquid with a high dielectric constant (52 at 25°C). It thus resembles water and ammonia in being a polar, highly ionizing, highly hydrogen-bonded protic solvent, although the chemical reactivity of hydrazine limits its application as a nonaqueous solvent. Hydrazine is endothermic (50 kJ/mol) but thermodynamically stable. Because of its endothermicity, however, hydrazine burns very exothermically (-622 kJ/mol) and is used as a rocket fuel. Aqueous hydrazine is a powerful reducing agent ($E° = -1.16$ V in basic solution), being oxidized to elemental nitrogen.

The preparation of hydrazine is somewhat delicate owing to the intermediate oxidation state of its nitrogen atoms. The classical method for hydrazine synthesis is the *Raschig synthesis*, which uses the hypochlorite oxidation of

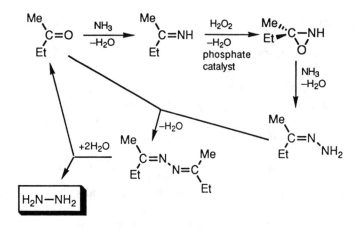

Figure 4.3.   The azine process for the synthesis of hydrazine.

ammonia with chloramine as an intermediate:

$$NH_3 + NaOCl \rightarrow NH_2Cl + NaOH \qquad (4.19a)$$

$$NH_3 + NH_2Cl + NaOH \rightarrow N_2H_4 + NaCl + H_2O \qquad (4.19b)$$

A difficulty with the Raschig synthesis of hydrazine is the following parasitic side reaction, which leads to destruction of the hydrazine with reduction of the yield:

$$2NH_2Cl + N_2H_4 \rightarrow 2NH_4Cl + N_2\uparrow \qquad (4.20)$$

This side reaction can be suppressed by the addition of gelatin, which apparently sequesters trace heavy metal catalysts for reaction 4.20. Some of the difficulties in the Raschig synthesis of hydrazine are avoided in the newer *azine process* for the synthesis of hydrazine (Figure 4.3). Waste is minimal because the methyl ethyl ketone used in the azine process can be recycled.

4.4.3 Diazene (Diimide)

Diazene or diimide, $HN{=}NH$, is a diamagnetic yellow compound that is unstable above $-180°C$. It can be generated in the pure state as the trans isomer by pyrolysis of sodium p-tolylsulfonylhydrazine:

$$p\text{-}CH_3C_6H_4SO_2N(Na)\text{---}NH_2 \xrightarrow{\Delta} trans\text{-}N_2H_2 + p\text{-}CH_3C_6H_4SO_2Na$$

$$(4.21)$$

Diazene may be regarded as the "hydronitrogen" analogue of ethylene, $H_2C{=}CH_2$, and is the parent of organic azo compounds such as azomethane, $CH_3N{=}NCH_3$, and azobenzene, $C_6H_5N{=}NC_6H_5$. Diazene generated in situ by hydrolysis of dipotassium azodicarboxylate, $KO_2C\text{---}N{=}N\text{---}CO_2K$, or by oxidation of hydrazine with two-electron oxidants (O_2, peroxides, etc.) is useful as a stereoselective reagent for the cis-hydrogenation of olefins, for example,[16]

$$RCH{=}CHR + N_2H_4 + H_2O_2 \xrightarrow{\text{Cu}^{2+}} RCH_2CH_2R + N_2{\uparrow} + 2H_2O$$

$$(4.22)$$

4.4.4 Hydrazoic Acid

Hydrazoic acid, HN_3, is a dangerously explosive, malodorous liquid (bp 37°C) that is toxic in concentrations below 1 ppm, although not a cumulative poison. It is a weak acid, $pK_a = 4.75$ at 25°C, forming stable salts called *azides*. Sodium azide can be prepared by the following reactions:

$$3NaNH_2 + NaNO_3 \xrightarrow{175°C} NaN_3 + 3NaOH + NH_3 \qquad (4.23a)$$

$$2NaNH_2 + N_2O \xrightarrow{190°C} NaN_3 + NaOH + NH_3 \qquad (4.23b)$$

Free hydrazoic acid can be generated in aqueous solution by the diazotization of hydrazine, analogous to the diazotization of an aromatic amine:

$$N_2H_5^+ + HNO_2 \rightarrow HN_3 + H_3O^+ + H_2O \qquad (4.24)$$

Azide salts may be classified as either ionic or covalent.[17] Ionic azides are formed by the most electropositive metals and contain the symmetric linear N_3^- anion with a nitrogen–nitrogen bond distance of 1.16 Å. Alkali metal azides, MN_3, are nonexplosive white solids, readily soluble in water; they undergo a smooth thermal decomposition with nitrogen evolution to give either the corresponding metal nitride (M = Li) or even the free alkali metal (M = Na, K, Rb, Cs):

$$3LiN_3 \xrightarrow{\Delta} Li_3N + 4N_2{\uparrow} \qquad (4.25a)$$

$$2NaN_3 \xrightarrow{\Delta} 2Na + 3N_2{\uparrow} \qquad (4.25b)$$

Pyrolysis of metal azides provides one of the few nonelectrolytic routes to the free alkali metals and also provides a source of small quantities of very pure nitrogen.

Covalent azides include the azides of heavy metals such as silver, lead, and mercury, which are some of the most dangerous known explosives. Thus lead azide, $Pb(N_3)_2$, is used as a detonator, and mercuric azide, $Hg(N_3)_2$, will explode even under water. Organic covalent azides, RN_3, are also known; they have structures with nonequivalent nitrogen–nitrogen distances and a bent R–N–N angle. Aliphatic azides, such as methyl azide. CH_3N_3, are dangerous explosives like diazomethane; trimethylsilyl azide, $(CH_3)_3SiN_3$, is much more stable and safer to handle.

4.4.5 Hydroxylamine

Hydroxylamine, H_2N—OH, can be derived from ammonia by replacement of one of the hydrogen atoms by a hydroxyl group; although not a binary compound of nitrogen with hydrogen, it is conveniently discussed here. Free hydroxylamine is a white solid (mp 32°C), which must be kept at 0°C to prevent decomposition. Hydroxylammonium salts such as $[NH_3OH]Cl$, $[NH_3OH]NO_3$, and $[NH_3OH]_2SO_4$ are much more stable and readily available white solids. Hydroxylamine is a weaker base than ammonia, $K_{25°C} = 6.6 \times 10^{-9}$, because of the electron-withdrawing effect of its hydroxyl group. Hydroxylammonium salts can be obtained by the following methods.

1. Reduction of nitrite ion with sulfur dioxide to give ultimately $[NH_3OH]_2SO_4$ through hydroxylamine sulfonic acid intermediates $[N(OH)(OSO_2)_2]^{2-}$ and $[NH(OH)(OSO_2)]^-$,
2. Hydrolysis of nitroalkanes with an α-CH_2 group by the reaction (R = H, etc.):

$$2RCH_2NO_2 + H_2SO_4 + 2H_2O \xrightarrow{\Delta} 2RCO_2H + [NH_3OH]_2SO_4 \quad (4.26)$$

3. Hydrogenation of nitric oxide in dilute aqueous sulfuric acid in the presence of a platinized active charcoal catalyst:[18]

$$3H_1 + 2NO \xrightarrow{Pt/C} 2NH_2OH \quad (4.27)$$

4. Electrolytic reduction of aqueous nitric acid between amalgamated lead electrodes in the presence of H_2SO_4/HCl.

Hydroxylammonium salts are useful reducing agents for antioxidants and photographic developers. In addition, hydroxylammonium salts are used to convert aldehydes and ketones to the corresponding oximes:

$$R_2C{=}O + NH_2OH \rightarrow R_2C{=}NOH + H_2O \quad (4.28)$$

4.5 Oxides of Nitrogen

4.5.1 Nitrous Oxide

Nitrous oxide, N_2O, is a colorless, odorless, tasteless gas (mp $-91°C$, bp $-89°C$). It has a linear structure, $:N{\equiv}N^+{-}\ddot{Q}:^-$, which is isoelectronic and isosteric with carbon dioxide (Figure 4.4). Nitrous oxide can be obtained by heating ammonium nitrate to 180–260°C. However, overheating ammonium nitrate above 300°C can cause an explosion.

   Nitrous oxide is relatively unreactive, being inert to the halogens, alkali metals, and ozone at ambient temperature. However, it oxidizes some low valent early transition metal derivatives such as $(C_5H_5)_2V$ and $(C_5H_5)_2Ti(CO)_2$ to

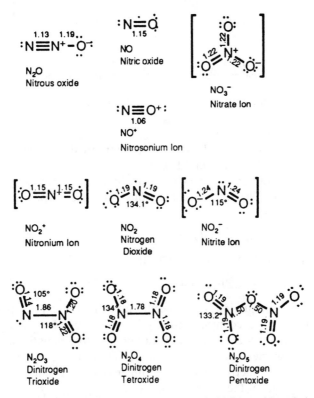

Figure 4.4. The structures of some molecular and ionic binary oxides of nitrogen; bond distances are given in angstrom units.

oxo complexes. Nitrous oxide forms a few metal complexes such as the octahedral ruthenium(II) complex $[Ru(NH_3)_5(N_2O)]^{2+}$. Nitrous oxide is used as an anesthetic ("laughing gas") and as an aerosol propellant.

4.5.2 Nitric Oxide and the Nitrosonium Ion

Nitric oxide, NO, is a colorless monomeric paramagnetic gas (mp $-164°C$, bp $-152°C$). It has a structure: $N\dot{=}\ddot{O}$: with a formal nitrogen–oxygen bond order of 2.5 and N=O distance of 1.15 Å. (Figure 4.4) Nitric oxide is frequently produced in the reduction of nitric acid, nitrate ion, or nitrite ion. Pure nitric oxide can be prepared by treatment of a mixture of sodium nitrite and ferrous sulfate with sulfuric acid according to the following equation:

$$2NaNO_2 + 2FeSO_4 + 3H_2SO_4 \rightarrow Fe_2(SO_4)_3 + 2NaHSO_4$$
$$+ 2H_2O + 2NO\uparrow \qquad\qquad (4.29)$$

The commercial preparation of nitric oxide uses the catalytic oxidation of

M—N≡O: M—N

linear NO   bent NO
3-electron donor   1-electron donor

Figure 4.5. Linear and bent NO groups in metal nitrosyls.

ammonia over a palladium/rhodium or platinum catalyst at temperatures in the range 500–900°C (equation 4.12).

Nitric oxide is rather reactive chemically, in accord with the presence of an unpaired electron. The following reactions thus occur spontaneously under ambient conditions:

$$2NO(g) + O_2(air) \rightarrow 2NO_2(g) \tag{4.30a}$$

$$2NO + X_2 \rightarrow 2XNO \quad (X = F, Cl, Br) \tag{4.30b}$$

$$2NO + 2CF_3I \rightarrow 2CF_3NO + I_2 \tag{4.30c}$$

The first reaction (equation 4.30a) occurs immediately upon exposure of NO to air and is readily recognized visually by the change from colorless NO to brown NO_2. The third reaction is also rather visually dramatic, since both NO and CF_3I are colorless gases, whereas CF_3NO is one of the few known blue gases. Nitric oxide dimerizes to a planar dimer N_2O_2 in condensed phases; both cis and trans forms of the dimer are known.

One of the important properties of nitric oxide is its ability to form transition metal complexes, commonly called *metal nitrosyls*.[19] Terminal nitrosyl groups in transition metal complexes can either be linear, in which a neutral NO ligand functions as a three-electron donor, or bent (M—N—O angles in the range 120–140°), in which a neutral NO ligand functions as a one-electron donor (Figure 4.5). The donation of three electrons by a neutral linear NO ligand can be dissected into the following two steps:

1. Donation to the metal of a single electron to give the nitrosonium cation, $:N{\equiv}O:^+$.
2. Donation to the metal of the nitrogen lone pair of the nitrosonium cation, $:N{\equiv}O:^+$, similar to donation to a transition metal of the carbon lone pair in carbon monoxide, $:C{\equiv}O:$.

The donation of one electron by a bent NO ligand consists of using the odd electron on a neutral NO molecule to form an M—N covalent bond; the lone pair on the nitrogen atom is not donated to the transition metal but remains stereochemically active, leading to the bent structure. Terminal linear nitrosyl groups can be recognized by strong infrared $\nu(NO)$ frequencies above 1610 cm^{-1}, while lower $\nu(NO)$ frequencies are found in terminal bent nitrosyl derivatives. Nitrosyl ligands can also bridge two (e.g., $[C_5H_5Mn(CO)(NO)]_2$) or three (e.g., $(C_5H_5)_3Mn_3(NO)_4$) metal atoms.

In general, the transition metal atom in metal nitrosyls has the usual 18-electron rare gas configuration. Thus the relatively unstable but nevertheless isolable $Co(NO)_3$ and $Cr(NO)_4$ are the only two known binary metal nitrosyls. Numerous metal carbonyl nitrosyls are known in which the transition metal has the favored rare gas configuration. An interesting complete series of neutral metal carbonyl nitrosyls consists of the isoelectronic tetrahedral molecules $Ni(CO)_4$, $Co(CO)_3NO$, $Fe(CO)_2(NO)_2$, $MnCO(NO)_3$, and $Cr(NO)_4$; all these molecules are relatively volatile. Among the earliest known metal nitrosyl derivatives are the so-called *nitroprussides* containing the red anion $Fe(CN)_5NO^{2-}$ in which the central iron atom has the favored 18-electron rare gas configuration. Nitroprussides were originally obtained from ferrocyanides (hexacyanoferrates(II)) and nitric acid.

The unpaired electron in nitric oxide is easily removed by one-electron oxidation to give the *nitrosonium* ion, $:N{\equiv}O:^+$, which is isoelectronic and isostructural with carbon monoxide (Section 2.5.1.1). The nitrosonium ion can be generated by dissolution of N_2O_3 or N_2O_4 (see Sections 4.5.3 and 4.5.4) in concentrated sulfuric acid, oxidation of NO with higher oxidation state halides such as MoF_6, or by treatment of nitrosyl halides, XNO, with Lewis acid halides. Nitrosonium hydrogen sulfate, $NO^+HSO_4^-$, is an important isolable intermediate in the lead chamber process for the manufacture of sulfuric acid, in which nitrogen oxides are used to catalyze the air oxidation of SO_2. Other isolable nitrosonium salts include $NO^+ClO_4^-$ and $NO^+BF_4^-$, which are isostructural with the corresponding ammonium salts. Nitrosonium salts readily hydrolyze to nitrous acid by the following reaction:

$$NO^+ + 2H_2O \rightarrow H_3O^+ + HNO_2 \qquad (4.31)$$

4.5.3 Dinitrogen Trioxide

Dinitrogen trioxide, N_2O_3 (mp $-100.6°$ C), is obtained under ambient conditions from stoichiometric quantities of NO and O_2 or of NO and N_2O_4; it forms an intensely blue liquid and a pale blue solid. The dissociation of blue N_2O_3 into colorless NO and brown NO_2 is significant about 30°C and is consistent with the long N—N bond (1.86 Å) in N_2O_3 (Figure 4.4). Liquid N_2O_3 undergoes some self-ionization according to the following equation:

$$N_2O_3 \rightleftarrows NO^+ + NO_2^- \qquad (4.32)$$

In addition, N_2O_3 is formally the anhydride of nitrous acid, which it forms upon reaction with water:

$$N_2O_3 + H_2O \rightleftarrows 2HNO_2 \qquad (4.33)$$

4.5.4 Nitrogen Dioxide/Dinitrogen Tetroxide

Nitrogen dioxide and dinitrogen tetroxide (Figure 4.4) exist in a strongly temperature-dependent equilibrium:

$$N_2O_4 \rightleftarrows 2NO_2 \qquad K = 1.4 \times 10^{-5}\ mol/L\ at\ 303\ K \qquad (4.34)$$

The presence of this equilibrium complicates the determination of these nitrogen oxides by gas chromatographic methods. The N—N bond in the dimer N_2O_4 is very long (1.78 Å); its dissociation energy in the gas phase is 75 kJ/mol. The monomer NO_2 is paramagnetic in accord with the presence of one unpaired electron The degree of dissociation of N_2O_4 can be recognized readily by its color: NO_2 is brown and N_2O_4 is colorless. In the gas phase above 140°C, complete dissociation of N_2O_4 to NO_2 occurs; at 100°C; the equilibrium mixture is 90% NO_2 and 10% N_2O_4. Liquid N_2O_4 at its boiling point (+21°C) is deep red-brown and contains 0.1% NO_2. Upon cooling to its freezing point (−101°C), liquid N_2O_4 becomes pale yellow and contains only 0.01% NO_2. Solid N_2O_4 is entirely colorless undissociated N_2O_4.

The N_2O_4/NO_2 equilibrium mixture is formed spontaneously from nitric oxide and oxygen by heating heavy metal nitrates such as $Pb(NO_3)_2$ at 400°C, or by reduction of nitric acid or nitrates. It is also produced as a by-product of atmospheric combustion and thus is a concern in atmospheric pollution.

The N_2O_4/NO_2 equilibrium mixture undergoes rapid disproportionation of the formal nitrogen(IV) during hydrolysis to give nitrogen(III) and nitrogen(V) in a mixture of nitrous and nitric acids:

$$N_2O_4 + H_2O \rightarrow HNO_2 + HNO_3 \qquad (4.35)$$

In a slower reaction, the nitrous acid hydrolysis product undergoes further disproportionation to give $NO + HNO_3$ and water (see equation 4.45). Pyrolysis of NO_2 to NO and O_2 (the reverse of one method of forming NO_2) begins at 150°C and is complete at 600°C. The N_2O_4/NO_2 equilibrium mixture is a fairly strong oxidizing agent in aqueous solution (comparable to Br_2):

$$N_2O_4 + H^+ + 2e^- \rightarrow 2HNO_2 \qquad E^\circ = +1.07\ V \qquad (4.36)$$

Liquid N_2O_4 is of interest as an aprotic nonionizing aqueous solvent, although its oxidizing power (equation 4.36) limits its application[20]; it is often used in conjunction with a nonreactive organic diluent such as ethyl acetate. Reactions of rather unreactive metals or inorganic salts with liquid N_2O_4 give the corresponding metal nitrates. Thus anhydrous copper nitrate, which is volatile at 250–200°C, can be prepared from copper metal and liquid N_2O_4 by the following sequence of reactions:

$$Cu + 3N_2O_4 \rightarrow 2NO\uparrow + Cu(NO_3)_2 \cdot N_2O_4 \qquad (4.37a)$$

$$Cu(NO_3)_2 \cdot N_2O_4 \rightarrow Cu(NO_3)_2 + N_2O_4 \qquad (4.37b)$$

In anhydrous acids N_2O_4 dissociates ionically as "nitrosonium nitrate" according to the following equation:

$$N_2O_4 \rightleftarrows NO^+ + NO_3^- \qquad (4.38)$$

This dissociation is complete in anhydrous nitric acid. This reaction can be used to prepare nitrosonium salts, for example,

$$N_2O_4 + H_2SO_4 \rightarrow NO^+HSO_4^- + HNO_3 \tag{4.39}$$

In addition, N_2O_4 is a powerful nitrosating agent for organic compounds and can be used to prepare nitrite esters, nitrosoamines, and so on:

$$ROH + N_2O_4 \rightarrow RONO + HNO_3 \tag{4.40a}$$

$$R_2NH + N_2O_4 \rightarrow R_2NNO + HNO_3 \tag{4.40b}$$

4.5.5 Dinitrogen Pentoxide[21]

Dinitrogen pentoxide (Figure 4.4), N_2O_5 (mp 30°C), is formally the anhydride of nitric acid and can be obtained by the dehydration of nitric acid with P_4O_{10}. Rehydration of N_2O_5 in water to regenerate HNO_3 is highly exothermic. In the solid phase, N_2O_5 has the ionic formulation $NO_2^+NO_3^-$. Dinitrogen pentoxide is not very stable and requires an atmosphere of ozonized oxygen to be distilled without decomposition (bp 47°C). Treatment of N_2O_5 with ozone gives a mixture in which NO_3 can be detected spectroscopically.

4.6 Oxoacids and Oxoanions of Nitrogen

4.6.1 Hyponitrous Acid, Hyponitrites, and Nitramide

Hyponitrous acid, $H_2N_2O_2$ or HON=NOH, is known as both the *cis* and *trans* isomers.[22] The sodium salt of the more readily available *trans* isomer is obtained by reduction of aqueous sodium nitrite with sodium amalgam, whereas the sodium salt of the *cis* isomer is obtained by treatment of nitric oxide with sodium in liquid ammonia. The yellow silver salt *trans*-$Ag_2N_2O_2$ is insoluble in water and can be used to prepare alkyl hyponitrites, *trans*-RON=NOR, by treatment with alkyl halides. Free hyponitrous acid is a weak dibasic acid, $pK_1 = 6.9$, $pK_2 = 11.6$, which can be obtained by treatment of $Ag_2N_2O_2$ with anhydrous HCl in diethyl ether. Hyponitrous acid is an explosive solid that decomposes irreversibly to water and N_2O.

An isomer of hyponitrous acid is nitramide, H_2N-NO_2, which is a weak acid, $K = 2.6 \times 10^{-7}$ at 25°C.

4.6.2 Trioxodinitrates

The trioxodinitrate or "α-oxyhyponitrite" ion, $N_2O_3^{2-}$, in which nitrogen has a formal $+2$ oxidation state as in nitric oxide, is obtained from hydroxylamine and alkyl nitrates by the following equation[23]:

$$RONO_2 + NH_2OH + 2NaOMe \xrightarrow[0°C]{MeOH} Na_2N_2O_3 + 2MeOH + ROH$$

$$(4.41)$$

However, the $N_2O_3^{2-}$ anion *cannot* be obtained directly upon treatment of NO with water or a base. The $N_2O_3^{2-}$ anion disproportionates through an unstable HNO ("nitroxyl") intermediate to give N_2O and nitrite in neutral or acid aqueous media according to the following sequence of reactions:

$$N_2O_3^{2-} + H^+ \rightarrow HNO + NO_2^- \qquad\qquad (4.42a)$$

$$2HNO \rightarrow N_2O + H_2O \qquad\qquad (4.42b)$$

4.6.3 Nitrous Acid, Nitrites, and Peroxonitrites

Nitrites of alkali and alkaline earth metals (Figure 4.4) are obtained by heating the corresponding nitrates, preferably in the presence of a mild reducing agent such as carbon, lead or iron, for example,

$$MNO_3 + C \rightarrow MNO_2 + CO\uparrow \qquad\qquad (4.43)$$

Free nitrous acid can be generated in aqueous solution by treatment of barium nitrite with the stoichiometric amount of sulfuric acid and filtering off the barium sulfate precipitate:

$$Ba(NO_2)_2 + H_2SO_4 \rightarrow BaSO_4\downarrow + 2HNO_2 \qquad\qquad (4.44)$$

However, concentration of aqueous nitrous acid results in its disproportionation to give NO and HNO_3:

$$3HNO_2 \rightarrow 2NO + HNO_3 + H_2O \qquad\qquad (4.45)$$

Metal nitrites in acid aqueous media or $[Ph_3P{=}N{=}PPh_3]NO_2$ in organic solvents are useful as nitrosating agents. Socium nitrite is mildly toxic (lethal dose ~ 100 mg/kg body weight) and is extensively used for curing meat.

The peroxonitrite ion, yellow $[O{-}O{-}N{=}O]^-$, an unstable isomer of nitrate ion (see next section), is obtained by a radical coupling reaction of NO with superoxide at pH 12–13[24]:

$$O_2 + NO \rightarrow [O{-}O{-}N{=}O]^- \qquad\qquad (4.46)$$

In addition, photolysis of solid potassium nitrate with 254 nm light at 42°C gives a yellow solid solution containing 30 μmol of $K[O{-}O{-}N{=}O]$ per gram of solid.[25]

4.6.4 Nitric Acid, Nitrates, Nitronium Ion, and Peroxonitrates

Aqueous nitric acid is obtained by catalytic ammonia oxidation to NO_2 followed by treatment with water and air for further oxidation (equation 4.13).

The aqueous acid is concentrated to a constant boiling mixture (bp 122°C) containing 68.4% HNO_3. Anhydrous nitric acid can be obtained either by treatment of potassium nitrate with concentrated sulfuric acid followed by vacuum distillation (bp 84°C) or by dehydration of constant boiling 68.4% nitric acid with H_2SO_4 or P_4O_{10}. Distillation or exposure to light decomposes nitric acid to give a yellow coloration from NO_2 produced by the following reaction:

$$4HNO_3 \rightarrow 4NO_2 + O_2\uparrow + 2H_2O \qquad (4.47)$$

"Fuming" nitric acid is a solution of N_2O_4 (see Section 4.5.4) in nitric acid.

Almost all metals form nitrates (Figure 4.4), which are frequently hydrated and almost always soluble in water. The nitrate ion is difficult to reduce in aqueous solution. Nevertheless, the nitrate ion is reduced to ammonia in concentrated aqueous alkali (e.g., NaOH) by treatment with metallic aluminum or zinc.

Concentrated nitric acid is a strong oxidizing agent. Copper and mercury but not gold and platinum dissolve in aqueous nitric acid, which is reduced to NO_2 or NO depending on its concentration. Certain base metals such as aluminum, iron, and chromium are not dissolved by concentrated nitric acid because of the formation of a dense adherent oxide layer, which *passivates* the metal by preventing further attack.

A mixture of concentrated nitric and sulfuric acids is used to nitrate organic compounds (e.g., in the conversion of benzene to nitrobenzene). A major component of this nitrating mixture is the *nitronium* ion NO_2^+, which is generated by reactions such as the following:

$$2HNO_3 \rightleftarrows NO_2^+ + NO_3^- + H_2O \qquad (4.48a)$$

$$HNO_3 + H_2SO_4 \rightleftarrows NO_2^+HSO_4^- + H_2O \qquad (4.48b)$$

Crystalline nitronium salts can be obtained by reactions such as the following:

$$N_2O_5 + HX \rightarrow NO_2^+X^- + HNO_3 \qquad (X = ClO_4 \text{ or } SO_3F) \quad (4.49a)$$

$$HNO_3 + 2SO_3 \rightarrow NO_2^+HS_2O_7^- \qquad (4.49b)$$

Nitronium salts are thermodynamically stable but chemically reactive; they are very strong nitrating agents and react violently with moisture to regenerate HNO_3.

Reaction of nitric acid or $NO_2^+BF_4^-$ with 90% H_2O_2 at 0°C gives unstable peroxonitric acid, HNO_4 or $HOOONO_2$, which can be removed from the reaction mixture by a stream of argon.[26]

4.6.5 Hydroxylamine Sulfonic Acid Derivatives Including Frémy's Salt, $K_2ON(SO_3)_2$

Replacement of N—H groups in hydroxylamine, NH_2OH, with N—SO_3H groups leads to hydroxylamine sulfonic acid derivatives. Such compounds are obtained by a redox reaction between the nitrite and bisulfite ions:

$$2HSO_3^- + HNO_2 \rightarrow HON(SO_3)_2^{2-} + H_2O \qquad (4.50a)$$

$$HON(SO_3)_2^{2-} + OH^- \rightarrow ON(SO_3)_2^{3-} + H_2O \qquad (4.50b)$$

Of particular interest is the one-electron oxidation (e.g., by air or PbO_2) of the trianion $ON(SO_3)_2^{3-}$ to give the dianion $ON(SO_3)_2^{2-}$, which can be isolated as its potassium salt, known as *Frémy's salt*. Formally Frémy's salt is an "inorganic" nitroxide similar to organic nitroxides of the general type R_2NO ($R = CF_3$, $(CH_3)_3C$, C_6H_5), which are paramagnetic molecules with one unpaired electron. However, Frémy's salt is dimorphic with a yellow monoclinic form and an orange-brown triclinic form. The orange-brown triclinic form is paramagnetic, but the yellow monoclinic form is nearly diamagnetic but with a thermally accessible triplet state. These unusual magnetic properties arise from magnetic interactions between the paramagnetic monomeric $ON(SO_3)_2^{2-}$ ions, which, in isolation, have a single unpaired electron. Frémy's salt is of interest as a one-electron oxidizing agent, being readily reduced back to $ON(SO_3)_2^{3-}$.

4.7 Nitrogen–Halogen Compounds

4.7.1 Nitrogen Fluorides[27]

Nitrogen trifluoride, NF_3 (mp $-206.8°C$, bp $-129°C$), is a very stable gas, normally reactive only above $\sim 250°C$. It is obtained by electrolysis of molten ammonium fluoride or by reaction of ammonia with fluorine over a copper catalyst. Nitrogen trifluoride is unaffected by water at room temperature. Its electron pair has no donor properties because of the strong electron-withdrawing properties of its three fluorine atoms.

Reaction of nitrogen trifluoride with elemental fluorine in the presence of a Lewis acid fluoride under pressure, a glow discharge, or ultraviolet irradiation gives salts of the *tetrafluoroammonium ion*, for example,

$$NF_3 + F_2 + BF_3 \xrightarrow{h\nu} NF_4^+ BF_4^- \qquad (4.51a)$$

$$NF_3 + F_2 + AsF_5 \longrightarrow NF_4^+ AsF_6^- \qquad (4.51b)$$

Tetrafluoroammonium salts, in contrast to ammonium salts, are readily hydrolyzed by water. The hydrolysis products of NF_4^+ are NF_3, H_2O_2, O_2, and HF.

Trifluoroamine oxide, NF_3O (mp $-160°C$, bp $-88°C$), is a toxic stable oxidizing gas that is resistant to hydrolysis. It is prepared from NF_3 and O_2 in an electric discharge or by photolysis of FNO (see Section 4.7.3) with F_2. Fluoride abstraction from NF_3O with SbF_5 gives the salt $[NF_2O][SbF_6]$ containing the NF_2O^+ cation. In addition, reduction of NF_3O with NO gives FNO.

Tetrafluorohydrazine or dinitrogen tetrafluoride, N_2F_4 (mp $-164.5°C$, bp $-73°C$), is obtained by the reaction of NF_3 with copper. In contrast to NF_3 it is very reactive because of its easy dissociation into NF_2 radicals[28,29]:

$$N_2F_4 \rightleftarrows 2NF_2 \qquad \Delta H = 78 \pm 6 \text{ kJ/mol at 298 K} \qquad (4.52)$$

Examples of chemical reactions of N_2F_4 through intermediate NF_2 radicals include the following:

$$N_2F_4 + 2NO \rightleftarrows 2ONNF_2 \text{ (unstable purple)} \qquad (4.53a)$$

$$N_2F_4 + Cl_2 \xrightarrow{hv} 2NF_2Cl \qquad (4.53b)$$

$$N_2F_4 + 2RI \xrightarrow{hv} 2RNF_2 + I_2 \qquad (4.53c)$$

$$N_2F_4 + 2RSH \rightarrow 2HNF_2 + RSSR \qquad (4.53d)$$

$$RCHO + N_2F_4 \rightarrow RC(O)NF_2 + HNF_2 \qquad (4.53e)$$

$$R_fCF{=}CF_2 + N_2F_4 \rightarrow R_fCF(NF_2)CF_2NF_2 \qquad (4.53f)$$

Treatment of N_2F_4 with SbF_5 results in fluoride abstraction to give the $N_2F_3^+$ cation:

$$N_2F_4 + SbF_5 \rightarrow N_2F_3^+ SbF_6^- \qquad (4.54)$$

Tetrafluorohydrazine is hydrolyzed by water after an induction period.

Difluoroamine, HNF_2, as obtained from N_2F_4 (equation 4.53d or 4.53e) or by the hydrolysis of the urea fluorination product H_2NCONF_2 with sulfuric acid, is a colorless explosive liquid (bp $23.6°C$), which, in contrast to NF_3, is a weak donor to strong Lewis acids such as BF_3.

Difluorodiazene, N_2F_2 or $FN{=}NF$, exists as either the *cis* isomer (bp $-106°C$) or the *trans* isomer (bp $-111°C$); the more reactive *cis* isomer of N_2F_2 predominates at room temperature. Difluorodiazene is obtained by the decomposition of fluorine azide (FN_3), the reaction of HNF_2 (equation 4.53d or 4.53e) with potassium fluoride at $-80°C$ followed by warming to $20°C$, or the treatment of N_2F_4 with aluminum chloride at lower temperatures. Abstraction of fluoride from *cis*-N_2F_2 with AsF_5 gives the salt $[N_2F][AsF_6]$; *trans*-N_2F_2 is unreactive toward AsF_5. The linear N_2F^+ cation in this salt is isoelectronic and isostructural with CO_2, N_2O, and NO_2^+.

4.7.2 Nitrogen Chlorides, Bromides, and Iodides[30]

All the nitrogen chlorides, bromides, and iodides are *dangerous explosives*.
Nitrogen trichloride, NCl_3, is a very reactive photosensitive pale yellow liquid (mp $-40°C$, bp $71°C$), which explodes violently upon impact or above its boiling point. The discoverer of NCl_3 (P. L. Dulong) lost three fingers and an

eye in handling NCl_3! The best preparation of NCl_3 uses the chlorination of ammonium chloride in acid aqueous solution according to the equation:

$$NH_4Cl + 3Cl_2 \rightarrow NCl_3 + 4HCl \tag{4.55}$$

The product can be extracted from the reaction mixture with carbon tetrachloride. The considerable hazards in handling NCl_3 can be minimized by handling it only in carbon tetrachloride solutions or in the dilute gas phase. Nitrogen trichloride is endothermic with an enthalpy of formation of 232 kJ/mol. The NCl_3 molecule is pyramidal like NH_3, with N—Cl = 1.753 Å and Cl–N–Cl = 107.47°. Hydrolysis of nitrogen trichloride gives hypochlorous acid (HOCl) and ammonia (NH_3) rather than nitrous acid (HONO) and hydrochloric acid (HCl), since nitrogen is more electronegative than chlorine in NCl_3. Reaction of NCl_3 with $SbCl_5$ and thionyl chloride gives the yellow salt $[ONCl_2]^+[SbF_6]^-$, which is stable to 145°C. Nitrogen trichloride vapor has been used to bleach flour.

Deep red *nitrogen tribromide*, NBr_3, is prepared by the following reaction:

$$(Me_3Si)_2NBr + 2BrCl \xrightarrow[-87°C]{pentane} NBr_3 + 3Me_3SiCl \tag{4.56}$$

It is a very temperature-sensitive explosive solid with properties similar to NCl_3.

Nitrogen triiodide is obtained as a black explosive solid polymeric ammoniate by reaction of concentrated aqueous ammonia with iodine according to the following equation:

$$2NH_3 + 3I_2 \rightarrow \frac{1}{n}[NI_3 \cdot NH_3]_n + 3HI \tag{4.57}$$

The structure of $[NI_3 \cdot NH_3]_n$ contains chains of vertex-sharing $NI_2I_{2/2}$ tetrahedra linked by NH_3 molecules.

4.7.3 Miscellaneous Nitrogen Halides

Chloramine, NH_2Cl (mp $-66°C$), is stable in the gas phase or in solution, but liquid and solid chloramine are explosive. It is obtained by chlorination of ammonia in the gas phase or by reaction of ammonia with hypochlorite at pH above 8.

Nitrosyl halides, X—N=O (X = F, Cl, Br) are formally acid halides of nitrous acid. They are obtained by treatment of nitric oxide with the corresponding halogen:

$$2NO + X_2 \rightarrow 2XNO \tag{4.58}$$

Nitrosyl chloride can also be obtained from N_2O_4 and KCl according to the following equation:

$$N_2O_4 + KCl \rightarrow ClNO + KNO_3 \tag{4.59}$$

The color deepens from colorless FNO (mp $-133°C$, bp $-60°C$) to yellow ClNO (mp $-60°C$, bp $-6°C$) and red BrNO (mp $-56°C$, bp $\sim 0°C$) with a concurrent decrease in stability. Nitrosyl halides are reactive and powerful oxidizing agents, which attack many metals. The ability of a 3:1 mixture of hydrochloric and nitric acids, called *aqua regia*, to dissolve noble metals such as gold and platinum is a consequence of the formation of nitrosyl chloride in this acid mixture. Nitrosyl halides are readily hydrolyzed to mixtures of HNO_3, HNO_2, NO, and HX.

The *nitryl halides*, FNO_2 (bp $-72°C$) and $ClNO_2$ (bp $-15°C$), are formally acid halides of nitric acid. They have planar structures and are obtained by reactions such as the following:

$$N_2O_4 + 2CoF_3 \xrightarrow{300°C} 2FNO_2 + 2CoF_2 \tag{4.60a}$$

$$ClSO_3H + HNO_3 \rightarrow ClNO_2 + H_2SO_4 \tag{4.60b}$$

The nitryl halides are readily decomposed by water to give HNO_3 and HX.

Nitrosyl hypofluorite, ONOF, is an isomer of nitryl fluoride that is obtained by treatment of nitrogen dioxide with elemental fluorine at $-30°C$.

Halogen nitrates, $FONO_2$ (bp $-46°C$); and $ClONO_2$ (bp $+22°C$), are highly reactive gases that are made by the reaction

$$XF + HNO_3 \rightarrow XONO_2 + HF \qquad (X = Cl, F) \tag{4.61}$$

Explosive $FONO_2$ readily decomposes into FNO and O_2.

References

1. Rozantsev, E. G., *Free Nitroxyl Radicals*, Plenum Press, New York, 1970

2. Latimer, W. M., *The Oxidation States of the Elements and Their Potentials in Aqueous Solutions*, Prentice-Hall, Englewood Cliffs, NJ, 1952, p. 104.

3. Randall, E. W.; Gillies, D. G., *Nitrogen Nuclear Magnetic Resonance*, in *Progress in Nuclear Magnetic Resonance Spectroscopy*, J. W. Emsley, J. Feeney, and L. H. Sutcliffe, Eds., Pergamon Press, Oxford, 1970.

4. von Philipsborn, W.; Müller, R., ^{15}N-NMR Spectroscopy—New Methods and Applications, *Angew. Chem. Int. Ed. Engl.*, 1986, **25**, 383–413.

5. Allen, A. D.; Harris, R. O.; Loescher, B. R.; Stevens, J. R.; Whiteley, R. M., Dinitrogen Complexes of the Transition Metals, *Chem. Rev.*, 1973, **73**, 11–20.

6. Sellmann, D., Dinitrogen–Transition Metal Complexes: Synthesis, Properties, and Significance, *Angew. Chem. Int. Ed. Engl.*, 1974, **13**, 639–649.

7. Allen, A. D.; Senoff, C. V., *J. Chem. Soc. Chem. Commun*, 1965, 621.

8. Al-Salih, T. I.; Pickett, C. J., *J. Chem. Soc. Dalton Trans.*, 1985, 1255–1264.

9. Wright, A. N.; Winkler, C. A., *Active Nitrogen*, Academic Press, New York 1968.

10. Jolly, W. L.; Hallada, C. J., *Liquid Ammonia*, in *Non-Aqueous Solvent Systems*, T. C. Waddington, Ed., Academic Press, London, 1965, pp. 1–45.

11. Fowles, G. W. A., *Inorganic Reactions in Liquid Ammonia*, in *Developments in Inorganic Nitrogen Chemistry*, C. B. Colburn, Ed., Elsevier, Amsterdam, 1966, pp. 522–576.

12. Bertie, J. E.; Shehata, N. R., *J. Chem. Phys.*, 1984, **81**, 27–30.

13. Garcia, E.; Cowley, A. H.; Bard, A. J., "Quaternary Ammonium Amalgams" as Zintl Ion Salts and Their Use in the Synthesis of Novel Quaternary Ammonium Salts, *J. Am. Chem. Soc.*, 1986, **108**, 6082–6083.

14. Eckhardt, E. W., *Hydrazine and Its Derivatives: Preparation, Properties, and Applications*, Wiley, New York, 1984.

15. Smith, P. A., *Derivatives of Hydrazine and Other Hydronitrogens with N—N Bonds*, Benjamin-Cummings, Menlo Park, CA, 1982.

16. Miller, C. E., Hydrogenation with Diimide, *J. Chem. Educ.*, 1965, **42**, 254–259.

17. Yoffe, A. D., *The Inorganic Azides*, in *Developments in Inorganic Nitrogen Chemistry*, Vol. 1, C. B. Colburn, Ed., Elsevier, Amsterdam, 1966, pp. 72–149.

18. Jauszik, R.; Crocetta, P., Production of Hydroxylamine from Nitrogen Oxide: A Short Review, *Appl. Catal.*, 1985, **7**, 1–21.

19. Richter-Adoo, G. B.; Legzdins, P., *Metal Nitrosyls*, Oxford University Press, New York, 1992.

20. Addison, C. C., Dinitrogen Tetroxide, Nitric Acid, and Their Mixtures as Media for Inorganic Reactions, *Chem. Rev.*, 1980, **80**, 21–39.

21. Addison, C. C.; Logan, N., *The Chemistry of Dinitrogen Pentoxide*, in *Developments in Inorganic Nitrogen Chemistry*, Vol. 2, C. B. Colburn, Ed., Elsevier, Amsterdam, 1973, pp. 27–69.

22. Hughes, M. N., Hyponitrites, *Q. Rev.*, 1968, **22**, 1–13.

23. Addison, C. C.; Gamlen, G. A.; Thompson, R., *J. Chem. Soc.*, 1952, 338–345.

24. Blough, N. V.; Zafirou, O. C., *Inorg. Chem.*, 1985, **24**, 3502–3504.

25. King, P. A.; Anderson, V. E.; Edwards, J. O.; Gustafson, G.; Plumb, R. C.; Suggs, J. W., *J. Am. Chem. Soc.*, 1992, **114**, 5430–5432.

26. Kenley, R. A.; Trevor, P. L.; Lan, B. Y., *J. Am. Chem. Soc.*, 1981, **103**, 2203–2206.

27. Colburn, C. B., Nitrogen Fluorides and Their Inorganic Derivatives, *Adv. Fluorine Chem.*, 1963, **3**, 92–116.

28. Johnson, F. H.; Colburn, C. B., *J. Am. Chem. Soc.*, 1961, **83**, 3043–3047.

29. Davies, P. B.; Kho, C. J.; Leong, W. K.; Lewis-Bevan, W., *J. Chem. Soc. Chem. Commun.*, 1982, 690–691.

30. Jander, J., Recent Chemistry and Structure Investigation of NI_3, NBr_3, NCl_3, and Related Compounds, *Adv. Inorg. Chem. Radiochem.*, 1976, **19**, 1–63.

CHAPTER

5

Phosphorus, Arsenic, Antimony, and Bismuth

5.1 General Aspects of the Chemistry of Phosphorus, Arsenic, Antimony, and Bismuth

The valence shells of phosphorus, arsenic, antimony, and bismuth are similar to those of nitrogen except for the presence of energetically available d orbitals. This similarity has a number of chemical consequences, as summarized in Table 5.1, which compares the chemistry of phosphorus with that of nitrogen (Chapter 4). Note that the presence of accessible d orbitals allows phosphorus and its heavier congeners to have coordination numbers larger than 4, in contrast to nitrogen, which is limited to the maximum coordination number of 4 provided by sp^3 hybridization.

The "typical" oxidation states of phosphorus and its heavier congeners are $+3$ and $+5$, in accord with their position in the periodic table. Examples of the stereochemistries of the $+3$ and $+5$ oxidation states of phosphorus are depicted in Figure 5.1.[1] The elements in their $+3$ oxidation states typically form three pyramidal single bonds plus a lone electron pair, as exemplified by the tertiary phosphines, R_3P: The tetrahedron in the $+3$ oxidation states can be completed by reactions involving the long pair such as the following in the case of tertiary phosphines:

Quaternization:
$$R_3P: + RX \rightarrow R_4P^+X^- \tag{5.1}$$

Example of metal complex formation
$$R_3P: + M(CO)_6 \rightarrow R_3P{\rightarrow}M(CO)_5 + CO\uparrow \ (M = Cr, Mo, W) \tag{5.2}$$

Table 5.1 A Comparison Between the Chemistry of Phosphorus and Nitrogen

Nitrogen	Phosphorus	Examples
No valence expansion	Valence expansion using the accessible d orbitals	P but not N forms derivatives with coordination numbers 5 (dsp^3 hybridization) and 6 (d^2sp^3 hybridization)
$p\pi$–$d\pi$ bonding not possible	Weak to moderate but important $d\pi$–$p\pi$ and even $d\pi$–$d\pi$ bonding	Back-bonding in R_3P but not R_3N metal complexes
Strong $p\pi$–$p\pi$ bonds	Weak $p\pi$–$p\pi$ bonds	N forms RON=O but P forms P(OR)$_3$

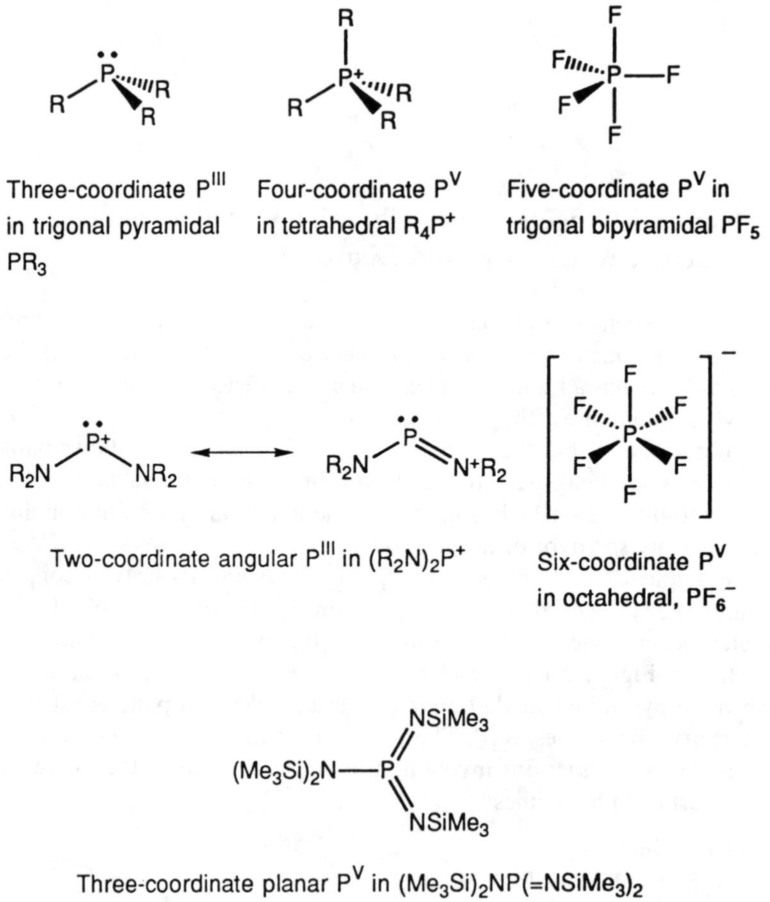

Three-coordinate P^{III} in trigonal pyramidal PR$_3$

Four-coordinate P^V in tetrahedral R$_4$P$^+$

Five-coordinate P^V in trigonal bipyramidal PF$_5$

Two-coordinate angular P^{III} in (R$_2$N)$_2$P$^+$

Six-coordinate P^V in octahedral, PF$_6^-$

Three-coordinate planar P^V in (Me$_3$Si)$_2$NP(=NSiMe$_3$)$_2$

Figure 5.1. Some examples of phosphorus stereochemistry in the $+3$ and $+5$ oxidation states.

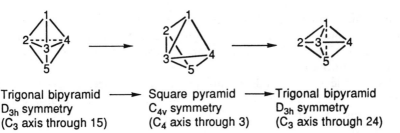

Trigonal bipyramid $\longrightarrow$ Square pyramid $\longrightarrow$ Trigonal bipyramid
D_{3h} symmetry $\qquad$ C_{4v} symmetry $\qquad$ D_{3h} symmetry
(C_3 axis through 15) $\quad$ (C_4 axis through 3) $\quad$ (C_3 axis through 24)

Figure 5.2. The Berry pseudorotation process converting a trigonal bipyramid to an isomeric trigonal bipyramid through a square pyramid intermediate.

The latter reaction is but one of literally thousands of examples of tertiary phosphine transition metal complexes involving donation of the phosphorus lone pair to the transition metal atom. Two-coordinate phosphorus(III) is found in anions of the type PH_2^- and PR_2^- (e.g., R = phenyl), the phosphaalkenes $R_2C{=}PR'$, and cations of the type $(R_2N)_2P^+$; the last are stabilized by N—P $p\pi$–$d\pi$ bonding and $p\pi$–$p\pi$ bonding.

Phosphorus and its heavier congeners in the +5 oxidation state typically form so-called *hypervalent derivatives* in which the d orbitals are involved in dsp^3 and d^2sp^3 hybrids. The latter is an octahedron, whereas dsp^3 hybrids can be either trigonal bipyramidal or square pyramidal depending on whether the d_{z^2} or $d_{x^2-y^2}$ orbital is used. The more electronegative substituents prefer axial sites in five-coordinate trigonal bipyramidal derivatives with mixed substituents. Phosphorus and its heavier congeners in the +5 oxidation state also form planar three-coordinate derivatives such as $(Me_3Si)_2NP({=}NSiMe_3)_2$ (Figure 5.1) in which the five valence electrons of phosphorus form only three bonds, two of which are double bonds.

An important property of five-coordinate complexes, such as those of phosphorus and its heavier congeners, is stereochemical nonrigidity.[2] Thus, trigonal bipyramids and square pyramids can undergo rapid interconversions through Berry pseudorotation processes (Figure 5.2). In the *Berry pseudorotation* process, an edge of the trigonal bipyramid (edge 24 in Figure 5.2) is broken, leading to a square pyramid after relatively small movement of the other edges to give a C_4 axis through the apical vertex 3. A new edge is then formed across a diagonal in the base of the square pyramid (vertices 1 and 5 in Figure 5.2), leading to a new trigonal bipyramid after slight movements of the other edges to give D_{3h} symmetry. Note that axial vertices of the original trigonal bipyramid (vertices 1 and 5 in Figure 5.2) become equatorial vertices in the new trigonal bipyramid, and two equatorial vertices in the original trigonal bipyramid (vertices 2 and 4 in Figure 5.2) become the two axial vertices in the new trigonal bipyramid. Thus a Berry pseudorotation can interchange axial and equatorial vertices in a trigonal bipyramid on a rapid time scale so that physical measurements with a sufficiently long time scale (e.g., NMR spectroscopy) show the axial and equatorial vertices to be equivalent despite their obvious topological difference. For example, the ^{19}F NMR spectrum of PF_5 under ambient conditions shows only a single sharp resonance, indicating equivalence of all five fluorine atoms on the NMR time scale because of a rapid Berry pseudorotation process.

Mononuclear compounds of phosphorus or its heavier congeners in oxidation states other than $+3$ and $+5$ are very rare. An exception is the stable phosphorus-centered free radical $[(Me_3Si)_2CH]_2P\cdot$, which can be obtained by the following reaction:

$$2[(Me_3Si)_2CH]_2PCl + (R_2N)_2C{=}C(NR_2)_2 \xrightarrow{h\nu}$$

$$2[(Me_3Si)_2CH]_2P\cdot + [(R_2N)_4C_2]Cl_2 \quad (5.3)$$

The stability of this phosphorus radical arises from the bulk of the bis(trimethylsilyl)methyl substituents.

The relative stabilities of the $+3$ and $+5$ oxidation states as well as their acid–base properties change markedly from phosphorus through arsenic and antimony to bismuth. The stability of the highest $(+5)$ oxidation state decreases with increasing atomic number. This is indicated by the strong reducing properties of many phosphorus(III) compounds such as tertiary phosphines, phosphites, and phosphorous acid (H_3PO_3), contrasted with the very strong oxidizing properties of all bismuth(V) compounds including BiF_5 and bismuth(V) oxoanions. Furthermore in a given oxidation state the metallic character of the elements, hence the basicity of the corresponding oxides, increases with increasing atomic number. Thus phosphorus(III) and arsenic(III) oxides are strictly acidic, antimony(III) oxide is amphoteric, and bismuth(III) oxide is strictly basic.

The only naturally occurring isotope of phosphorus is ^{31}P. This isotope, with a spin of $\frac{1}{2}$ and an NMR sensitivity 0.066 that of the proton and about 377 times that of ^{13}C, is a very favorable nucleus for NMR studies.[3,4] Phosphorus-31 NMR data therefore have played an important role in the elucidation of the nature of phosphorus compounds in solution. All naturally occurring isotopes of arsenic, antimony, and bismuth have significant quadrupole moments; the resulting quadrupole broadening makes NMR spectroscopy for these elements. The antimony isotope ^{121}Sb, of 57.25% natural abundance, is suitable for Mössbauer studies.

5.2 Elemental Phosphorus, Arsenic, Antimony, and Bismuth

Except for minute amounts of elemental arsenic and bismuth, phosphorus, arsenic, antimony, and bismuth are found in nature only in combined form. The abundance of these elements in the earth's crust decreases with increasing atomic number. Thus phosphorus is much more abundant than arsenic, antimony, and bismuth combined. Important phosphorus minerals include *collophanite*, $Ca_3(PO_4)_2 \cdot H_2O$, and *apatite*, $Ca_5X(PO_4)_3$ (X = F, Cl, OH), with the latter being the most important phosphorus source. Arsenic and antimony are associated in nature with sulfide minerals, particularly those of copper, silver, and lead. Bismuth minerals include the oxide Bi_2O_3 as well as sulfides.

Elemental phosphorus occurs in three main forms: white phosphorus, black phosphorus, and red phosphorus. White phosphorus (mp 44°C, bp 280°C), consists of tetrahedral P_4 molecules with P—P distances of 2.21 Å. It can be prepared in large quantities by reduction of phosphate rock with coke in the presence of silica in an electric furnace according to the following reaction:

$$2Ca_3(PO_4)_2 + 6SiO_2 + 10C \xrightarrow{\Delta} P_4\uparrow + 6CaSiO_3 + 10CO\uparrow \qquad (5.4)$$

Under these conditions the white phosphorus distills out of the hot reaction mixture and is condensed in the cooler parts of the reactor.

White phosphorus is a spontaneously flammable solid ($> \sim 35°C$) that is often kept as sticks under water, in which it is insoluble. In accord with its molecular structure, it is soluble in many nonaqueous solvents, notably carbon disulfide (880 g P_4/100 g CS_2 at 10°C) but also liquid ammonia, phosphorus trichloride, liquid sulfur dioxide, amines, and so on. Controlled slow oxidation of white phosphorus leads to emission of light, which is the origin of the term *phosphorescence*. White phosphorus is highly toxic, with a fatal human dose of ~ 50 mg.

The remaining forms of elemental phosphorus, namely black and red phosphorus, are insoluble polymers, which are much less reactive chemically than white phosphorus. Black phosphorus is the most thermodynamically stable and least reactive form of the element; it can be obtained by heating white phosphorus under pressure. The initially produced form of black phosphorus is the orthorhombic form, which is graphitic in appearance and conductance. Its structure consists of a double layer of three-coordinate phosphorus atoms. Application of pressure above 12 kbar converts ortho-rhombic black phosphorus successively into rhombohedral black phosphorus, with a structure consisting of hexagonal nets of three-coordinate phosphorus atoms, and finally into cubic black phosphorus. Red phosphorus, in contrast to black phosphorus, is not crystalline but amorphous. It is readily obtained by heating white phosphorus to $\sim 400°C$ in a sealed container. Red phosphorus is believed to consist of polymers obtained by partial opening of the P_4 tetrahedra of white phosphorus, but its structure is not definitively known because of its amorphous nature.

The heavier congeners of phosphorus have fewer allotropic forms. They commonly occur as α-rhombohedral allotropes similar to rhombohedral black phosphorus discussed previously. Unstable molecular allotropes of arsenic and antimony, E_4 (E = As, Sb), analogous to white phosphorus, can be obtained by rapid condensation of the vapors.

5.3 Phosphides, Arsenides, Antimonides, and Bismuthides

Direct combination of phosphorus, arsenic, antimony, and bismuth with metallic elements, frequently at elevated temperatures, leads to phosphides, arsenides, antimonides, and bismuthides, respectively. Important compounds of this type include those discussed in Section 5.3.1.

5.3.1 III–V Compounds

Compounds such as gallium arsenide, GaAs, and indium phosphide, InP (i.e., the so-called *III–V compounds*), are isoelectronic with the semiconducting elements silicon and germanium. They are therefore important semiconductors for the fabrication of microelectronic and optoelectronic devices including solid state lasers, light-emitting diodes, and photodetectors. For such applications the III-V compounds are generally made under milder conditions than those required for direct combination of the elements such as the vapor phase thermal decomposition of a volatile mixture or a single compound containing both elements (e.g., a mixture of Me_3Ga and AsH_3 or single source precursors such as $[Bu_2^tGa(\mu\text{-}AsH_2)]_3$ or $[Me_2Ga(\mu\text{-}AsBu_2^t)]_2$ for GaAs).[5] Thus the pyrolysis of $[Bu_2^tGa(\mu\text{-}AsH_2)]_3$ to give GaAs proceeds cleanly with the evolution of isobutane according to the following equation[6]:

$$[Bu_2^tGa(\mu\text{-}AsH_2)]_3 \xrightarrow{\Delta} 6Bu^iH \uparrow + 3GaAs \qquad (5.5)$$

5.3.2 Transition Metal Phosphides and Arsenides

Solid state transition metal phosphides can be classified into two types on the basis of the coordination numbers of the component atoms.[7] In the transition metal phosphides with high coordination numbers, the metal atoms are surrounded by more than six phosphorus atoms and at least some phosphorus atoms have more then four nearest neighbors. The bonding in such transition metal phosphides cannot be rationalized simply by two-center, two-electron bonds, and such compounds have metallic properties like other intermetallic compounds. In the transition metal phosphides with low coordination numbers, the metal atoms have six (octahedral), four (tetrahedral or square planar), or even as few as two (linear) phosphorus neighbors, and the phosphorus atoms have tetrahedral coordination if their lone pairs are counted. The transition metal phosphides with low coordination numbers are richer in phosphorus than those with high coordination numbers, and their chemical bonding can be rationalized simply by two-center, two-electron bonds.[8] Examples of transition metal phosphides with low coordination numbers include MP_4 (M = Fe, P, Os, Mn, Tc, Re, Cr, Mo, W, V), MP_3 (M = Co, Tc), MP_2 (M = Co, Rh, Ir, Fe, Ru, Os), Re_2P_5, and Re_6P_{13}.

Transition metal arsenides include $CoAs_3$ (*skutteridite*), which has planar

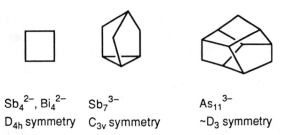

Sb_4^{2-}, Bi_4^{2-} Sb_7^{3-} As_{11}^{3-}

D_{4h} symmetry C_{3v} symmetry ~D_3 symmetry

Figure 5.3. The frameworks of Zintl ions of arsenic, antimony, and bismuth, which have been characterized by X-ray crystallography.

As_4 rings, $NiAs_2$ (*marcasite*), which has distinct As_2 pairs, and NiAs, which is the prototype for many MX structures.

5.3.3 Zintl Phases

The elements phosphorus, arsenic, antimony, and bismuth all can be reduced by the most electropositive metals (alkali and alkaline earth metals) to give a considerable variety of derivatives containing E_n^{m-} ions of various types.[9] Such ions are generically called *Zintl* phases after the chemist who first characterized these systems by potentiometric titration in the early 1930s.[10] Single crystals of many of these alkali metal phosphides, arsenides, or bismuthides suitable for structure determination by X-ray crystallography can be obtained by dissolving the solid reduction products in liquid ammonia or ethylenediamine and adding a cryptate to complex the alkali metal cation. In general these derivatives are very air-sensitive. Polyatomic anions of arsenic, antimony, and bismuth found in Zintl phases that have been characterized by X-ray crytallography (Figure 5.3) include the square anions Sb_4^{2-} and Bi_4^{2-}, which may be considered to have six delocalized π electrons such as $C_4H_4^{2-}$ (or even benzene, C_6H_6), as well as the more complicated species Sb_7^{3-} and As_{11}^{3-} with threefold symmetry axes passing through one or two vertices, respectively. The bonding in Sb_7^{3-} and As_{11}^{3-} can be rationalized by simple two-electron, two-center bonds with single negative charges on the two-coordinate atoms, analogous to the negative charge on the PH_2^- phosphorus atom. The structure of Sb_7^{3-} is closely related to the structure of the isoelectronic P_4S_3 (Section 5.7).

5.4 Hydrides of Phosphorus, Arsenic, Antimony, and Bismuth

The hydrides EH_3 (E = P, As, Sb, Bi) may be obtained by treatment of phosphides or arsenides with acid or by reduction of arsenic, antimony, or bismuth derivatives either electrolytically or with an electropositive metal. Their stabilities decrease rapidly in the sequence $PH_3 > AsH_3 > SbH_3 > BiH_3$, so that SbH_3 and BiH_3 decompose at room temperature and $-45°C$ respectively.

Phosphine, PH_3, is a very toxic gas that can be prepared by treatment of Ca_3P_2 or AlP with dilute acid, pyrolysis of H_3PO_3 at 200°C, treatment of white phosphorus with aqueous NaOH or KOH, or reduction of PCl_3 with $LiAlH_4$ or LiH. Phosphine oxidizes readily in air, but when pure is not spontaneously flammable. However, the presence of a P_2H_4 impurity, such as occurs when PH_3 is prepared from Ca_3P_2, can make PH_3 spontaneously inflammable. The boiling point of PH_3 ($-87.7°C$) is much lower than that of NH_3 ($-33.5°C$), in accord with the absence of significant hydrogen bonding in PH_3, in contrast to NH_3 (see Chapter 1), leading to lack of association in liquid PH_3. Phosphine is only sparingly soluble in water but more soluble in organic liquids such as benzene, carbon disulfide, and trichloroacetic acid. Phosphine can lose a proton upon treatment with an alkali metal in liquid ammonia to give the phosphide ion, PH_2^-:

$$2PH_3 + 2Na \xrightarrow{\text{liq } NH_3} 2NaPH_2 + H_2\uparrow \qquad (5.6)$$

Similarly, phosphine can add a proton upon treatment with hydrogen iodide to give phosphonium iodide:

$$PH_3(g) + HI(g) \rightarrow PH_4^+I^-(s) \qquad (5.7)$$

In addition, phosphine is protonated by very strong acids such as $BF_3 \cdot H_2O$. Since, however, both the acidic ($K_a = 1.6 \times 10^{-29}$) and the basic ($K_b = 4 \times 10^{-28}$) properties of phosphine are minimal,[11] "salts" containing either PH_2^- or PH_4^+ are rapidly hydrolyzed by water, regenerating PH_3, and PH_3 is neutral in aqueous solution. However, formation of $PH_4^+I^-$ followed by its decomposition in aqueous alkali is a useful way to obtain very small amounts of very pure PH_3.

The P—H bonds in phosphine are reactive toward addition to unsaturated organic compounds. An important reaction of this type is the addition of phosphine to the C=O bonds in four equivalents of formaldehyde according to the following equation:

$$PH_3 + 4H_2C{=}O + HCl \rightarrow [P(CH_2OH)_4]^+Cl^- \qquad (5.8)$$

The resulting product, tetrakis(hydroxymethyl)phosphonium chloride, is used as the major ingredient with urea or melamine–formaldehyde resins for the permanent flameproofing of cotton cloth.

Phosphine has a trigonal pyramidal structure with H–P–H angles of 93.7°. The inversion barrier of the PH_3 pyramid (155 kJ/mol) is much higher than that of the NH_3 pyramid (24 kJ/mol). There are similar differences between the inversion barriers of organic phosphines RR'R"P and organic amines RR'R"N making it possible to resolve asymmetric phosphines RR'R"P but not the corresponding asymmetric amines RR'R"N into pure enantiomers.

Higher phosphorus hydrides with phosphorus–phosphorus bonds are much less stable than PH_3. However, *diphosphine*, P_2H_4 (H_2P—PH_2), which is an analogue of hydrazine (Section 4.4.2) or ethane, can be isolated as a by-product from the hydrolysis of Ca_3P_2 as a spontaneously flammable yellow liquid (mp $-99°C$, bp $63.5°C$, extrap), which decomposes to yellow insoluble polymers when stored at room temperature.

Arsine, AsH_3 (mp $-116°C$, bp $-62°C$), is a very toxic gas that is formed when many arsenic-containing compounds are reduced by nascent hydrogen. The facile decomposition of arsine to an arsenic mirror is the basis of the Marsh test for arsenic. Arsine has recently become important as an arsenic source in the chemical vapor deposition of the important III-V semiconductor gallium arsenide. Arsenic, antimony, and bismuth analogues of the phosphonium ion PH_4^+ are unknown.

5.5 Halides of Phosphorus, Arsenic, Antimony, and Bismuth

5.5.1 Trihalides

All 16 possible trihalides EX_3 (E = P, As, Sb, Bi; X = F, Cl, Br, I) are known. In general they are readily hydrolyzed by water and rather volatile. This class of compounds exhibits the following structural features:

1. *Gaseous EX_3 derivatives.* Discrete molecular C_{3v} trigonal pyramidal structures.
2. *Iodides EI_3 (E = As, Sb, Bi).* Layer lattices with no discrete molecules; their structures may be regarded as close-packed arrays of iodine atoms with the As, Sb, or Bi atoms in the octahedral interstices.
3. *BiF_3.* Ionic lattice with nine-coordinate tricapped trigonal prismatic bismuth atoms.
4. *SbF_3.* SbF_3 molecules linked through fluorine bridges.

Phosphorus trifluoride (trifluorophosphine), PF_3 (mp $-151.5°C$, bp $-101.8°C$), is a colorless, odorless, toxic gas, which can be generated from phosphorus trichloride and ZnF_2 or AsF_3. Unlike most other EX_3 trihalides, PF_3 is unreactive toward pure water under ambient conditions and can be freed from more reactive impurities by passing through a water trap. Phosphorus trifluoride functions as a strong π acceptor toward transition metals similar to carbon monoxide (Section 2.5.1.1)[12,13] and, like carbon monoxide, is toxic because of complex formation with the iron in hemoglobin. The strong π-acceptor properties of PF_3 arise from the presence of three strongly electronegative fluorine atoms to withdraw much of the electron density from low-valent transition metal atoms through $d\pi$–$d\pi$ bonding (Figure 5.4). Zerovalent PF_3 transition metal complexes, such as $M(PF_3)_6$ (M = Cr, Mo, and W), $Fe(PF_3)_5$, and $Ni(PF_3)_4$ are volatile air-stable solids or liquids much like the corresponding zerovalent metal carbonyls. In some cases stable PF_3 derivatives (e.g., $Pt(PF_3)_4$ and $Rh_2(PF_3)_8$) can be prepared which have no stable carbonyl analogues. Since, however, PF_3 has a much lower tendency to function as a bridging ligand than CO, no PF_3 analogues of $Fe_2(CO)_9$ and $Co_2(CO)_8$, which have bridging carbonyl groups, are known.

Figure 5.4. The $d\pi$–$d\pi$ $M \to P$ back-bonding in transition metal complexes of PF_3.

Arsenic trifluoride, AsF_3 (mp $-6°C$, bp $62.8'C$), and *antimony trifluoride*, SbF_3 (mp $292°C$), do not form extensive series of transition metal complexes like PF_3, probably because As and Sb are weaker σ donors than P. However, these fluorides are useful fluorinating agents, as indicating by the use of AsF_3 for the conversion of PCl_3 to PF_3 mentioned earlier. The relatively nonvolatile antimony trifluoride is useful not only for the conversion of reactive nonmetallic inorganic chlorides to the corresponding fluorides but also for the conversion of sufficiently reactive C—Cl bonds to C—F bonds in the synthesis of fluorocarbons.

Phosphorus trichloride (mp $-93.5°C$, bp $+76.1°C$) is a readily hydrolyzed liquid that is made industrially on a large scale by the direct chlorination of phosphorus suspended in previously prepared PCl_3. Phosphorus trichloride is used as a precursor for the manufacture of number of organic phosphorus derivatives of the types PR_3, PR_nCl_{3-n}, $PR_n(OR)_{3-n}$, $(RO)_3P{=}O$, and $(RO)_3P{=}S$ by combinations of halogen substitution and $P(III) \to P(V)$ oxidation reactions. Organic phosphorus derivatives of these types are used in large quantities as oil additives, plasticizers, flame retardants, fuel additives, and intermediates in the manufacture of insecticides.

Arsenic trichloride (mp $-16.2°C$, bp $103.2°C$, viscosity 1.23 cPs, dielectric constant 12.8, conductivity 1.4×10^{-7} $\Omega \cdot cm^{-1}$),[14] and *antimony trichloride* (mp $73°C$, bp $223°C$, viscosity 2.58 cPs, dielectric constant 33.2, conductivity 1.4×10^{-6} $(\Omega \cdot cm)^{-1}$),[15] and good nonaqueous solvents for chloride transfer reactions. However, their self-ionization to give ECl_2^+ and ECl_4^- is very low.

Phosphorus triiodide[16] (mp $61.2°C$), which can be synthesized from elemental phosphorus and iodine in the correct mole ratio under mild conditions, is useful as a mild deoxygenating agent in dichloromethane solution.[17] Thus PI_3 converts R_2SO to R_2S and RCH_2NO_2 or $RCH{=}NOH$ to $RC{\equiv}N$ all in high yield (75–95%).

5.5.2 Pentahalides

All the pentafluorides EF_5 (E = P, As, Sb, Bi) are known. *Phosphorus pentafluoride* (bp $-101.5°C$), and *arsenic pentafluoride* (bp $-53°C$), are monomeric molecular trigonal bipyramidal species. Their ^{19}F NMR spectra exhibit single resonances, even though the axial and equatorial fluorine atoms are different

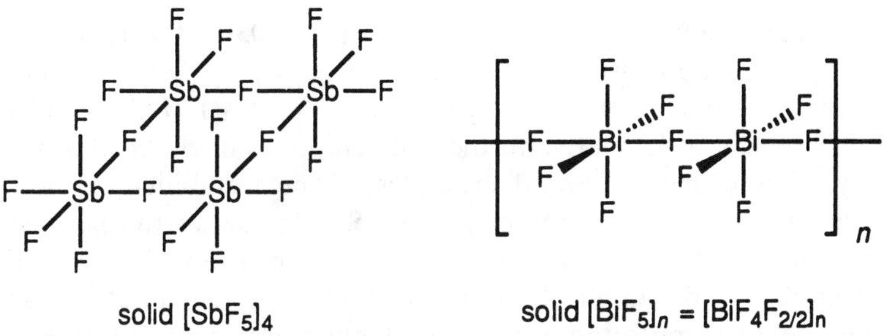

Figure 5.5. Association through fluorine bridging in SbF_5 and BiF_5.

owing to Berry pseudorotation (Figure 5.2), discussed in Section 5.1. *Antimony pentafluoride* (mp 7°C, bp 149.5°C), is markedly different from PF_5 and AsF_5 because it is associated through fluorine bridges even in the gas phase.[18] Thus, liquid SbF_5 is very viscous owing to the formation of liquid polymers by fluorine bridging (Figure 5.5). Solid SbF_5 forms cyclic tetramers wth bridging fluorine atoms. *Bismuth pentafluoride* also has a polymeric structure consisting of linear chains of BiF_6 octahedra linked by *trans*-fluoride bridges (Figure 5.5); it thus can be represented as $BiF_4F_{2/2}$.

The pentafluorides EF_5 (E = P, As, Sb, Bi) are all potent fluoride acceptors, readily forming the octahedral EF_6^- anions. The *hexafluorophosphate anion*, PF_6^-, is commonly used when a reasonably large spherical, noncomplexing anion is needed: it is readily available as ammonium and alkali metal salts, which are soluble and stable in water. Bismuth pentafluoride reacts explosively with water to evolve a mixture of ozone and OF_2 with the formation of a voluminous brown precipitate.

Phosphorus pentachloride (mp 167°C, bp 160°C, subl) is molecular trigonal bipyramidal PCl_5 in the gas and liquid phases but ionic $[PCl_4]^+[PCl_6]^-$ in the solid phase. It is hydrolyzed by water to give first $O{=}PCl_3$ and finally phosphoric acid. Reactions of alcohols (ROH) and carboxylic acids (RCO_2H) with PCl_5 provide useful syntheses of alkyl chlorides (RCl) and acyl chlorides (RCOCl), respectively. *Phosphorus pentabromide* is dissociated completely to PBr_3 and Br_2 in the gas phase and is normally ionic $[PBr_4]^+Br^-$ in the solid phase, although another ionic phase $[PBr_4]^+[Br_3^-]\cdot PBr_3$, containing both phosphorus(III) and phosphorus(V) as well as the tribromide ion, can be obtained by rapidly cooling PBr_5 vapor at 15 K. *Phosphorus pentaiodide* has not been well characterized but is probably $[PI_4]^+I^-$ in solution.

Many of the pentahalides of arsenic, antimony, and bismuth containing halogens heavier than fluorine, including $BiCl_5$ and all the pentabromides and pentaiodides EX_5 (E = As, Sb, Bi; X = Br, I), cannot be isolated or observed because internal redox reactions give the elemental halogen and the corresponding trihalide. Pale yellow *arsenic pentachloride* has been prepared by the ultraviolet irradiation of $AsCl_3$ with chlorine at $-105°C$, but it is stable only below $-50°C.$[19] *Antimony pentachloride* (bp $2.8°C$, bp $79°C/22$ mm Hg), is considerably more stable and decomposes only above $+140°C$. The instability of arsenic pentachloride relative to the pentachlorides of either phosphorus or antimony is only one of several examples of the instability of the highest oxidation states of the elements of the As, Se, Br row of the periodic table relative to those of either the row above (P, S, Cl) or below (Sb, Te, I); this has been attributed to the stabilization of the $4s^2$ electron pair in all the elements following the first transition series. Solid antimony pentachloride has the dimeric structure $Cl_4(\mu\text{-Cl})_2SbCl_4$ (**5.1**).

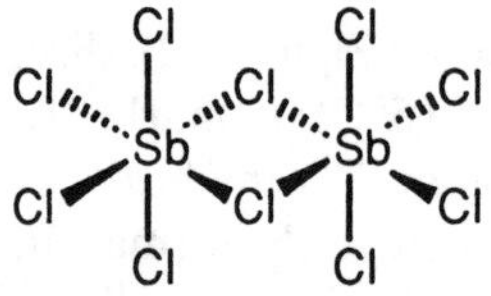

5.1

5.5.3 Lower Halides

5.5.3.1 The Dimers $X_2E—EX_2$

Phosphorus halides of the type $X_2P—PX_2$ (X = F, Cl, I) can be obtained by the following reactions:

$$2PF_2I + 2Hg \rightarrow F_2P—PF_2 + Hg_2I_2 \tag{5.9a}$$

$$2PCl_3 + H_2 \xrightarrow{\text{silent discharge}} Cl_2P—PCl_2 + 2HCl \tag{5.9b}$$

$$P_4 + 4I_2 \xrightarrow{CS_2} 2I_2P—PI_2 \tag{5.9c}$$

Other related halides include the iodide As_2I_4 and Sb_2I_4, which readily disproportionate into EI_3 and the free element.

5.5.3.2 Bismuth Cluster Halides

Dinuclear bismuth halides of the type Bi_2X_4 are unknown. An attempt to prepare such a halide by the reaction of bismuth trichloride with elemental bismuth leads instead to a material of stoichiometry $Bi_{24}Cl_{28}$, shown by X-ray crystallography[20] to be $[Bi_9^{5+}]_2[BiCl_5^{2-}]_4[Bi_2Cl_8^{2-}]$ containing a tricapped trigonal prismatic Bi_9^{5+} cluster cation as well as $BiCl_5^{2-}$ and $Bi_2Cl_8^{2-}$ anions containing square pyramidal bismuth(III), which can alternatively be regarded as ψ-octahedral with a stereochemically active lone pair (Figure 5.6).

Bi_9^{5+} $BiCl_5^{2-}$ $Bi_2Cl_8^{2-}$

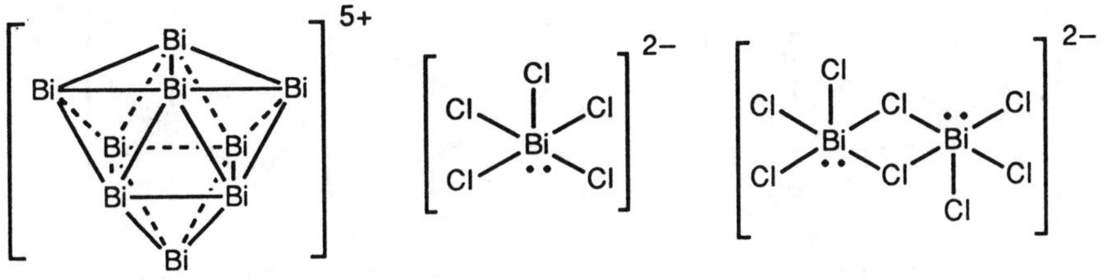

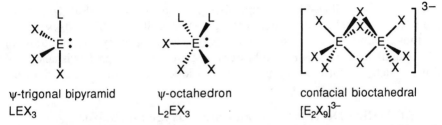

Figure 5.6. The species present in $Bi_{24}Cl_{28} = [Bi_9^{5+}]^2[BiCl_5^{2-}]_4[Bi_2Cl_8^{2-}]$. Similar $As_2X_8^{2-}$ (X = Cl, Br) anions are also known.

5.5.4 Complexes of Trivalent Halides: Stereochemically Active Lone Electron Pairs

The phosphorus trihalides exhibit essentially no Lewis acidity. Arsenic trihalides AsX_3 (X = Cl, Br) display a limited Lewis acidity and form the anions $As_2X_8^{2-} = X_3As(\mu\text{-}X)_2AsX_3^{2-}$ with structures similar to the $Bi_2Cl_8^{2-}$ depicted in Figure 5.6; the polymeric infinite chain $[AsBr_4^-]_n = [AsBr_3Br_{2/2}]_n^{n-}$ is also known. Antimony and bismuth trihalides are considerably stronger Lewis acids and form complexes with a variety of Lewis bases. Thus reactions of $SbCl_3$ with nitrogen donor ligands give either ψ-trigonal bipyramidal $LSbCl_3$ or ψ-octahedral $LSbCl_4$ derivatives with stereochemically active lone pairs (Figure 5.7). Similarly, anions of the type EX_4^- obtained from antimony or bismuth halides may be ψ-trigonal bipyramidal, but sometimes they form more complicated structures from association through bridging halides. Hexacoordinate anions of the type EX_6^{3-} (E = Sb, Bi) are frequently regular octahedra, indicating that the lone pair is *not* stereochemically active. In such compounds the lone pair is assumed to occupy the *s* orbital leaving the *p* and *d* orbitals to form the octahedral hybrid. The confacial bioctahedral binuclear anions $E_2Cl_9^{3-}$ (E = Sb, Bi) also have regular octahedral coordination with three bridging halogen atoms (Figure 5.7).

ψ-trigonal bipyramid
LEX_3

ψ-octahedron
L_2EX_3

confacial bioctahedral
$[E_2X_9]^{3-}$

Figure 5.7. Some complexes of the halides (X = Cl, Br) of antimony and bismuth (E = Sb, Bi) showing the stereochemically active lone electron pairs in LEX_3 and L_2EX_3 and the confacial bioctahedral $[E_2X_9]^{3-}$.

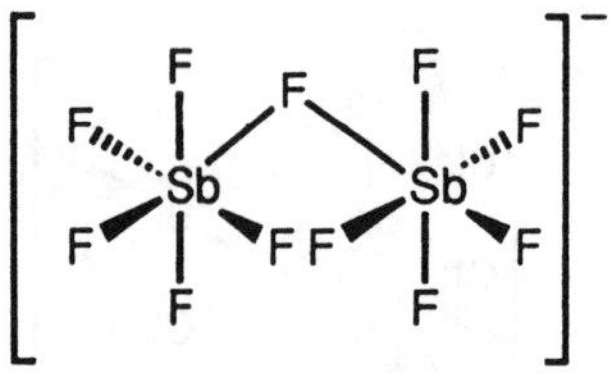

$Sb_2F_{11}^- = F_5Sb(\mu\text{-}F)SbF_5^-$ 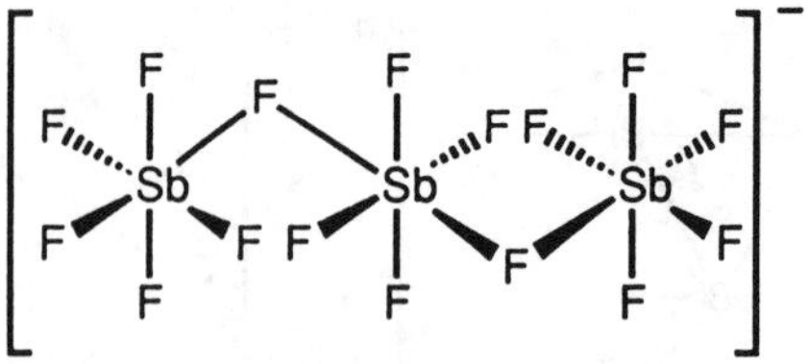 $Sb_3F_{16}^- = trans\text{-}F_5Sb(\mu\text{-}F)SbF_4(\mu\text{-}F)SbF_5^-$

Figure 5.8. Fluorine bridging in $Sb_2F_{11}^-$ and $Sb_3F_{16}^-$.

Many of the complexes of the trivalent halides shown in Figure 5.7 exhibit stereochemically active lone electron pairs. The stereochemical activity of a lone pair in a complex of trivalent arsenic, antimony, or bismuth is shown to *decrease* upon an increase in the coordination number, an increase in the atomic number of the halogen, and an increase in the atomic number of the central element.[21] In most six-coordinate complexes of arsenic, antimony, and bismuth, the lone pair is not stereochemically active, thus the coordination polyhedron is a regular octahedron.

5.5.5 Complexes of Pentavalent Halides: Fluorine Bridges in Antimony(V) Derivatives

The pentahalides, EX_5, in general are very strong Lewis acids with an overwhelming tendency to interact with a ligand (L) to give octahedral derivatives LEX_5. Thus PF_5 reacts readily with fluoride ion to give the octahedral hexafluorophosphate ion, PF_6^-. The hexafluorophosphate anion is a useful counteranion for the isolation of cations when an anion of low coordinating power is desired. In this sense it is a safe substitute for the perchlorate ion, ClO_4^-, which sometimes gives dangerously explosive salts. The related hexahaloantimonates(V), SbX_6^- (X = F, Cl), are analogously obtained by reactions of the corresponding antimony pentahalide and halide ion; they are useful counteranions for the isolation of relatively stable salts of unusual cations such as O_2^+ (Section 6.3.2), ICl_2^+, and IBr_2^+. The polynuclear fluoroantimonate(V) ions $Sb_2F_{11}^- = F_5Sb(\mu\text{-}F)SbF_5^-$ and $Sb_3F_{16}^- = trans\text{-}F_5Sb(\mu\text{-}F)SbF_4(\mu\text{-}F)SbF_5$ are also known. Their structures (Figure 5.8) have Sb—F—Sb bridges similar to those found in SbF_5 (see Figure 5.5). The propensity of antimony(V) compounds to form Sb—F—Sb bridges is related to the high Lewis acidity of SbF_5, which can form strong bonds with a lone pair of an adjacent fluorine.

5.5.6 Halide Derivatives Containing Oxygen or Sulfur

The most important oxohalides are the so-called *phosphoryl halides*, $O{=}PX_3$ (X = F, Cl, Br, I). The phosphoryl halides have a pyramidal PX_3 group with the oxygen atom occupying the fourth position of a distorted tetrahedron; the phosphorus–oxygen bond distance of 1.55 Å, suggests double bonding. The most common phosphoryl halide is phosphoryl chloride, $O{=}PCl_3$ (bp 105.3°C), which can be obtained by heating PCl_3 in oxygen or by the reaction of P_4O_{10} with PCl_5 according to the following equation:

$$P_4O_{10} + 6PCl_5 \rightarrow 10O{=}PCl_3 \tag{5.10}$$

The chlorine atoms in $O{=}PCl_3$ are very reactive toward substitution. In addition, the oxygen atom in $O{=}PCl_3$ has donor properties toward Lewis acid metal ions. Thus aluminum chloride can be removed from Friedel–Crafts reaction products by formation of a very strong $Cl_3P{=}O \rightarrow AlCl_3$ complex; this is particularly useful when addition of water to the reaction mixture has to be avoided. Some phosphoryl chloride complexes are quite volatile. Thus zirconium and hafnium can be separated by fractional vacuum distillation of their $Cl_3P{=}O{\rightarrow}MCl_4$ (M = Zr, Hf) complexes. Other related compounds include *thiophosphoryl chloride*, $S{=}PCl_3$ (bp 125°C), which is readily obtained by treatment of PCl_3 with elemental sulfur or treatment of P_4S_{10} with PCl_5. The binuclear *pyrophosphoryl halides*, $X_2P({=}O){-}O{-}P({=}O)X_2$ (X = Cl, F), are also known.

Mononuclear arsenic analogues of the phosphoryl halides appear to be unstable under ambient conditions. The only $O{=}AsX_3$ derivative to have been isolated is $O{=}AsCl_3$, and it is unstable above $-25°C$, decomposing to $AsCl_3$, Cl_2, and a polymer of stoichiometry $[As_2O_3Cl_4]_n$. The instability of $O{=}AsCl_3$ is another manifestation of the instability of arsenic(V) relative to phosphorus(V) and antimony(V) (Section 5.5.2).

Simple mononuclear oxohalides of antimony(V) and bismuth(V) are not known. However, dilution of concentrated hydrochloric acid solutions of Sb(III) and Bi(III) precipitates the polymeric metal(III) oxohalides $[OECl]_n$ (E = Sb, Bi), which have complicated layer structures.

5.6 Oxides of Phosphorus, Arsenic, Antimony, and Bismuth

5.6.1 Oxides of Phosphorus

The well-characterized oxides of phosphorus have the general formula P_4O_{6+n} ($0 \leq n \leq 4$), containing a P_4O_6 tetrahedral core with n exotetrahedral oxygen atoms (Figure 5.9). The most readily available member of this series is the most readily available member of this series is the most highly oxidized member, *phosphorus(V) oxide*, P_4O_{10}, commonly but incorrectly known as "phosphorus pentoxide," in which all four phosphorus atoms have the maximum +5 oxidation state. This oxide, a volatile white solid subliming at 360°C and 1 atmos, is the main product obtained by burning elemental phosphorus:

$$P_4 + 5O_2 \rightarrow P_4O_{10} \qquad \Delta H° = -1493 \text{ kJ/mol } P_4O_{10} \tag{5.11}$$

Reactions of P_4O_{10} with basic metal oxides leads to diverse solid metal phosphates.

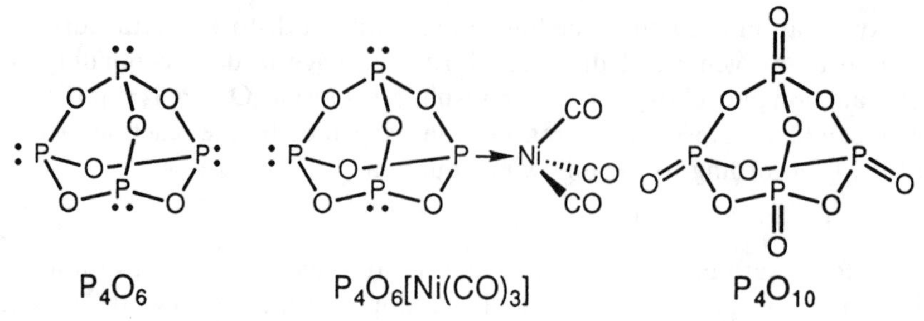

P_4O_6 $P_4O_6[Ni(CO)_3]$ P_4O_{10}

Figure 5.9. The P_4O_6 framework in the phosphorus oxides and in the nickel complex $P_4O_6[Ni(CO)_3]$.

The most important property of P_4O_{10} is its high affinity for water, with which it reacts exothermically to form a syrupy mixture of polyphosphoric acids. For this reason it is a very useful and powerful dehydrating agent. It can therefore be used to dehydrate nitric, sulfuric, perchloric, and malonic acid to the corresponding anhydrides N_2O_5 (Section 4.5.5), SO_3 (Section 6.7.3), Cl_2O_7 (Section 7.4.1.4), and C_3O_2 (Section 2.5.1.3), respectively.

The least oxidized binary oxide of phosphorus is *phosphorus(III) oxide*, P_4O_6 (mp 24°C, bp 170°C), which has a structure similar to that of P_4O_{10} except for the lack of exocyclic oxygen atoms; this oxide is obtained by burning white phosphorus in an oxygen-deficient atmosphere (e.g., 75% O_2/25% N_2 at 50°C/90 mm Hg).[22] Hydrolysis of P_4O_6 in cold water readily gives phosphorus acid, $HP(O)(OH)_2$. This process is irreversible; that is, $HP(O)(OH)_2$ cannot be dehydrated back to P_4O_6. Reaction of P_4O_6 with hot water leads to a complicated disproportionation reaction with PH_3, H_3PO_4, and elemental phosphorus as major products. Transition metal complexes such as $P_4O_6[Ni(CO)_3]$ (Figure 5.9) can be obtained by complexation of the phosphorus lone pairs in P_4O_6 with suitable transition metal moieties.

5.6.2 Oxides of Arsenic

5.6.2.1 Arsenic(III) Oxide

Arsenic(III) oxide exists in both molecular and polymeric modifications. Cubic arsenic(III) oxide (*arsenolite*) contains discrete As_4O_6 molecules similar to P_4O_6 already discussed and is the major product obtained by burning arsenic in air. Monoclinic arsenic(III) oxide (*claudetite*) consists of sheets of $AsO_{3/2}$ units in which each oxygen atom bridges two arsenic atoms. The monoclinic form of arsenic(III) oxide is 8.7% more dense than the cubic form. Reaction of arsenic(III) oxide with water gives solutions of "arsenious acid," which is presumed to be $As(OH)_3$ but has not been well characterized. Treatment of arsenic(III) oxide with bases gives arsenite ions formed by ionization of As(OH)$_3$, namely AsO(OH)$_2^-$, AsO$_2$OH^{2-}, and AsO$_3^{3-}$.

5.6.2.2 Arsenic(V) Oxide

Arsenic(V) oxide, As$_2$O$_5$, *cannot* be obtained by the direct reaction of elemental arsenic with oxygen, in contrast to phosphorus(V) oxide, P$_4$O$_{10}$; this is another consequence of the lower stability of the highest oxidation states of the elements following the 4d transition metals relative to those of the corresponding elements in the rows immediately above or below this row of the periodic table. However, As$_2$O$_5$ can be obtained by the oxidation of elemental arsenic with nitric acid followed by dehydration. Pyrolysis of As$_2$O$_5$ readily leads to loss of oxygen to form As$_2$O$_3$. Water dissolves As$_2$O$_5$ to give arsenic acid, H$_3$AsO$_4$. The structure of As$_2$O$_5$ contains equal numbers of AsO$_{6/2}$ octahedra and AsO$_{4/2}$ tetrahedra with each oxygen bridging two arsenic atoms: As$_2$O$_5$ can thus more precisely be represented as (AsO$_{6/2}$)(AsO$_{4/2}$) = As$_2$O$_5$.

5.6.3 Oxides of Antimony

5.6.3.1 Antimony(III) Oxide

Antimony(III) oxide, like arsenic(III) oxide, exists in two forms: cubic *senarmontite*, containing discrete tetrahedral Sb$_4$O$_6$ molecules, and orthorhombic *valentinite*, containing polymeric layers consisting of SbO$_{3/2}$ units in which each oxygen atom bridges two antimony atoms. The orthorhombic form of antimony(III) oxide is 11.3% more dense than the cubic form. Antimony(III) oxide can be obtained by direct reaction of antimony with oxygen. The form with discrete Sb$_4$O$_6$ molecules exists in the vapor phase and in the solid phase below 570°C; the orthorhombic polymeric form occurs in the solid state at high temperatures. Antimony(III) oxide is insoluble in water or dilute nitric or sulfuric acid. However, it is soluble in strong acids that form strong complexes with antimony, such as hydrochloric acid. In addition, antimony(III) oxide dissolves in bases to form solutions of antimonates(III). Since antimony(III) oxide is soluble in both certain acids and strong bases, it can be regarded as amphoteric. Antimony(III) oxide is used extensively as a flame retardant in fabrics, paper, paints, plastics, epoxy resins, adhesives, and rubbers.

5.6.3.2 Antimony(V) Oxide

Antimony(V) oxide can be obtained by treatment of antimony(III) oxide with elemental oxygen at high pressures and temperatures or by the hydrolysis of antimony pentachloride with aqueous ammonia followed by heating to 275°C, to remove water and ammonia from the initial precipitate. The antimony atoms are all octahedrally coordinated in the structure of antimony(V) oxide.

5.6.3.3 Antimony(III,V) Oxides

Heating antimony(III) oxide in air at 900°C gives a yellow product having composition SbO_2 and containing equal amounts of antimony(III) and antimony(V) rather than antimony(IV) as shown, for example, by its diamagnetism. In the orthorhombic form of SbO_2, the antimony(III) is coordinated to four oxygen atoms and the antimony(V) is coordinated to six oxygen atoms.

5.6.4 Oxides of Bismuth

The only well-characterized oxide of bismuth is bismuth(III) oxide, Bi_2O_3. Bismuth(V) oxide is extremely unstable and has not been obtained in a pure state. Bismuth(III) oxide is strictly basic; it is thus insoluble in aqueous alkalis but dissolves in acids to form the corresponding bismuth(III) salts.

5.7 Sulfides and Other Chalcogenides of Phosphorus, Arsenic, Antimony, and Bismuth

Phosphorus sulfides[23] have the general formula P_4S_n ($3 \leq n \leq 10$, but $n \neq 6$); their structures, like the phosphorus oxides, are based on P_4 tetrahedra. However, the tetrahedral frameworks of phosphorus sulfides can contain direct P—P bonds as well as P—S—P units, whereas the tetrahedral frameworks of the phosphorus oxides (see Section 5.6.1) contain only P—O—P units. The phosphorus sulfides P_4S_3, P_4S_7, P_4S_9, and P_4S_{10} can be obtained by heating red phosphorus and sulfur in the indicated formula ratios. Melts of phosphorus and sulfur in P/S mole ratios intermediate between 4/3 and 4/7 contain complicated mixtures, as indicated by phosphorus-31 NMR results; since these complicated mixtures cannot be efficiently separated, P_4S_4 and P_4S_5 are made by indirect methods. Phosphorus sulfides are readily hydrolyzed to H_2S and various oxoacids of phosphorus. Yellow crystalline P_4S_3 (mp 174°C, bp 408°C), has a structure analogous to that of Sb_7^{3-} (Figure 5.3). A mixture of P_4S_3 with the oxidizer $KClO_3$ provides an igniter for "strike anywhere" matches. The phosphorus sulfide, P_4S_{10}, has a structure similar to that of P_4O_{10} (Figure 5.9) and is useful in organic chemistry to convert oxygen functionalities (e.g.,

$$\text{C}{=}\text{O}, -\text{C}({=}\text{O})\text{OH}, -\text{C}({=}\text{O}){-}\text{N}(\text{H}){-}, \text{C}{-}\text{OH})$$ to their sulfur analogues.

Treatment of P_4S_{10} with alcohols gives phosphorus thioacids of the type $(RO)_2P(S)SH$, which are useful as antioxidants and corrosion inhibitors. In general phosphorus selenides are rather similar to phosphorus sulfides.

Arsenic sulfides, like phosphorus sulfides, have molecular structures containing four arsenic atoms; the most important arsenic sulfides are yellow As_2S_3 (*orpiment*), with a layer structure analogous to As_2O_3; orange-yellow As_4S_3 (*dimorphite*), which exists in two different forms[24]; and orange-red As_4S_4 (*realgar*),[25] containing an eight-membered As_4S_4 ring with alternating arsenic and sulfur atoms in which the four sulfur atoms are coplanar. Treatment of Sb(III) and Bi(III) solutions with H_2S precipitates E_2S_3 (E = Sb, Bi), which form ribbonlike polymers $ES_{3/2}$ with interlocking ES_3 and SE_3 pyramids in which each sulfur atom bridges two metal atoms; these sulfides exhibit semiconductor properties. Antimony sulfide, Sb_2S_3, occurs as the gray to black mineral *stibnite* and is made industrially on a moderately large scale for use in the manufacture of safety matches, military ammunition, and pyrotechnic products because of its vigorous reactions with oxidizing agents.

5.8 Oxoacids of Phosphorus and Their Derivatives

All phosphorus atoms in oxoacids of phosphorus are tetracoordinate. The hydrogen atoms in P—OH groups are ionizable, whereas those in P—H groups are not. Thus hypophosphorous acid, $H_3PO_2 = H_2P(O)OH$, is a monobasic acid rather than a dibasic or tribasic acid, and phosphorous acid, $H_3PO_3 = HP(O)(OH)_2$ is a dibasic rather than a tribasic acid. Also all the mononuclear phosphorous oxyacids, namely H_3PO_2, H_3PO_3, and H_3PO_4, have roughly the same pK_1 values, corresponding to those expected for an oxoacid H_nXO_m with $m - n = 1$ (compare Section 1.4 and equation 1.21b).

5.8.1 Hypophosphorous Acid (Phosphinic Acid)

The hypophosphite ion can be generated by heating white phosphorus with an alkali or alkaline earth hydroxide according to the equation

$$P_4 + 4OH^- + 4H_2O \rightarrow 4H_2PO_2^- + 2H_2\uparrow \tag{5.12}$$

The free acid can be obtained by treatment of white phosphorus with $Ba(OH)_2$ according to equation 5.12 followed by precipitation of the Ba^{2+} counterion as $BaSO_4$ with sulfuric acid. Because of oxidation and disproportionation reactions, pure hypophosphorous acid cannot be obtained simply by evaporation of its concentrated aqueous solution. However, hypophosphorous acid can be isolated from its concentrated aqueous solution by extraction with diethyl ether. White crystalline hypophosphorous acid, $H_3PO_2 = H_2P(O)OH$ (mp 26.5°C), is a monobasic acid, $pK = 1.2$. Hypophosphorous acid, unlike most of the phosphorous oxoacids, forms a water-soluble calcium salt. Hypophosphorous acid and its salts are strong reducing agents; the sodium salt $NaH_2PO_2 \cdot H_2O$ is used as an industrial reducing agent in the electroless plating of nickel.[26] Numerous organic derivatives of the type $R_2P(O)(OH)$ are known in which the P—H bonds of hypophosphorus acid are replaced by P—R bonds; these derivatives are called dialkylphosphinic acids. Strictly speaking, hypophosphorus acid should be called phosphinic acid, but this more exact name is rarely used.

5.8.2 Phosphorous Acid (Phosphonic Acid) and Its Esters

White crystalline deliquescent phosphorous acid, $H_3PO_3 = HP(O)(OH)_2$ (mp 70.1°C), is obtained by treatment of PCl_3 or P_4O_6 with water; it is a dibasic rather than a tribasic acid ($pK_1 = 1.3$, $pK_2 = 6.7$). The oxidation of phosphorous acid to phosphate by reagents such as sulfur dioxide and halogens is a slow and complicated reaction. Numerous organic derivatives of the type $RP(O)(OH)_2$ are known in which the P—H bond of phosphorous acid is replaced by a P—R bond; these derivatives are called alkylphosphonic acids. Strictly speaking, phosphorous acid should be called phosphonic acid, but this more exact name is rarely used.

The esters of phosphorous acid, namely the trialkyl- and triarylphosphites, $(RO)_3P:$, contain a three-coordinate phosphorus atom with a lone pair in contrast to phosphorous acid and its salts with four-coordinate phosphorus atoms. These esters can be readily synthesized by reactions of phosphorus trichloride with alcohols or phenols in the presence of a base such as a tertiary amine:

$$PCl_3 + 3ROH + 3R_3'N \rightarrow (RO)_3P + 3[R_3'NH]Cl \qquad (5.13)$$

As indicated in the following examples, the chemical reactivity of trialkylphosphites relates largely to the presence of a three-coordinate phosphorus atom and a lone electron pair.

1. Trialkylphosphites form numerous stable transition metal complexes in which the phosphorus lone pair is coordinated to the transition metal (e.g., $(MeO)_3P \rightarrow Cr(CO)_5$). Trialkylphosphites are rather strong π-acceptor ligands. Thus in many of their transition metal complexes the transition metal has a relatively low formula oxidation state, which is stabilized by removal of electron density through $d\pi$–$d\pi$ bonding similar to that in PF_3 transition metal complex (see Figure 5.4). However, trialkylphosphites are not as strong π acceptors as PF_3, since alkoxy groups are not as strongly electron-withdrawing as fluorine atoms.

2. Trialkylphosphites react readily with alkyl halides to give dialkyl alkylphosphonates through an intermediate quaternary salt in the *Michaelis–Arbusov reaction*[27]:

$$(RO)_3P + R'X \rightarrow \{[(RO)_3PR']X\} \rightarrow (RO)_2P(O)R' + RX \qquad (5.14)$$

A related reaction is the spontaneous isomerization of trimethylphosphite, $(MeO)_3P:$, which has a three-coordinate phosphorus atom, to dimethyl methylphosphonate, $MeP(O)(OMe)_2$, which has a four-coordinate phosphorus atom.

3. Although trialkylphosphites are relatively stable toward air oxidation, stronger oxidizing agents can convert them to trialkylphosphates by the following reaction:

$$(RO)_3(\text{trialkylphosphite}) + [O] \rightarrow (RO)_3P{=}O \ (\text{trialkylphosphate})$$

$$(5.15)$$

5.8.3 Hypophosphoric Acid

Hypophosphoric acid, $(HO)_2P(O)$—$P(O)(OH)_2$, is the simplest and longest known example of a phosphorus oxoacid containing a direct phosphorus–phosphorus bond. The hypophosphate ion can be isolated from the mixture obtained by the oxidation of red phosphorus with hypochlorite or chlorite ion, for example,

$$2P + 2NaClO_2 + 8H_2O \rightarrow Na_2H_2P_2O_6 \cdot 6H_2O + 2HCl \qquad (5.16)$$

This sodium salt can be crystallized from this reaction mixture at pH 2; its diamagnetic $H_2P_2O_6^{2-}$ anion has a P—P distance of 2.17 Å, corresponding to a direct phosphorus–phosphorus single bond. Free hypophosphoric acid can be obtained from the sodium salt by use of an acidic ion exchange resin; it is a tetrabasic acid with successive pK values of 2.2, 2.8, 7.3, and 10.0. The hypophosphate ion is very stable in alkali, even 80% NaOH at 200°C. However, hypophosphoric acid is readily hydrolyzed in acid to phosphorous and phosphoric acids:

$$(HO)_2P(O)\text{—}P(O)(OH)_2 + H_2O \rightarrow HP(O)(OH)_2 + OP(OH)_3 \qquad (5.17)$$

The half-life of this reaction is 180 days in 1 M hydrochloric acid but less than an hour in 4 M hydrochloric acid.

5.8.4 Phosphorus(V) Acids and Their Salts

The traditional nomenclature of phosphorus(V) oxyacids recognizes three types of phosphorus oxoacid: *orthophosphoric acid* of the stoichiometry H_3PO_4, *pyrophosphoric acid* of the stoichiometry $H_4P_2O_7$, and *metaphosphoric acid* of the stoichiometry HPO_3. Orthophosphoric acid and pyrophosphoric acid (diphosphoric acid) and their anions are now known to be monomers, as written, but monomeric metaphosphoric acid, which would necessarily have planar tricoordinate phosphorus, is not stable. Instead the metaphosphoric acid stoichiometry $[HPO_3]_n$ is known in cyclic oligomers and polymers. However, a few formal derivatives of monomeric metaphosphoric acid are known in which the P—O and P=O bonds have been replaced by corresponding P—NR and P=NR with bulky R groups (e.g., $pMe_3Si)_2NP$ ($=NSiMe_3$ in Figure 5.1). Monomeric metaphosphate esters such as $MeOPO_2$ can also be generated as unstable intermediates in pyrolysis reactions.[28]

5.8.4.1 Orthophosphoric Acid and Its Salts

Orthophosphoric acid, H_3PO_4, namely $O=P(OH)_3$, is obtained in large quantities from phosphate rock either by treatment with sulfuric acid or by reduction to elemental phosphorus (equation 5.4) followed by oxidation and hydration. Orthophosphoric acid is normally marketed as the syrupy 85% aqueous acid. Pure orthophosphoric acid, a deliquescent solid (mp 42.3°C), can be obtained by vacuum evaporation of the concentrated aqueous acid at 80°C. It is very stable and has no oxidizing properties below 350–400°C, although it undergoes partial dehydration to pyrophosphoric acid, $H_4P_2O_7$. Fused orthophosphoric acid has appreciable ionic conductivity, suggesting the auto-protolysis

$$2H_3PO_4 \rightleftarrows H_4PO_4^+ + H_2PO_4^- \tag{5.18}$$

Orthophosphoric acid is a tribasic acid ($pK_1 = 2.15$, $pK_2 = 7.1$, $pK_3 = 12.4$) and forms three series of salts. Pure orthophosphoric acid and its crystalline hydrates have tetrahedral PO_4 groups connected by hydrogen bonds. Hydrogen bonding accounts for the relatively high viscosity of concentrated aqueous H_3PO_4.

Orthophosphates of most metals and other cations are known. In most cases orthophosphates of monovalent metals and other monovalent cations are soluble in water, whereas orthophosphates of divalent and trivalent metals are insoluble in water. Ammonium phosphates are used as fertilizers and alkali metal phosphates are used as buffers. Natural phosphate minerals are all orthophosphates; fluoroapatite, $Ca_5F(PO_4)_3$, is the most important major natural phosphate. The mineral part of teeth consists of hydroxyapatites (e.g., $Ca_5(OH)(PO_4)_3$.

5.8.4.2 Condensed Phosphates

Condensed phosphates are polynuclear phosphates(V) containing more than one phosphorus atom and P—O—P bonds.[29,30] They are constructed by combining monofunctional "end" units, $\mathscr{E} = P(O)O_2O_{1/2}^{2-}$, bifunctional "middle" units, $\mathscr{M} = P(O)OO_{2/2}^-$, and trifunctional "branching" units, $\mathscr{B} = P(O)O_{3/2}$ (see Figure 5.10). Phosphorus-31 NMR spectroscopy (Section 5.1) has proven to be very important for the characterization of condensed phosphates.

Polyphosphates may be classified into the following principal types.

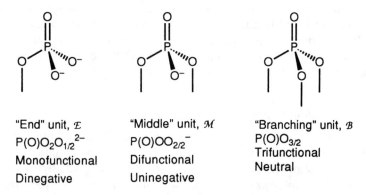

"End" unit, $\mathscr{E}$
$P(O)O_2O_{1/2}^{2-}$
Monofunctional
Dinegative

"Middle" unit, $\mathscr{M}$
$P(O)OO_{2/2}^-$
Difunctional
Uninegative

"Branching" unit, $\mathscr{B}$
$P(O)O_{3/2}$
Trifunctional
Neutral

Figure 5.10. The building blocks for condensed phosphate structures.

1. *Chain polyphosphates* of the type $[P_nO_{3n+1}]^{(n+2)-}$ or $\mathscr{E}-\mathscr{M}_{n-2}-\mathscr{E}$, which are terminated by the "end" units, $\mathscr{E}$ (Figure 5.10). The simplest chain phosphates are the pyrophosphates or, more systematically, the diphosphates, $P_2O_7^{4-}$ ($n = 2$ in $[P_nO_{3n+1}]^{(n+2)-}$). Tripolyphosphates $M_5^IP_3O_{13}$ ($\mathscr{E}-\mathscr{M}-\mathscr{E}$; i.e., $n = 3$) are useful as detergents and act as water softeners by sequestering Ca^{2+} and Mg^{2+} ions.

2. *Polymeric infinite chain metaphosphates* $[PO_3]_n^{n-}$ or $\mathscr{M}_n^{n-}$.

3. *Cyclic metaphosphates* $[PO_3]_n^{n-}$ or $\mathscr{M}_n^{n-}$ ($3 \leq n \leq 10$ or more); the cyclic metaphosphates with $n = 3$ and 4 are the most common.[31] The six-membered P_3O_3 rings in the trimetaphosphate ion $P_3O_9^{3-}$ have the chair conformation.

4. *Ultraphosphates*, which contain "branching" units $\mathscr{B}$. The ultimate ultra-phosphate is the neutral P_4O_{10}—that is, $\mathscr{B}_4$ (Figure 5.9)—since branching units $\mathscr{B}$ bear no negative charge (Figure 5.10). *Ultraphosphonic acid* is $H_2P_4O_{11}$ (i.e., $H_2\mathscr{M}_2\mathscr{B}_2$). Note that $\mathscr{B}$ in condensed phosphates, "branching units," are much more readily hydrolyzed than "end" units $\mathscr{E}$ or "middle" units $\mathscr{M}$.

Condensed phosphates can be made by the following general methods.

1. Dehydration of $H_2PO_4^-$ and HPO_4^{2-} salts at temperatures in the range 300–1200°C for example,

$$\text{Pyrophosphate: } 2M_2^IHPO_4 \xrightarrow{\Delta} M_4^IP_2O_7 + H_2O\uparrow \tag{5.19a}$$

polymeric or cyclic metaphosphates:

$$nNaH_2PO_4 \xrightarrow{\Delta} Na_nP_nO_{3n} + nH_2O\uparrow \tag{5.19b}$$

2. Controlled addition of water to P_4O_{10}, for example,

$$\text{ultraphosphonic acid: } P_4O_{10} + H_2O \rightarrow H_2P_4O_{11} \tag{5.20}$$

Further addition of water to ultraphosphonic acid leads to the so-called *polyphosphoric acid*, which contains various mixtures of the free acids $H_{n+2}P_nO_{3n-1}$ ($2 \leq n \leq {\sim}8$). Equilibrium in these systems is achieved relatively slowly, and in the absence of catalysts, half-lives for the formation or hydrolysis of P—O—P linkages can be years.

3. Elimination of AgCl from the reactions of chlorophosphates with silver phosphates.

In addition, a specific method for the preparation of the cyclometaphosphate ion, $P_3O_9^{3-}$ uses the reaction

$$2(MeO)_3PO + (Me_2N)_3PO \rightarrow [Me_4N]_3P_3O_9 \tag{5.21}$$

In this reaction the dimethylamino groups of the hexamethylphosphoramide, $(Me_2N)_3PO$, are methylated by trimethyl phosphate, $(MeO)_3PO$, to give the tetramethylammonium cation.

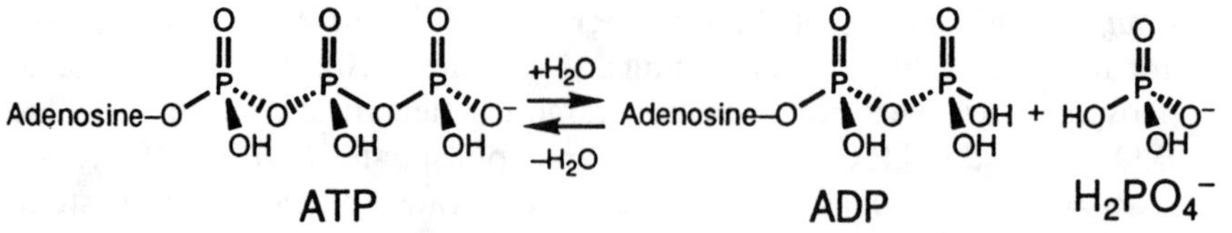

Figure 5.11. The hydrolysis of ATP to ADP as an example of a biologically significant phosphate ester hydrolysis.

5.8.4.3 Phosphate Esters and Their Significance in Biology

Phosphate esters, $(RO)_3P{=}O$, can be made by the oxidation of phosphite esters (equation 5.13) or by reaction of $O{=}PCl_3$ with an alcohol in the presence of a base, for example,

$$O{=}PCl_3 + 3ROH + 3R_3N \rightarrow O{=}P(OR)_3 + 3[R_3NH]Cl \qquad (5.22)$$

Tri-*n*-butyl phosphate, $O{=}P(OBu^n)_3$, is used in the Purex process for the extraction of uranium and plutonium from spent nuclear fuel.[32]

Many of the most essential chemicals in life processes, including DNA and RNA, are esters of polyphosphoric acids with P—O—P linkages. The hydrolysis of adenosine triphosphate (ATP) to adenosine diphosphate (ADP) is depicted schematically in Figure 5.11.

Because of the importance of phosphate esters in biological processes, their hydrolysis has been studied in detail. Phosphate triesters of the type $O{=}P(OR)_3$ are attacked by OH^- at phosphorus and by H_2O at carbon as shown by ^{18}O labeling studies (Figure 5.12). Phosphate diesters of the type $O = P(OR)_2(OH)$ occur completely in the anionic form $O{=}P(OR)_2O^-$ at normal and physiological pH values:

$$O{=}P(OR)_2(OH) \rightleftarrows O{=}P(OR)_2O^- + H^+ \qquad pK_a \approx 1.6 \qquad (5.23)$$

Because of the negative charge on the $O{=}P(OR)^2O^-$ anion, phosphate diesters are relatively resistant to nucleophilic attack by OH^- or H_2O; their hydrolysis is therefore slow in the absence of suitable enzymes (i.e., "biological" catalysts). Hydrolysis of phosphate esters can proceed either by a bimolecular S_N2 reaction involving a five-coordinate intermediate or by a unimolecular reaction invol-

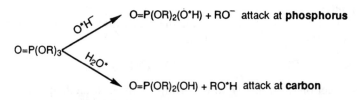

Figure 5.12. Oxygen-18 labeling studies ($O^* = {}^{18}O$) indicating attack at phosphorus and at carbon in the hydrolysis of phosphate triesters $O{=}P(OR)_3$.

ing a three-coordinate metaphosphate intermediate analogous to the bimolecular S_N2 and unimolecular S_N1 reactions at a tetrahedral carbon atom, respectively. Pseudorotation (see Figure 5.2) has been shown to occur in the five-coordinate intermediate in the S_N2 hydrolysis of phosphate esters.

5.9 Oxoacids and Oxoanions of Arsenic, Antimony, and Bismuth

5.9.1 Arsenic

White crystalline pentavalent *arsenic acid*, H_3AsO_4, can be obtained by treatment of elemental arsenic with nitric acid, by the catalyzed treatment of As_2O_3 with air and water under pressure, or by the reaction of As_2O_5 with water. Arsenic acid, unlike phosphoric acid, is a moderately strong oxidizing agent in acid solution ($E° = 0.559$ V). Orthoarsenates generally resemble orthophosphates and are often isomorphous to them. Condensed arsenates are much less stable than condensed phosphates and hydrolyze rapidly in aqueous solution; therefore condensed arsenates are known in much less detail than condensed phosphates. Metaarsenates of the stoichiometry $[M^+AsO_3^-)]_n$ have been shown to have either infinite chain structures (e.g., $M^+ = Li^+$ or Na^+) or cyclic structures (e.g., $M^+ = K^+$, $n = 3$).

Trivalent *arsenous acid* is generated by dissolving As_4O_6 in acid aqueous solution. Raman spectra indicate that arsenous acid, H_3AsO_3, is $As(OH)_3$ with three-coordinate pyramidal arsenic, in contrast to phosphorous acid, H_3PO_3, which is $HP(O)(OH)_2$ with four-coordinate phosphorus (Section 5.8.2). The arsenite ion in solid salts generally occurs as AsO_3^{3-}, although all three ions $As(OH)_2O^-$, $As(OH)O_2^{2-}$, and AsO_3^{3-} can be detected spectroscopically when solutions of $As(OH)_3$ are treated with hydroxide ion; all these arsenite ions contain three-coordinate pyramidal arsenic. Arsenites of the alkali metals are very soluble in water, those of the alkaline earth metals are less so, and heavy metal arsenites are essentially insoluble in water. "Metaarsenites" such as $[NaAsO_2]_n$ are known containing polymeric chain anions constructed from $AsOO_{2/2}^-$ units containing the usual three-coordinate pyramidal As(III).

5.9.2 Antimony

The four-coordinate tetrahedral pentavalent antimony oxoanion SbO_4^{3-} analogous to PO_4^{3-} and AsO_4^{3-} is unknown. Instead the anion $Sb(OH)_6^-$, with six-coordinate octahedral antimony(V), is generated in basic solutions of antimony(V)—(for example, by treating $SbCl_5$ with aqueous sodium hydroxide. The sodium salt $NaSb(OH)_6$ is the least soluble of the alkali metal salts and one of the least soluble sodium salts; it can be precipitated from aqueous solution. However, the aqueous solubility of $NaSb(OH)_6$ is not low enough for it to be useful for the gravimetric determination of sodium. The $Sb(OH)_6^-$ anion is one member of an isoelectronic and isostructural series $Sn(OH)_6^{2-}$ (Section 3.6.2), $Sb(OH)_6^-$, $Te(OH)_6$ (Section 6,8,8), and $I(OH)_6^+$ (Section 7.5.4).

No well-defined antimony(III) oxoacid is known, but the hydrous oxide $Sb_2O_3 \cdot xH_2O$ dissolves in bases to form antimonite salts, most of which (e.g., $NaSbO_2$) appear to have polymeric structures.

5.9.3 Bismuth

Bismuth(III) has no acidic properties and thus forms no oxyanions stable in aqueous solution. However, "bismuthates" containing bismuth(V) can be obtained in low purity, either by "wet" methods such as the treatment of $Bi(OH)_3$ with chlorine in alkaline solution or by "dry" methods such as heating Na_2O_2 with Bi_2O_3. "Sodium bismuthate," nominally $NaBi^VO_3$ but never obtained pure, is a yellow-brown, poorly characterized solid that is a useful potent oxidizing agent ($E° = +2.03$ V) in aqueous solution. "Sodium bismuth-ate" is thus capable of oxidizing manganese(II) to permanganate, which provides the basis for a method for the analysis of manganese in steel.

5.10 Phosphorus–Nitrogen Compounds

5.10.1 Cyclophosphazenes

Cyclophosphazenes contain even-membered $(PN)_n$ rings with alternating phosphorus(V) and nitrogen(III) atoms.[33,34] Dimeric cyclophosphazene derivatives containing P_2N_2 four-membered rings are rather rare but can be made by the following reaction sequence if R is a relatively bulky group such as isopropyl:

$$R_2PN_3 \xrightarrow{\Delta} \{R_2PN\} + N_2 \qquad (5.24a)$$

$$2\{R_2PN\} \rightarrow [R_2PN]_2 \qquad (5.24b)$$

Reaction 5.24a involves dinitrogen elimination from azide to give a reactive dialkylphosphinonitrene intermediate, which can then undergo dimerization. The $[R_2PN]_2$ dimers are planar, with equal P—N distances suggestive of delocalized π-bonding. They are thermodynamically stable and not air-sensitive.

Trimeric and tetrametric cyclophosphazenes are much more abundant and better known than the dimeric cyclophosphazene derivatives. Larger rings are also known. The cyclophosphazenes $[Me_2PN]_n$ ($3 \leq n \leq 12$) have been characterized by X-ray crystallography. The standard preparation of trimeric and tetrameric cyclophosphazenes uses the reaction of phosphorus pentachloride with ammonium chloride in an inert high-boiling solvent such as chlorobenzene or 1,2,3,2-tetrachloroethane; this reaction proceeds at 120–150°C according to the following general equation:

$$nPCl_5 + nNH_4Cl \xrightarrow{\Delta} [NPCl_2]_n + 4nHCl\uparrow \qquad (5.25)$$

The chlorine atoms in $[NPCl_2]_n$ are reactive toward nucleophilic substitution,[35] which can be used to introduce other substituents into the cyclophosphazene ring, for example,

$$[NPCl_2]_3 + 6NaF \xrightarrow{\text{MeCN}} [NPF_2]_3 + 6NaCl \tag{5.26a}$$

$$[NPCl_2]_2 + 6NaOR \rightarrow [NP(OR)_2]_3 + 6NaCl \tag{5.26b}$$

In many cases such nucleophilic substitution reactions can be controlled to give partially substituted derivatives.

Delocalization within the cyclohosphazene rings has been a matter of some controversy. The six-membered rings $[NPX_2]_3$ are planar or nearly so. However, except for the fluoride $[NPF_2]_4$ the larger rings $[NPX_2]_n$ ($n \geq 4$) are nonplanar, with the angles N–P–N $\approx 120°$ and P–N–P $\approx 132°$ in all cases. In $[NPX_2]_n$ the P—M distances are nearly equal (1.55–1.61 Å), and shorter than the expected P—N single bond lengths of ~ 1.75–1.80 Å, indicative of P—N π-bonding.

5.10.2 Linear Polyphosphazenes

Linear polyphosphazenes are examples of purely "inorganic" polymers with no carbon in their backbones, which consist of chains of alternating phosphorus and nitrogen atoms. They can be obtained by ring-opening polymerization of certain cyclophosphazenes, for example,

$$\frac{n}{3}[NPCl_2]_3 \xrightarrow{230°C} [-N{=}PCl_2{-}]_n \tag{5.27}$$

Soluble polyphosphazenes with $n > 15,000$ can be obtained by using pure $[NPCl_2]_3$ as a precursor. However, heating $[-N{=}PCl_2{-}]_n$ above $350°C$ results in regeneration of $[NPCl_2]_3$ and other $[NPCl_2]_n$ cyclophosphazenes. The chlorine atoms in $[-N{=}PCl_2{-}]_n$ can be replaced by other groups using nucleophilic substitution reactions (with NaOR, RNH_2, R_2NH, etc.), leading to useful polymers exhibiting diverse properties.[36] Replacement of the chlorine atoms in $[-N{=}PCl_2{-}]_n$ results in improved stability toward hydrolysis. A variety of physical properties can be obtained in the linear polyphosphazene polymers using different combinations of substituents on the phosphorus–nitrogen backbone; these include glasses, rubbers, tough flexible solids, fibers, and biodegradable polymers. Poly[2-(2-methoxyethoxy)ethoxy]phosphazene doped with lithium salts is a useful electrolyte for "solid state" lithium batteries.[37]

A material closely related to the linear polyphosphazenes in *phospham*, $[PN_2H]_n$, presumed to have the polymeric structure **5.2** with considerable cross-linking. Phospham is made by heating ammonia with red phosphorus in a sealed reaction vessel at $500°C$.

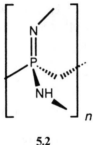

5.2

5.10.3 Other Phosphorus–Nitrogen Compounds

The halogen atoms in various phosphorus halides can be replaced by dialkyl-amino groups upon treatment with secondary amines under relatively mild conditions, for example,

$$PCl_3 + 6R_2NH \rightarrow (R_2N)_3P + 3[R_2NH_2]Cl \tag{5.28a}$$

$$O{=}PCl_3 + 6R_2NH \rightarrow (R_2N)_3P{=}O + 3[R_2NH_2]Cl \tag{5.28b}$$

Tris(dialkylamino)phosphines such as $(Me_2N)_3P$ undergo quaternization and metal complexation reactions selectively at the phosphorus atom rather than at one of the nitrogen atoms, for example,

$$(Me_2N)_3P + MeI \rightarrow [(Me_2N)_3PMe]I \tag{5.29a}$$

$$(Me_2N)_3P + Cr(CO)_6 \rightarrow (Me_2N)_3P \rightarrow Cr(CO)_5 + CO{\uparrow} \tag{5.29b}$$

Hexamethylphosphoramide, $(Me_2N)_3P{=}O$ (mp 7°C, bp 230–232°C and 740 mm Hg), is a good polar aprotic solvent, which gives blue solutions of alkali metals similar to those of alkali metals in liquid ammonia.

Reaction of PCl_3 with four equivalents of a secondary amine, R_2NH, give the halides $(R_2N)_2PCl$, which undergo halogen abstraction with the strong Lewis acid $AlCl_3$ to give the stable two-coordinate bis(dialkylamino)phosphenium cations, $[R_2N{-}P^+{-}NR_2]$.[38] Such cations are stabilized by resonance structures of the type $[R_2N^+{=}P{-}NR_2]$; for this reason, analogous two-coordinate dialkylphosphenium or diarylphosphenium ions, R_2P^+, without at least one dialkylamino substituent are not stable.

The very large bis(triphenylphosphine)iminium or "PPN" cation, $[Ph_3P{=}N{=}PPh_3]^+$, has been widely used to isolate large anions in crystalline form or to give reagents soluble in organic solvents for use in organic synthesis. This cation can be synthesized from triphenylphosphine, chlorine, and hydroxylamine hydrochloride by the following sequence of reactions[39]:

$$2Ph_3P + 2Cl_2 \rightarrow 2Ph_3PCl_2 \tag{5.30a}$$

$$2Ph_3PCl_2 + Ph_3P + NH_3OH^+Cl^- \rightarrow$$

$$[Ph_3P{=}N{=}PPh_3]Cl + Ph_3PO + 4HCl \tag{5.30b}$$

5.11 Organic Derivatives of Phosphorus, Arsenic, Antimony, and Bismuth

5.11.1 Trivalent Derivatives R_3E (E = P, As, Sb, Bi) and the Corresponding Oxides $R_3E{=}O$ and Quaternary Cations R_4E^+

Alkylations or arylations of halides of phosphorus, arsenic, antimony, and bismuth with reactive organometallic compounds lead to extensive series of organic derivatives containing carbon–element bonds, for example,

$$EX_3 + 3RMgX \rightarrow R_3E + 3MgX_2 \text{ (E = P, As, Sb, Bi)} \qquad (5.31a)$$

$$O{=}PCl_3 + 6RMgX \rightarrow 2R_3P{=}O + 3MgCl_2 + 3MgX_2 \qquad (5.31b)$$

Alkylations and arylations of the trihalides EX_3 using a limited amount of the alkylating or arylating reagent can be controlled to result in partial substitution of the halogen atoms of the trihalides EX_3 to give R_nEX_{3-n} ($n = 1, 2$), particularly when bulky groups such as *tert*-butyl or neopentyl are used.

The trivalent tricoordinate phosphorus derivatives, R_3P, known as *phosphines* in the English literature and *phosphanes* in the German literature (even when translated into English!), are readily oxidized to the pentavalent tetracoordinate phosphine oxides, $R_3P{=}O$. Trimethylphosphine, Me_3P (bp 39°C), is a spontaneously flammable liquid, but higher trialkylphosphines, especially those with bulky groups, are less air-sensitive. Triphenylphosphine, Ph_3P, is an air-stable white solid that requires an oxidizing agent such as hydrogen peroxide for conversion to the corresponding phosphine oxide, $Ph_3P{=}O$. Phosphine oxides are very stable compounds with a very short phosphorus–oxygen bond (e.g., 1.483 Å in $Ph_3P{=}O$), indicative of a bond order greater than 2. The phosphorus–halogen bonds in halogen derivatives of the types R_2PX and RPX_2 are very sensitive toward hydrolysis, similar to the phosphorus trihalides PX_3.

Optically active phosphines and phosphine oxides of the type $RR'R''P{:}$ and $RR'R''P{=}O$ can be prepared in which R, R', and R'' are different alkyl or aryl groups. The optically active phosphine oxides $RR'R''P{=}O$ can be deoxygenated upon treatment with $HSiCl_3$ in the presence of a tertiary amine or with Si_2Cl_6 in the absence of a base to give the corresponding optically active phosphine $RR'R''P{:}$. In the case of $HSiCl_3$, deoxygenation occurs with either retention or inversion depending on the base, whereas deoxygenation with Si_2Cl_6 always results in inversion.

An important property of tertiary phosphines is their ability to form a very large variety of transition metal complexes in which the phosphines function as Lewis bases through donation of the phosphorus lone pair to the metal atom.[40] A variety of polyphosphines forming chelate rings with transition metals have also been synthesized (Figure 5.13). Phosphine complexes of transition metals are important homogeneous catalysts for a variety of reactions. Thus triphenylphosphine rhodium complexes are used industrially as catalysts for hydroformylation reactions.

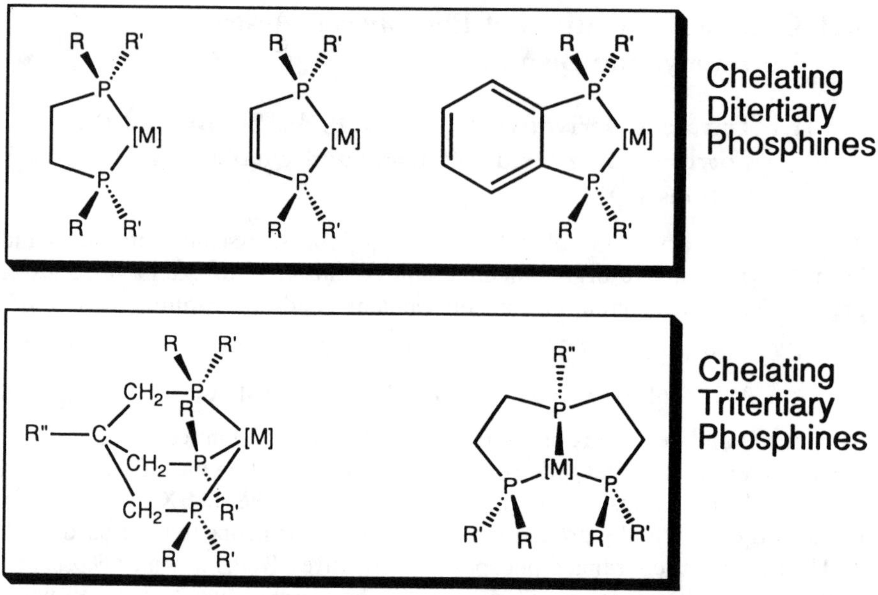

Chelating Ditertiary Phosphines

Chelating Tritertiary Phosphines

Figure 5.13. Some structural types of chelating ditertiary and tritertiary phosphines frequently used in transition metal coordination chemistry: [M] refers to the central transition metal atom with the other associated ligands. Ligands with R = R' = phenyl are the most readily available and in general are air-stable. Ligands with R ≠ R' can be resolved into pure enantiomers and then used as ligands in homogeneous asymmetric catalysis.

Another important property of the trivalent derivatives R_3E is their alkylation with alkyl halides to give quaternary "onium" cations:

$$R_3E + R'X \rightarrow [R_3ER']^+X^- \tag{5.32}$$

In most cases halide salts of these quaternary cations are soluble in water, although unlike PH_4^+, discussed earlier they do not react with it. Quaternary cations such as the tetraphenylphosphonium cation, Ph_4P^+, and the tetraphenylarsonium cation, Ph_4As^+, are large univalent cations that frequently are useful for isolating large anions as readily crystallized stable salts.

5.11.2 Compounds with Phosphorus–Carbon Multiple Bonds

Alkylidenephosphoranes[41] are derivatives of four-coordinate phosphorus(V) of the type $R_3P=CR'_2$ containing a phosphorus(V)–carbon double bond. The alternative resonance structure $R_3P^+-C^-R'_2$ leads to the alternative name "phosphonium ylide" for this type of compound. Alkylidenephosphoranes can be obtained by the deprotonation of tetraalkylphosphonium salts having hydrogen atoms on carbon atoms α to the phosphorus atom. Thus methyl-enetriphenylphosphorane, $Ph_3P{=}CH_2$, can be prepared by the following sequence of reactions:

$$Ph_3P + MeI \rightarrow [Ph_3PMe]^+I^- \tag{5.33a}$$

$$[Ph_3PMe]^+I^- + Bu^nLi \rightarrow Ph_3P{=}CH_2 + LiI + C_4H_{10} \tag{5.33b}$$

Alkylidenephosphoranes are very air-sensitive substances, which are volatile enough to be purified by vacuum distillation in the case of derivatives such as $Me_3P{=}CH_2$ containing low molecular weight alkyl groups. Their phosphorus–carbon bond distances range from 1.66 to 1.74 Å, indicative of considerable if not full double-bond character; this suggests the dominance of the resonance structure $R_3P{=}CR_2'$ relative to $R_3P^+{-}C^-R_2'$.

The most important application of alkylidenephosphoranes is their use in the *Wittig reaction*[42] for the conversion of carbonyl compounds, such as aldehydes and ketones, to the corresponding olefins, for example,

$$RR'C{=}O + Ph_3P{=}CH_2 \rightarrow RR'C{=}CH_2 + Ph_3P{=}O \tag{5.34}$$

The first step of this reaction can involve attack of the nucleophilic methylene carbon atom in $Ph_3P{=}CH_2 \leftrightarrow Ph_3P^+{-}C^-H_2$ on the carbonyl carbon of the aldehyde or ketone, similar to the reactions of aldehydes and ketones with the carbanions in organolithium and organomagnesium compounds. The driving force of the elimination of the phosphine oxide ($Ph_3P{=}O$ in equation 5.34) is the higher strength of the P—O bond relative to the P—C bond. For use in the Wittig reaction, the alkylidenephosphoranes need not be isolated in the pure state but can be generated *in situ* from the corresponding phosphonium salt and a strong base, such as an alkyllithium derivative or sodium amide.

The methods for the preparation of alkylidenephosphoranes containing phosphorus(V)–carbon double bonds (e.g., equations 5.32) can be applied in slightly modified form to the preparation of the "carbon complex" carbon-bis(triphenylphosphorane),[43] $(Ph_3P)_2C$, by the following sequence of reactions:

$$2Ph_3P + CH_2Br_2 \rightarrow [Ph_3PCH_2PPh_3]Br_2 \tag{5.35a}$$

$$[Ph_3PCH_2PPh_3]Br_2 + 2K \rightarrow Ph_3P{=}C{=}PPh_3 + H_2\uparrow + 2KBr \tag{5.35b}$$

The P–C–P angle in this cumulenelike derivative is 137°, indicative of appreciable contribution of resonance structures of the type $Ph_3P^+{-}\ddot{C}^-{=}PPh_3$.

Compounds of the type $R_2C{=}PX$, containing two-coordinate phosphorus(III) and with a double bond between phosphorus(III) and carbon ("*phosphaalkenes*"), are also known.[44] However, such compounds generally are isolable only when the phosphorus and carbon atoms forming the double bond are sterically protected by large substituents (e.g., trimethylsilyl or $2,4,6{-}R_3'C_6H_2$ ($R' = Me$ or Bu^t). Olefin chemistry is a useful model for the preparation and chemical reactivity of phosphaalkenes. Thus phosphaalkenes can be prepared by elimination reactions similar to the synthesis of olefins, for example,

$$R_2CH\ddot{P}Cl_2 \rightarrow R_2C{=}\ddot{P}Cl + HCl \tag{5.36}$$

Phosphaalkenes in which the phosphorus atom is bonded to two carbon atoms can be obtained by reactions of $R_2C{=}PCl$ derivatives with carbon nucleophiles, for example,

$$R_2C{=}\ddot{P}Cl + LiR' \rightarrow R_2C{=}\ddot{P}R' + LiCl \qquad (5.37)$$

The C–P–C angle in phosphaalkenes is bent ($\sim 110°$), similar to the corresponding C–C–C angle in olefins and indicative of stereochemical activity of the phosphorus lone pair. The short P=C distance of ~ 1.68 Å is suggestive of a double bond. Phosphaalkenes form transition metal complexes (e.g., $(R_2C{=}\ddot{P}R')Ni(PMe_3)_2$) with structures analogous to those of corresponding metal–olefin complexes. A cumulene-type phosphorus(III) derivative with two-coordinate phosphorus atoms of the type $RP{=}C{=}PR$ ($R{=}2,4,6\text{-}Bu^t_3C_6H_2$) is also known[45]; the P–C–P angle in this derivative is nearly linear ($173°$), in contrast to the bent P–C–P angle of $137°$ in the phosphorus(V) cumulene $Ph_3P{=}C{=}PPh_3$ discussed in connection with equations 5.35.

Phosphaalkynes of the type $RC{\equiv}P\colon$ containing a triple bond between phosphorus(III) and carbon are also known[46,47]; phosphaalkynes are examples of one-coordinate phosphorus(III). Phosphaalkynes, like phosphaalkenes, require a bulky R group to be isolable. The phosphaalkyne that has been studied in the greatest detail is the t-butyl derivative $Bu^tC{\equiv}P\colon$, a reasonably stable distillable liquid, which can be prepared by the following sequence of reactions:

$$Bu^tC({=}O)Cl + (Me_3Si)_3P \rightarrow Bu^tC({=}O)P(SiMe_3)_2 + Me_3SiCl \qquad (5.38a)$$

$$Bu^tC({=}O)P(SiMe_3)_2 \rightarrow (Me_3SiO)(Bu^t)C{=}P{-}SiMe_3 \qquad (5.38b)$$

$$(Me_3SiO)(Bu^t)C{=}P{-}SiMe_3 \xrightarrow{\text{NaOH}} Bu^tC{\equiv}P\colon + (Me_3Si)_2O \qquad (5.38c)$$

The driving force behind this synthesis is the stability of Si—Cl and Si—O bonds relative to Si—P bonds, and the overall result of this synthetic sequence is the replacement of both the oxygen and chlorine atoms in the acid chloride $Bu^tC({=}O)Cl$ with a single phosphorus atom by use of $(Me_3Si)_3P$. Phosphaalkynes, like alkynes, form a variety of complexes with transition metals.

5.11.3 Pentavalent Derivatives, R_5E

Five-coordinate pentavalent compounds[48] of the type R_5E tend to be more stable with the heavier elements, particularly antimony. Thus reaction with methyllithium of tetramethylstibonium halides, $[Me_4Sb]^+X^-$, gives pentamethylantimony, a colorless distillable liquid, whereas the corresponding reaction with methyllithium of tetramethylphosphonium halides, $[Me_4P]^+X^-$, results in methane elimination to five methylenetrimethylphosphorane (see preceding section):

$$[Me_4Sb]^+X^- + LiMe \rightarrow Me_5Sb + LiX \qquad (5.39a)$$

$$[Me_4P]^+X^- + LiMe \rightarrow Me_3P{=}CH_2 + CH_4{\uparrow} + LiX \qquad (5.39b)$$

Pentaphenylphosphorus, Ph_5P, however, is a stable crystalline solid (mp 124°C, dec), since it has no hydrogen atoms bonded to an α-carbon atom.[49,50] In most R_5E derivatives the central five-coordinate atom has trigonal bipyramidal coordination; an exception is crystalline Ph_5Sb (mp 171°C), in which the antimony coordination is closer to square pyramidal.[51] Pentaarylbismuth derivatives, R_5Bi, typically have square pyramidal bismuth coordination and are frequently red or violet.[52].

5.11.4 Binuclear and Polynuclear Organometallic Derivatives of Phosphorus, Arsenic, Antimony, and Bismuth

Compounds of the type R_2E—ER_2, with E—E single bonds, can be obtained by Wurtz-like coupling reactions of R_2ECl derivatives, for example,

$$2Ph_2BiCl + 2Na \xrightarrow{\text{liq } NH_3} Ph_2Bi\text{—}BiPh_2 + 2NaCl \tag{5.40}$$

The prior preparation and isolation of R_2PCl (R = alkyl) derivatives for the synthesis of the tetraalkylbiphosphines, R_2P—PR_2, can be avoided by using the anomalous reaction of the readily prepared $S{=}PCl_3$ (Section (5.5.6)) with alkylmagnesium compounds leading directly to the formation of a phosphorus—phosphorus bond, for example,

$$2S{=}PCl_3 + 6MeMgI \rightarrow$$

$$Me_2P({=}S)P({=}S)Me_2 + 3MgCl_2 + 3MgI_2 + C_2H_6\uparrow \tag{5.41a}$$

$$Me_2P({=}S)P({=}S)Me_2 + 2Fe \xrightarrow{\Delta} 2FeS + Me_2P\text{—}PMe_2 \tag{5.14b}$$

The intermediate $Me_2P({=}S)P({=}S)Me_2$ is an air-stable crystalline solid, whereas Me_2P—PMe_2 is an evil-smelling volatile liquid, which is spontaneously flammable in air. The arsenic analogue Me_2As—$AsMe_2$ has similar properties; it was one of the first organometallic derivatives of any type to be prepared and was originally called "cacodyl" because of its bad odor. Some distibines, R_2Sb—SbR_2, and dibismuthines, R_2Bi—BiR_2, exhibit thermochromic properties; that is, their colors are strongly temperature-dependent.[53–55] Thus all distibines and dibismuthines are yellow to orange in the liquid phase. However, derivatives with small alkyl groups (e.g., methyl) are orange, red, blue, or violet in the solid state. The presence of E—E$\cdots$E—E$\cdots$ chains with alternating short bonding E—E distances (~ 2.8 Å for E = Sb) and long semibonding E$\cdots$E distances (~ 3.6 Å for E = Sb) in the solid phase is postulated to explain these thermochromic properties.

Cyclopolyphosphines and cyclopolyarsines with the general formula $(RE)_n$ (E=P, As; $3 \leq n \leq 6$) are known.[56,57] In general they can be prepared by treatment of $RECl_2$ derivatives with a strong reducing agent such as magnesium metal. The arsenic compounds can also be obtained by reduction of the salts $RAsO_3Na_2$ with excess hypophosphorous acid. The preferred ring size depends on the size and/or the electronegativity of the R group with an increase in the size of the R group leading to a smaller ring size. Thus reaction with magnesium of Bu^tPCl_2 containing the relatively bulky t-butyl group can be controlled to give either the four-membered ring $Bu^t_4P_4$ or even the three-membered ring $Bu^t_3P_3$.

The ultimate ring size limit of the cyclopolyphosphines $(RE)_n$ is the dimer $(RE)_2 = RE{=}ER$, in which the $E{=}E$ double bond can be interpreted as a "two-membered ring." Until relatively recently, stable compounds of the type $RP{=}PR$, analogous to organic azo compounds, $RN{=}NR$, were believed to be impossible. One relevant argument for the instability of $RP{=}PR$ derivatives is the much lower π-bond strength of a $P{=}P$ bond (156 kJ/mol) relative to an $N{=}N$ bond (256 kJ/mol). Attempts to make a stable phosphorus analogue of azobenzene, namely $PhP{=}PPh$, always lead instead to a cyclopolyphosphine, $(PhP)_n$ ($n = 4, 5$). However, recently it has been shown that use of very bulky R groups (e.g., $R = 2,4,6-Bu^t_3C_6H_2$ or $(Me_3Si)_3C$) can lead to stable *diphosphenes*,[58,59] $RP{=}PR$, by reactions such as the following:

$$2RPCl_2 + 2Mg \rightarrow RP{=}PR + 2MgCl_2 \tag{5.42a}$$

$$2RPCl_3 + 4NaC_{10}H_8 \rightarrow RP{=}PR + 4NaCl + 4C_{10}H_8 \tag{5.42b}$$

$$RPCl_2 + R'PH_2 \xrightarrow{\text{base}} RP{=}PR' + 2HCl \tag{5.42c}$$

The last method can be used to prepare unsymmetrical diphosphenes such as $2,4,6-Bu^t_3C_6H_2P{=}PC(SiMe_3)_3$. Related methods can be used to prepare analogous derivatives with $P{=}As$, $As{=}As$, and $P{=}Sb$ double bonds. Compounds with a $P{=}P$ bond are colored (yellow to red-orange) owing to $n \rightarrow \pi^*$ and/or $\pi \rightarrow \pi^*$ transitions of the $P{=}P$ chromophore in the visible region. Some $RP{=}PR$ derivatives with sufficiently large R groups are stable in air, notably $2,4,6-Bu^t_3C_6H_2{=}PC_6H_2Bu^t_3-2,4,6$. X-ray crystallography on stable crystalline $RP{=}PR$ derivatives indicate $P{=}P$ bond distances in the range 2.00–2.03 Å which is ~ 0.2 Å shorter than similar $P{-}P$ single bonds. Diphosphenes, $RP{=}PR$, can form a variety of transition metal complexes involving coordination of the phosphorus lone pairs and/or the $P{=}P$ double bond.

5.12 Cationic Chemistry of Phosphorus, Arsenic, Antimony, and Bismuth in Aqueous Solution

There is little evidence for cationic derivatives of phosphorus or arsenic in aqueous solution. Antimony(III) gives SbO^+ derivatives in aqueous solution, but the species observed vary markedly with the acid concentration and the anions present. In aqueous sulfuric acid solutions, the cations SbO^+ and $Sb(OH)_2^+$ are found in concentrations below 1.5 M and sulfato complexes such as $SbOSO_4^-$ and $Sb(SO_4)_2^-$ in more concentrated sulfuric acid solutions. The presence of organic complexing agents in aqueous antimony(III) solutions

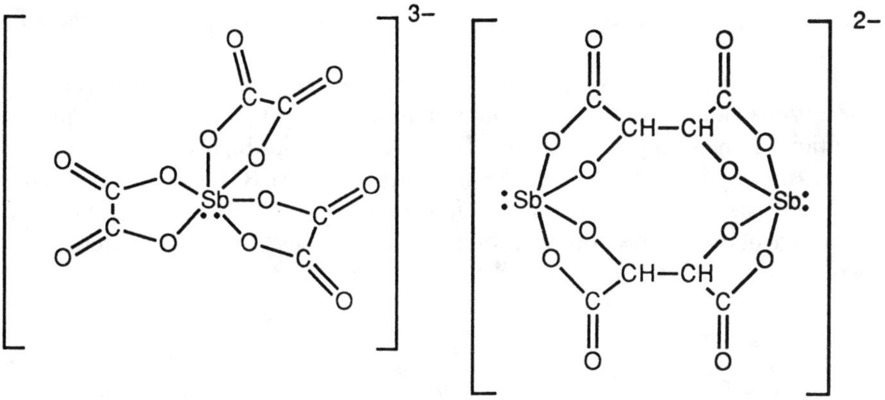

ψ-pentagonal bipyramidal
$[Sb(C_2O_4)_3]^{3-}$

ψ-trigonal bipyramidal
$[Sb_2(C_4H_2O_6)_2]^{2-}$

Figure 5.14. The antimony coordination in the $[Sb(C_2O_4)_3]^{3-}$ and $[Sb_2(C_4H_2O_6)_2]^{2-}$ complex atoms showing the stereochemically active lone pairs.

leads to complexes in which the lone pair of antimony(III) is stereochemically active (Figure 5.14). In the oxalate complex $[Sb(C_2O_4)_3]^{3-}$ the six-coordinate antimony(III) is ψ-pentagonal bipyramidal with the lone pair in an axial position. The tartrate complex $K_2[Sb_2(d\text{-}C_4H_2O_6)_2] \cdot 3H_2O$, which is called *tartar emetic*, has been known for more than 300 years; it is used in medicine to treat the parasitic diseases schistosomiasis and leishmaniasis. The binuclear anion in tartar emetic (Figure 5.14) contains two four-coordinate ψ-trigonal bipyramidal antimony(III) atoms with a stereochemically active lone pair; the

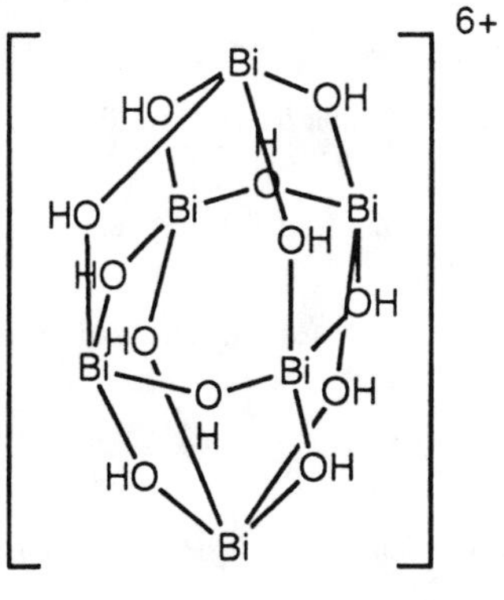

Figure 5.15. The framework of the $Bi_6(OH)_{12}^{6+}$ cation.

tetradeprotonated tartaric acid in tartar emetic acts as a tetradentate bimetallic ligand.

Bismuth(III) forms well-defined cationic species in aqueous solution, but there is no evidence for the simple mononuclear $Bi(H_2O)_n^{3+}$ ion in such aqueous solutions. In neutral perchlorate the hexanuclear species $Bi_6O_6^{6+}$ or its hydrated form, $Bi_6(OH)_{12}^{6+}$ is produced. The structure of the $Bi_6(OH)_{12}^{6+}$ cation (Figure 5.15) consists of an octahedron of bismuth ions with a hydroxy group bridging each edge of the octahedron; vibrational spectra suggest some weak interactions between the bismuth atoms.

References

1. Corbridge, D. E. C., *The Structural Chemistry of Phosphorus*, Elsevier, Amsterdam, 1974.

2. Luckenbach, R., *Dynamic Stereochemistry of Pentacoordinated Phosphorus and Related Elements*, Thieme, Stuttgart, 1973.

3. Crutchfield, M. M.; Dungan, C. H.; Letcher, J. H.; Mark, V.; Van Wazer, J. R., Compilation of ^{31}P NMR Data, in *Topics in Phosphorus Chemistry*, M. Grayson and E. J. Griffiths, Eds., Wiley-Interscience, New York, 1967, Chapter 4.

4. Verkade, J. G.; Quin, L. D., Eds., *Phosphorus-31 NMR Spectroscopy in Stereochemical Analysis*, VCH Publishers, Deerfield Beach, FL, 1987.

5. Zanella, P.; Rossetto, G.; Brianese, N.; Ossola, F.; Porchia, M.; Williams, J. O., Organometallic Precursors in the Growth of Epitaxial Thin Films of Group III–V Semiconductors by Metal–Organic Chemical Vapor Deposition, *Chem. Mater*, 1991, **3**, 225–242.

6. Cowley, A. H., Jones, R. A., Single Source III/V Precursors: A New Approach to Gallium Arsenide and Related Semiconductors, *Angew. Chem. Int. Ed. Engl.*, 1989, **28**, 1208–1215.

7. Jeitschko, W.; Flörke, M.; Möller, M. H.; Rühl, R., Preparation, Crystal Structures, and Properties of Transition Metal Polyphosphides, *Ann. Chim. (Paris)*, 1982, **7**, 525–537.

8. King, R. B., *Inorg. Chem.*, 1989, **28**, 3048–3051.

9. Corbett, J. D., Polyatomic Zintl Anions of the Post-Transition Elements, *Chem. Rev.*, 1985, **85**, 383–397.

10. Zintl, E.; Goubeau, J.; Dullenkopf, W., *Z. Phys. Chem. A*, 1931, **154**, 1–96.

11. Weston, R. E.; Bigeleisen, J., *J. Am. Chem. Soc.*, 1954, **76**, 3074–3078.

12. Kruck, T., Trifluorophosphine Complexes of Transition Metals, *Angew. Chem. Int. Ed. Engl.*, 1967, **6**, 53–67.

13. Nixon, J. F., Recent Progress in the Chemistry of Fluorophosphines, *Adv. Inorg. Chem. Radiochem.*, 1970, **13**, 363–469.

14. Payne, D. S., in *Non-aqueous Solvent Systems*, T. C. Waddington, Ed., Academic Press, London, 1976, pp. 301–321.

15. Baughan, E. C., in *The Chemistry of Non-aqueous Solvents*, Vol. 4, J. J. Lagowski, Ed., Academic Press, London, 1976, pp. 129–165.

16. Kirsanov, A. V.; Gorbatenko, Zh. K.; Feshchenko, Chemistry of Phosphorus Iodides, *Pure Appl. Chem.*, 1975, **44**, 125–139.

17. Denis, J. N.; Krief, A., *J. Chem. Soc. Chem. Commun.*, 1980, 544–545.

18. Passmore, J., *J. Chem. Soc. Dalton Trans.*, 1985, 9–16.

19. Seppelt, K., Arsenic Pentachloride, $AsCl_5$, *Angew. Chem. Int. Ed. Engl.*, 1976, **15**, 377–378.

20. Hershaft, A.; Corbett, J. D., *Inorg. Chem.*, 1963, **2**, 979–985.

21. Shustorovich, E.; Dobosh, P. A., *J. Am. Chem. Soc.*, 1979, **101**, 4090–4095.

22. Heinze, D., Zur Chemie von Phosphor(III) Oxid, *Pure Appl. Chem.*, 1975, **44**, 141–172.

23. Meisel, M., Über Darstellung und Konstitution von Phosphorchalkogeniden mit Adamantanstruktur, *Z. Chem.*, 1985, **23**, 117–125.

24. Whitfield, H. J., *J. Chem. Soc. A*, 1970, 1800–1803; 1973, 1737–1738.

25. Lauer, W.; Becke-Goehring, M.; Sommer, K., *Z. Anorg. Chem.*, 1969, **371**, 193–200.

26. Niederprüm, H., Chemical Nickel Plating, *Angew. Chem. Int. Ed. Engl.*, 1975, **14**, 614–620.

27. Bhattacharyya, A. K.; Thyagarajan, G., The Michaelis–Arbusov Rearrangement, *Chem. Rev.*, 1981, **81**, 415–430.

28. Westheimer, F. H., Monomeric Metaphosphates, *Chem. Rev.*, 1981, **81**, 313–326.

29. Griffith, E. J., The Chemical and Physical Properties of Condensed Phosphates, *Pure Appl. Chem.*, 1975, **44**, 173–200.

30. Majling, J.; Hunic, F., Phase Chemistry of Condensed Phosphates, *Top. Phosphorus Chem.*, 1980, **10**, 341–502.

31. Kalliney, S. Y., Cyclophosphates. *Top. Phosphorus Chem.*, 1972, **7**, 255–309.

32. Choppin, G. R.; Rydberg, J., *Nuclear Chemistry: Theory and Applications*, Pergamon Press, Oxford, 1980, Chapter 20.

33. Allcock, H. R., Recent Advances in Phosphazene (Phosphonitrilic) Chemistry, *Chem. Rev.*, 1972, **72**, 315–356.

34. Krishnamurthy, S. S.; Sau, A. C.; Woods, M., Cyclophosphazenes, *Adv. Inorg. Chem. Radiochem.*, 1978, **21**, 41–112.

35. Allen, C. W., Radio- and Stereochemical Control in Substitution Reactions of Cyclophosphazene, *Chem. Rev.*, 1991, **91**, 119–135.

36. Neilson, R. H.; Wisian-Neilson, P., Poly(alkyl/arylphosphazenes) and Their Precursors, *Chem. Rev.*, 1988, **88**, 541–562.

37. Nazri, G.; MacArthur, D. M.; Ogara, J. F., *Chem. Mater.*, 1989, **1**, 370–374.

38. Cowley, A. H.; Kemp, R. A., Synthesis and Reaction Chemistry of Stable Two-Coordinate Phosphorus Cations (Phosphenium Ions), *Chem. Rev.*, 1985, **85**, 367–382.

39. Ruff, J. K.; Schlientz, W. J., *Inorg. Synth.*, 1974, **15**, 84.

40. McAuliffe, C. A., Ed., *Transition Metal Complexes of Phosphorus, Arsenic and Antimony Ligands*, Macmillan, London, 1973.

41. Schmidbaur, H., Inorganic Ylides, *Acc. Chem. Res.*, 1970, **8**, 62–70.

42. Maryanoff, B. E.; Reiz, A. B., The Wittig Olefination Reaction and Modifications involving Phosphoryl-Stabilized Carbanions. Stereochemistry, Mechanism, and Selected Synthetic Aspects, *Chem. Rev.*, 1989, **89**, 863–927.

43. Vincent, A. T.; Wheatley, P. J., *J. Chem. Soc. Commun.*, 1971, 592.

44. Appel, R.; Knoll, F.; Ruppert, I., Phospha-Alkenes and Phospha-Alkynes, Genesis and Properties of the $(p-p)\pi$-Multiple Bond. *Angew. Chem. Int. Ed. Engl.*, 1981, **20**, 731–744.

45. Karsch. H. H.; Reisacher, H.-U.; Müller, G., *Angew. Chem. Int. Ed. Engl.*, 1984, **23**, 638–639.

46. Regitz, M.; Binger, P., Phosphaalkynes—Synthesis, Reactions, Coordination Behavior, *Angew. Chem. Int. Ed. Engl.*, 1988, **27**, 1484–1508.

47. Regitz, M., Phosphaalkynes: New Building Blocks in Synthetic Chemistry, *Chem. Rev.*, 1990, **90**, 191–213.

48. Hellwinkel, D., Penta- and Hexaorganyl Derivatives of the Main Group Five Elements, *Top. Curr. Chem.*, 1983, **109**, 1–63.

49. Wittig, G.; Rieber, M., *Liebigs Ann. Chem.*, 1949, **562**, 187–192.

50. Wittig, G.; Geissler, G., *Liebigs Ann. Chem.*, 1953, **580**, 44–57.

51. Wheatley, P. J., *J. Chem. Soc.*, 1964, 3718.

52. Seppelt, K., Structure, Color, and Chemistry of Pentaaryl Bismuth Compounds, *Adv. Organomet. Chem.*, 1992, **34**, 207–217.

53. Ashe, A. J., III; Butler, W.; Diephouse, T. R., *J. Am. Chem. Soc.*, 1981, **103**, 207–209.

54. Hughbanks, T.; Hoffmann, R.; Whangbo, M.-H.; Stewart, K. R.; Eisenstein, O.; Canadell, E., *J. Am. Chem.*, 1982, **104**, 3876–3879.

55. Ashe, A. J., III, Thermochromic Distibines and Dibismuthines, *Adv. Organometl. Chem.*, 1990, **30**, 77–97.

56. Baudler, M., Chain and Ring Phosphorus Compounds—Analogies Between Phosphorus and Carbon Chemistry, *Angew. Chem. Int. Ed. Engl.*, 1982, **21**, 492–512.

57. Baudler, M., Polyphosphorus Compounds—New Results and Perspectives, *Angew. Chem. Int. Ed. Engl.*, 1987, **26**, 419–441.

58. Cowley, A. H., Stable Compounds with Double Bonding Between the Heavier Main-Group Elements, *Acc. Chem. Res.*, 1984, **17**, 386–392.

59. Cowley, A. H.; Norman, N. C., The Synthesis, Properties, and Reactivities of Stable Compounds Featuring Double Bonding Between Heavier Group 14 and 15 Elements, *Prog. Inorg. Chem.*, 1986, **34**, 1–63.

6

The Chalcogens

6.1 General Aspects of Chalcogen Chemistry

The elements oxygen, sulfur, selenium, tellurium, and polonium of group VIA (old designation) or group 16 (new designation) of the periodic table are called the *chalcogens.* The chemistry of oxygen is distinct from that of the other chalcogens and, therefore, is discussed separately.

6.1.1 Oxygen and Oxides

The oxygen atoms has six valence electrons. It can, therefore, complete its octet in the following ways:

1. Gain of two electrons to form the oxide ion, O^{2-}.
2. Formation of two single covalent bonds (e.g., R—O—R) or one double covalent bond ($R_2C{=}O$, $O{=}C{=}O$, etc.).
3. Gain of one electron and formation of one single bond (OH^-, OR^-, etc.).
4. Formation of three or four covalent bonds with a compensating positive charge (e.g., the oxonium ions R_3O^+).

6.1.1.1 Oxides

The formation of the oxide ion O^{2-} from elemental oxygen, O_2, requires the expenditure of considerable energy (~ 1000 kJ/mol) by the following sequence of reactions:

$$O_2(g) = 2O(g) \qquad\qquad \Delta H = 248 \text{ kJ/mol} \qquad\qquad (6.1a)$$

$$O(g) + 2e^- = O^{2-}(g) \qquad \Delta H = 752 \text{ kJ/mol} \qquad\qquad (6.1b)$$

In addition, formation of an ionic metal oxide requires energy in vaporizing and ionizing the metal atoms. Despite this considerable energy requirement for the formation of metal oxides, many essentially ionic oxides (e.g., CaO) are very stable because their lattice energies containing the relatively small oxide ion, O^{2-} (1.40 Å), are very high. Furthermore, the high lattice energies of metal oxides contribute to the stabilization of a number of unusual high metal oxidation states in metal oxides relative to aqueous solutions; some examples are manganese(IV) in MnO_2, silver(II) in AgO, and praseodymium(IV) in PrO_2.

Oxides can be classified according to their acid–base behavior in aqueous solution into the following four types:

1. *Basic oxides.* Basic oxides are formed by the most electropositive metals. Soluble basic metal oxides such as Na_2O form the metal cation upon dissolving in water; insoluble basic metal oxides such as MgO and La_2O_3 usually dissolve in dilute aqueous nonoxidizing acids to form the corresponding metal cation. Such basic metal oxides are *basic anhydrides.*
2. *Acidic oxides.* Most covalent oxides of the nonmetals, particularly in their highest oxidation states, are acidic (e.g., CO_2, N_2O_5, P_4O_{10}, SO_3, Cl_2O_7, etc.) and dissolve in water to form acids; they, therefore, function as *acid anhydrides.* A few high oxidation state metal oxides such as CrO_3, Mn_2O_7, and Re_2O_7 are also acidic oxides.
3. *Amphoteric oxides.* A few metal oxides such as BeO, ZnO, Al_2O_3, and Cr_2O_3 are *amphoteric;* they behave as acids toward strong bases and as bases toward strong acids.
4. *Inert oxides.* A few oxides such as N_2O, CO, MnO_2, and PbO_2 are inert and dissolve in neither acids nor bases to form cations or anions, respectively. Some inert oxides (e.g., N_2O and CO) are dehydration products of acids (e.g., hyponitrous and formic acids, respectively) but do not regenerate the acid upon addition of water. Other inert oxides (e.g., MnO_2 and PbO_2) contain metals in relatively high oxidation states and will dissolve in some reducing acids, particularly HX ($X = Cl$, Br, and I), by redox reactions to form a lower metal oxidation state.

For a given metal, the acidity of its oxides increases with increasing oxidation state. This point is illustrated especially well with chromium, where $Cr^{II}O$ is strictly basic, $Cr_2^{III}O_3$ is amphoteric, and $Cr^{VI}O_3$ is strongly acidic. Many oxides of metals with multiple oxidation states are nonstoichiometric; these typically consist of lattices of oxide ions with the interstices partially filled by metal ions. A good example of a nonstoichiometric metal oxide is iron(II) oxide, which has a composition range from $FeO_{0.90}$ to $FeO_{0.95}$ depending on the conditions used for its preparation.

Specific oxides are discussed under the element forming the oxide in question.

6.1.1.2 Hydroxides

Discrete OH^- ions exist only in the hydroxides of the most electropositive elements, namely the alkali and alkaline earth metals. Other metal ions, particularly those in relatively high oxidation states such as Al^{3+} and Fe^{3+}, generate acidic solutions (Section 1.4) by hydrolysis according to the following general scheme:

$$[M(H_2O)_n]^{m+} \rightarrow [M(H_2O)_{n-1}(OH)]^{(m-1)+} + H^+ \tag{6.2a}$$

$$[M(H_2O)_{n-1}(OH)]^{(m-1)+} \rightarrow [M(H_2O)_{n-2}(OH)_2]^{(m-2)+} + H^+, \text{ etc.} \tag{6.2b}$$

Such hydrolyses lead ultimately to the precipitation of hydrous metal oxides, commonly called "metal hydroxides," although they do not contain discrete OH^- ions. Hydrous metal oxides are precipitated upon treatment of aqueous solutions of most metals other than the alkali and alkaline earth metals with a solution of an alkali metal hydroxide. In the case of amphoteric metals such as zinc and aluminum, the initially precipitated hydrous oxide will redissolve upon addition of excess alkali metal hydroxide, forming hydroxo anions such as $Zn(OH)_4^{2-}$ and $Al(OH)_4^-$.

6.1.1.3 Covalent Oxygen Compounds

The most common coordination number of oxygen is 2, which is found in $X—\ddot{O}—X$ derivatives. In the absence of π-bonding, the X–O–X angle is bent relating to the stereochemical activity of the two lone electron pairs on the oxygen atom. Thus the X–O–X angles in H_2O, Me_2O, and OF_2 are 104.5°, 110°, and 103°, respectively. When the X atoms forming the X—O—X bond have π orbitals or d orbitals capable of interacting with the oxygen lone pair orbitals, the X—O bonds acquire some π character and become shorter, with concurrent widening of the X–O–X bond angle. Examples of such widened X—O—X bond angles include the C–O–C bond angle of 124° in diphenyl ether, $(C_6H_5)_2O$, and the Si–O–Si bond angle of 142° in quartz. The limiting case of such π interaction in X—O—X systems leads to two linear sp hybrids on the oxygen atom with an X–O–X angle of 180°. This occurs in a few oxo-bridged metal complexes such as $[Cl_5Ru—O—RuCl_5]^{4-}$.

Oxygen can also have a coordination numbers of 1 in multiply bonded derivatives, although the actual bond order to one-coordinate oxygen can depend drastically on the charge distribution. Thus the nitrogen–oxygen bond order is essentially unity in trimethylamine N-oxide, which can be described by the polar structure $Me_3N^+—O^-$, whereas the carbon–oxygen bond order in aldehydes and ketones, $RR'C=O$, is essentially 2, with the carbon–oxygen double bond consisting of a simple σ bond and a simple π bond. In most one-coordinate oxygen derivatives of heavier elements such as R_3PO, R_3AsO, PO_4^{3-}, SO_4^{2-}, ClO_4^-, CrO_4^{2-}, MnO_4^-, and OsO_4, two orthogonal π- interactions are possible between X and O in mutually perpendicular planes that intersect along the X—O axis.

Coordination numbers higher than 2 are occasionally found for oxygen. Thus three-coordinate oxygen is found in the oxonium ion H_3O^+ (Section 1.4) as well as its organic derivatives such as protonated ethers, R_2OH^+, trialkyloxonium ions, R_3O^+, and ether adducts of Lewis acids such as $Et_2O \rightarrow BF_3$. Three-coordinate oxygen atoms also occur in metal complexes in which $(\mu_3\text{-O})M_3$ units can be either pyramidal as in OHg_3^+, $Os_4O_4(CO)_{12}$, and so on, or planar as in trinuclear basic carboxylates of trivalent metals of the type $[M_3O(O_2CR)_3L_6]^+$ (M = V, Cr, Mn, Fe, Co, Rh, Ir, etc.; L = H_2O, pyridine, etc.). Four-coordinate oxygen atoms are found as $(\mu_4\text{-O})M$ units in a few rather ionic oxides such as PbO and tetranuclear metal complexes such as $Mg_4OBr_6 \cdot 4\ Et_2O$, $Cu_4OCl_6(OPPh_3)_4$, and $M_4O(O_2CR)_6$ (M = Be, Zn). A six-coordinate $(\mu_6\text{-O})$ oxygen atom is found in the center of octahedral isopolyoxometallates of the type $M_6O_{19}^{n-}$ (M = Nb, Ta, $n = 8$; M = Mo, W; $n = 2$).

Catenation is very limited in oxygen compounds. Chains of oxygen atoms longer than two are found only in ozone, O_3; the ozonide ion, O_3^-; a few highly fluorinated trioxides of the type $(R_f)_2O_3 = R_fO\text{---}O\text{---}OR_f$; and the very unstable $O_4F_2 = FO\text{---}O\text{---}O\text{---}OF$. The weakness of the oxygen–oxygen bonds in H_2O_2, O_2, and O_2^{2-}, like the limited catenation ability of oxygen, is attributed to repulsions between lone electron pairs on adjacent oxygen atoms.

Oxygen has the following three stable isotopes:

^{16}O: 99.759% natural abundance and nonmagnetic
^{17}O: 0.0374% natural abundance with a spin of $\frac{5}{2}$
^{18}O: 0.2039% natural abundance and nonmagnetic

Oxygen-17 can be used as an NMR nucleus despite its quadrupole moment $(I = \frac{5}{2})^1$; because of the low natural abundance of this isotope, however, enriched samples are frequently required for satisfactory ^{17}O NMR spectra. The heavier oxygen isotopes can be enriched by the fractional distillation of water. Since, it is difficult to separate the pure middle isotope, ^{17}O, of interest in NMR studies, this isotope is very rare and costly.

6.1.2 Sulfur, Selenium, Tellurium, and Polonium Chemistry

The chemistry of the heavier chalcogens differs from that of oxygen in the following ways.

1. The lower electronegativities of the elements from sulfur to polonium lessen the ionic character of their compounds analogous to oxygen compounds and thus decrease drastically the importance of hydrogen bonding (Section 1.3).

2. The availability of d orbitals for chemical bonding in the elements from sulfur to polonium means that their maximum valencies are no longer restricted to 2 nor their maximum coordination numbers to 4, as is the case with oxygen. The involvement of d orbitals in the chemical bonding of sulfur and its heavier congeners leads to so-called *hypervalent compounds*, which

can be defined loosely as compounds in which d orbitals are involved in the primary bonding. Thus sulfur forms several hexacoordinate compounds (e.g., SF_6), and the characteristic coordination number of tellurium is 6. The d orbitals of sulfur and its heavier congeners can also in involved in multiple bonding such as in the sulfur–oxygen bonds in sulfates, sulfones, and sulfoxides.

3. In contrast to oxygen, sulfur has a strong tendency toward catenation, as exhibited in the S_n rings in free sulfur, diverse polysulfides S_n^{2-} in ionic salts, and chelating ligands in transition metal chemistry, *sulfanes* of the general formula XS_nX ($X = H$, halogen, CN, NR_2, etc.), and *polythionic* acids $HO_3SS_nSO_3H$ and their salts.

The following general trends can be observed in the sequence $S \rightarrow Se \rightarrow Te \rightarrow Po$:

1. Increasing metallic character of the elements.
2. The emergence of cationic properties for polonium and to a lesser extent for tellurium.
3. A decrease in the thermal stability of the hydrogen compounds H_2E so that H_2Te is significantly endothermic.
4. An increasing tendency to form hypervalent anionic complexes such as $SeBr_6^{2-}$, $TeBr_6^{2-}$, and PoI_6^{2-}.

The stable nuclei ^{77}Se (7.6% natural abundance) and ^{125}Te (7.0% natural abundance) both have spins of $\frac{1}{2}$ and are suitable nuclei for NMR measurements in selenium and tellurium derivatives, respectively.[2] Sulfur isotopes are much less suitable for NMR studies than these selenium and tellurium isotopes. However, sulfur NMR spectra have been obtained using the quadrupolar sulfur isotope ^{33}S with a spin of $\frac{3}{2}$ and a natural abundance of 0.75%.

6.2 The Elemental Chalcogens

6.2.1 Oxygen

The earth's crust contains $\sim 50\%$ by weight of oxygen and the earth's atmosphere is $\sim 20\%$ O_2. The second allotrope of oxygen, namely ozone, O_3, is also present in small amounts in the earth's atmosphere and is partially concentrated at altitudes of 15–25 km.[3] The small concentrations of ozone in the earth's atmosphere exercise an important function by absorbing much of the ultraviolet energy of wavelengths of 200–360 nm.

6.2.1.1 Dioxygen

The stable and most readily available allotrope of elemental oxygen is dioxygen, O_2, a colorless, odorless, tasteless, paramagnetic gas (mp $-218.8°C$, bp $-183.0°C$). The ground state of O_2 is the triplet $^3\Sigma_g^-$ state with its unpaired electrons in two orthogonal π orbitals leading to an oxygen–oxygen bond order

Table 6.1 Properties of Dioxygen Species and Their π-Orbital Occupancies

Species	O—O Distance (Å)	Stretching Frequency, v(O—O)(cm^{-1})	π^*-Orbital Occupancy		Unpaired Electrons
			π_a^*	π_b^*	
O_2^+	1.12	1905	↑		1
$O_2(^3\Sigma_g^-)$	1.21	1580	↑	↑	2
$O_2(^1\Delta_g)$			↑↓		0
$O_2(^3\Sigma_g^+)$			↑	↓	2
O_2^-	1.33	1097	↑↓	↑	1
O_2^{2-}	1.49	802	↑↓	↑↓	0

of $1 + \frac{2}{3}$ (Table 6.1). Electrons can be either added to these π orbitals or removed from them giving the ionic species O_2^- (superoxide), O_2^{2-} (peroxide), and O_2^+ (dioxygenyl cation); the properties of these dioxygen species, including the occupancies of the π^*-antibonding orbitals of their oxygen–oxygen bonds are summarized in Table 6.1. Note that as the oxygen–oxygen bond distance increases, indicating a lower bond order, the oxygen–oxygen stretching frequency decreases, indicating a weaker bond.

The presence of the unpaired electrons in the antibonding π^* orbitals of dioxygen makes dioxygen a fairly good oxidizing agent; that is, in aqueous solution,

$$O_2 + 2H_2O + 4e^- = 4OH^- \qquad E^\circ = +0.401 \text{ V} \qquad (6.3)$$

Dioxygen is readily soluble in organic solvents. For this reason degassing is necessary before organic solvents are used to handle compounds that are sensitive to air oxidation, such as many organometallic compounds. In addition, some organic solvents, notably many ethers, form explosive peroxides upon long standing in air. Nontoxic fluorocarbons, such as perfluorodecahydro-naphthalene, dissolve large amounts of oxygen and can, therefore, serve as blood substitutes.[4]

Photolysis of normal triplet dioxygen in the presence of a sensitizer such as methylene blue or fluorescein can lead to the excited singlet $^1\Delta_g$ state with an energy 94.7 kJ/mol above that of the ground state but with a sufficient lifetime to use as a reactive chemical intermediate (Table 6.1).[5] Singlet oxygen can also be generated by chemical methods such as the reaction

$$H_2O_2 + ClO^- \rightarrow Cl^- + H_2O + {}^1\Delta_g O_2 \qquad (6.4)$$

Reversion of singlet $^1\Delta_g$ oxygen to the ground state triplet $^3\Sigma_g^-$ oxygen can lead to a red chemiluminescence. The oxygen–oxygen bond in singlet oxygen contains a single σ bond and a single π bond similar to the carbon–carbon double bonds in olefins such as ethylene.[6] For this reason the chemical reactivity of singlet oxygen is similar to that of an olefin with electronegative substituents such as tetracyanoethylene. Thus singlet oxygen, like electronegatively substituted olefins, undergoes Diels–Alder-like [4 + 2] cycloadditions to 1,3-dienes, for example,

Such reactions are sometimes called *photooxidation reactions*, since the singlet oxygen is usually generated by bubbling oxygen into an irradiated solution of the substrate containing a suitable dye for the photosensitized conversion of ground state triplet oxygen to excited singlet oxygen such as methylene blue or fluorescein. Singlet oxygen plays an important role in the photodegradation of polymers in air.

6.2.1.2 Ozone[7]

Ozone, O_3 (bp $-112°C$, mp $-193°C$), has a bent structure with an O–O–O angle of $116.8 \pm 0.5°$ and an oxygen–oxygen distance of 1.278 ± 0.003 Å, indicative of partial double-bond character. Unlike dioxygen, ozone is diamagnetic, indicating the absence of unpaired electrons. The structure of ozone can be described by the resonance hybrids **6.1a** and **6.1b**.

6.1a   6.1b

Ozone is obtained by passing a silent electric discharge ($10–20$ kV/$50–500$ Hz) through dioxygen and thus arises naturally in the atmosphere from lightning discharges in air. This process can lead to the production of dioxygen containing up to $\sim 10\%$ of ozone. Pure ozone can be obtained by the fractional liquefaction of oxygen–ozone mixtures. Ozone forms a deep blue liquid and a black-violet solid. Ozone in the condensed phase is a dangerous explosive, since the reaction $2O_3 \rightarrow 3O_2$ is exothermic by -142 kJ/mol.

Ozone is chemically very reactive and a much more powerful oxidant than dioxygen. Thus the oxidation potential of ozone is $+1.24$ V in neutral aqueous solution and $+2.07$ V in acidic aqueous solution. Ozone is useful for the purification of drinking water because of its powerful oxidizing ability and its innocuous reduction products. Ozone is destroyed by the NO generated in high temperature combustion processes (e.g., from supersonic aircraft), to give NO_2, and by chlorine atoms generated from chlorofluorocarbons, to generate reactive ClO radicals. These processes are of considerable environmental concern

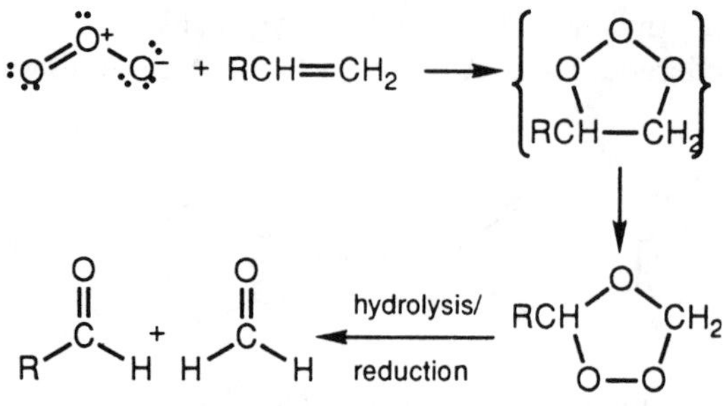

Figure 6.1. Cleavage of a double bond by ozonization.

because of their potential to destroy the protective ozone layer in the atmosphere, which screens out harmful ultraviolet irradiation from sunlight.

Ozone is reactive toward unsaturated organic compounds, since it can undergo [2 + 3] dipolar cycloaddition reactions (Figure 6.1) to give products call *ozonides*.[8] Rearrangement of the originally produced [2 + 3] cycloadduct (Figure 6.1) to reduce the number of direct oxygen–oxygen bonds from two to one leads concurrently to rupture of the carbon–carbon double bond of the olefin. Pure ozonides are generally unstable explosive compounds. Generation of an ozonide followed by mild reduction of the ozonide in situ produces aldehydes and ketones (Figure 6.1). A consequence of the reactivity of carbon–carbon double bonds toward ozone is the onset, upon exposure to ozone, of brittleness and fracture in rubbers and other organic polymers, with residual unsaturation.

6.2.2 Sulfur

Sulfur occurs widely in nature as elemental sulfur (S_8), sulfide minerals such as pyrites (FeS_2), sulfate minerals such as anhydrite ($CaSO_4$), and in crude oils, coal, natural gas (H_2S), and other substances. Sulfur is used in large quantities as a source of sulfuric acid and for the vulcanization of rubber.

The allotropy of sulfur is far more complicated than that of any other element.[9] This property arises from a delicate balance between the catenation ability of sulfur and the reactivity of sulfur–sulfur bonds. Thus sulfur in the solid state can consist of either discrete molecular S_n rings or polymeric S_n chains. However, the sulfur–sulfur bonds in these structures undergo rupture upon melting.

6.2.2.1 Molecular Forms of Solid Sulfur

Sulfur rings, S_n (n = 6–12, 18, and 20) have been characterized.[10] In all these sulfur rings, the S—S distances (~ 2.05 Å) and S–S–S angles ($\sim 106°$) are very similar, but the S–S–S–S torsion angles vary considerably. The S—S bond energies in elemental sulfur are ~ 265 kJ/mol and are the highest homonuclear bond energies except for H—H and C—C bonds.

The most thermodynamically stable form of solid sulfur is the eight-membered ring *cyclo-octasulfur*, S_8, which has the crown conformation (**6.2**). This form of sulfur is obtained from chemical reactions liberating elemental sulfur except for reactions directed toward other S_n ring sizes, for example,

$$H_2S_4 + S_2Cl_2 \rightarrow cyclo\text{-}S_6 + 2HCl \qquad (6.5a)$$

$$H_2S_8 + S_4Cl_2 \rightarrow cyclo\text{-}S_{12} + 2HCl \qquad (6.5b)$$

$$(C_5H_5)_2TiS_5 + S_2Cl_2 \rightarrow cyclo\text{-}S_7 + (C_5H_5)_2TiCl_2 \qquad (6.5c)$$

The common form of sulfur, namely *cyclo*-octasulfur, S_8, is a yellow solid (mp 119°C), which has a density of 2.069 g/cm^3 and is a good electrical insulator. All solid molecular forms of sulfur are readily soluble in carbon disulfide.

6.2

6.2.2.2 Polymeric Solid Sulfur (Catenasulfur)

Solid "plastic sulfur" is obtained upon rapid quenching of molten sulfur by pouring it into ice water. This form of sulfur can be obtained as fibers consisting of polymeric helical chains of S—S bonds with ~ 3.5 atoms per turn of the helix. Polymeric solid sulfur is insoluble in carbon disulfide and transforms slowly to the stable cyclic molecular form of sulfur, *cyclo-S_8*.

6.2.2.3 Liquid Sulfur

cyclo-Octasulfur, S_8, first gives a yellow, transparent, mobile liquid upon melting at ~ 119°C. This liquid becomes brown and increasingly viscous upon further heating above ~ 160°C. The viscosity reaches a maximum at ~ 200°C and, thereafter, falls until the boiling point (~ 445°C), where sulfur is a mobile, dark red liquid. The changes in viscosity of liquid sulfur can be related to the length of the sulfur chains, with an increase in viscosity corresponding to an increase in the average chain length in accord with standard relationships between polymer viscosity and molecular weight. Thus the sulfur chains reach their greatest length, namely 500,000–800,000 atoms, at ~ 200°C, the temperature of highest viscosity. The color changes upon melting arise from an increase in the intensity and the shift of an absorption band to the red. These color changes are associated with the formation of red S_3 and S_4, comprising 1–3% of sulfur at its boiling point.

6.2.2.4 Sulfur Vapor

The vapor of sulfur contains the species S_n ($2 \leq n \leq 10$) with S_8 dominating at $\sim 600°C$. However, upon further heating the equilibrium shifts in favor of S_2, which dominates above $\sim 720°C$. The sulfur–sulfur bond in S_2 is short (1.89 Å) and strong (422 kJ/mol), suggesting multiple bonding. Furthermore, S_2 is a triplet analogous to O_2 (Section 6.2.1.1).

6.2.3 Selenium and Tellurium

Selenium and tellurium are much less abundant than sulfur. They are recovered from anode slime deposited in the electrolytic purification of copper arising from impurities in the copper sulfide ore. Selenium and tellurium are also by-products from the aqueous processing of other sulfide ores.[11]

Elemental selenium, like elemental sulfur, occurs in both molecular and polymeric forms. However, unlike elemental sulfur, the polymeric form of elemental selenium, namely "gray" selenium, is more thermodynamically stable than the molecular form of elemental selenium. Gray selenium, (mp 217°C, bp 685°C); has metallic properties and can be obtained crystalline from molten selenium or from hot solutions of selenium in aniline. The structure of gray selenium consists of infinite spiral —Se—Se—Se—Se— chains with strong Se—Se bonds between adjacent selenium atoms in the chain and weak metallic interactions between the neighboring atoms of different chains.

Gray selenium is of interest because of its photoconductive properties. In the dark the electrical conductivity of gray selenium (2×10^{11} $\mu\Omega \cdot cm$) is far less than that of true metals. However, upon exposure to light the electrical conductivity of elemental selenium increases drastically. Elemental selenium is, therefore, used in photoelectric devices and xerography.[12]

Metastable molecular forms of selenium Se_n ($n = 6, 7, 8$) are known, and Se_6 and Se_8 have been isolated in the solid state. However, they readily revert to gray selenium upon application of heat or pressure.

The single known crystalline form of elemental tellurium is isomorphic with gray selenium.

6.2.4 Polonium

Polonium has no stable isotopes. The isotope ^{210}Po, an α-emitter with a half-life of 138.4 days, occurs in trace quantities in uranium and thorium minerals as a steady state intermediate (3×10^{-10} ppm) in the radioactive decay series. Since because of its short half-life, only trace quantities of polonium are found in nature, ^{210}Po is more readily obtained in gram quantities from nuclear reactors by the process

$$^{209}Bi + n \rightarrow {}^{210}Bi + \gamma \tag{6.6a}$$

$$^{210}Bi \rightarrow {}^{210}Po + \beta^- \tag{6.6b}$$

The study of polonium chemisty is complicated immensely by the intense alpha radiation arising from its relatively short half-life; this intense radiation causes damage to solutions and solids, generates much heat, and requires special safety precautions.[13]

Elemental polonium has two allotropes, both with metallic structures. α-Polonium is the only example of a simple cubic structure in a pure element; the polonium atoms have octahedral six-coordination in this structure. Heating α-polonium above 36°C gives β-polonium (mp 254°C), with a rhombohedral structure. Both forms of polonium exhibit metallic resistivity with a positive temperature coefficient.

6.2.5 Reactions of the Elemental Chalcogens

Elemental sulfur, selenium, tellurium, and polonium burn in air to form the corresponding dioxides, EO_2, and react when heated with halogens, most metals, and many nonmetals to form the corresponding binary compounds. Elemental sulfur, selenium, and tellurium are not affected by nonoxidizing acids. However, polonium metal dissolves even in strong nonoxidizing acids such as hydrochloric acid to give saltlike pink Po^{2+} and yellow Po^{4+}.

The molecular forms of sulfur and selenium, namely cyclic E_n (E = S, Se), dissolve readily in carbon disulfide and other nonpolar solvents. Such solutions become cloudy upon exposure to light, possibly owing to small amounts of ring opening to form polymeric E_x chains; this insoluble material reverts to soluble E_8 (E = S, Se), slowly in the dark or rapidly in the presence of a tertiary amine such as triethylamine. Secondary amines are capable of opening sulfur rings to form diaminopolysulfanes, $R_2N-(S_{n-2})-NR_2$, with the evolution of H_2S according to the following equation:

$$2R_2NH + S_n \rightarrow R_2N-(S_{n-2})-NR_2 + H_2S\uparrow \qquad (6.7)$$

For this reason, many reactions of elemental sulfur are catalyzed by amines. Electron paramagnetic resonance (EPR) studies on such reaction mixtures suggest the presence of one reactive free radical per 10,000 sulfur atoms. Sulfur and selenium are also capable of dehydrogenating cyclohexane derivatives to the corresponding benzenoid derivatives (e.g., Figure 6.2). The vulcanization of natural and synthetic rubbers is based on the cross-linking of the polymer chains by additon of sulfur to residual carbon–carbon double bonds at elevated temperatures.

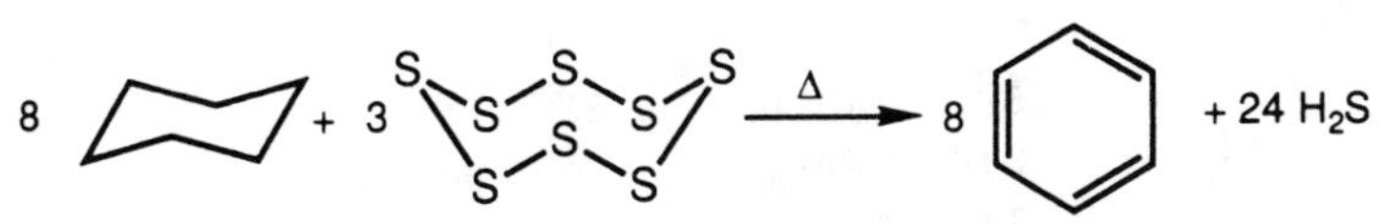

Figure 6.2. Dehydrogenation of cyclohexane to benzene by heating with elemental sulfur (S_8).

6.2.6 Elemental Chalcogen Cations

Elemental sulfur, selenium, and tellurium have long been known to dissolve in fuming sulfuric acid (H_2SO_4 plus excess SO_3) to give blue, green, or red solutions. The nature of these solutions has long been uncertain because of their instability and the inability of chemists to isolate pure compounds from them. Early reports, now known to be incorrect, claimed unusual chalcogen oxides, such as "S_2O_3" in the blue solutions of elemental sulfur in fuming sulfuric acid.

Modern work[14] has shown that elemental chalcogen cations with the chalcogen in a fraction formal positive oxidation state are present in such solutions of elemental chalcogens in fuming sulfuric acid. Pure crystalline products cannot be obtained from the fuming sulfuric acid solutions of elemental chalcogens, but by using other strong Lewis acid oxidants, pure crystalline salts of these cations suitable for their structure determinations by X-ray diffraction can be obtained by reactions such as the following:

$$S_8 + 3SbF_5 \xrightarrow{SO_2} [S_8]^{2+}[SbF_6^-]_2 + SbF_3 \tag{6.8a}$$

$$4Se + S_2O_6F_6 \rightarrow [Se_4]^{2+}[SO_3F^-]_2 \tag{6.8b}$$

$$7Te + TeCl_4 + 4AlCl_3 \rightarrow 2[Te_4]^{2+}[AlCl_4^-]_2 \tag{6.8c}$$

The structures of such chalcogen cations have been established by X-ray crystallography as well as by ^{77}Se and ^{125}Te NMR spectroscopy.

The most frequently occurring elemental chalcogen cations are the square E_4^{2+} and the bicyclic E_8^{2+} cations (Figure 6.3), although other related cations such as Te_6^{2+}, Te_6^{4+}, Se_{10}^{2+}, and even S_{19}^{2+} are also known. The square E_4^{2+} dications correspond to delocalized 6π-electron systems related to the square Bi_4^{2-} anion (Section 5.3.3) and more distantly to the cyclobutadiene dianion, $C_4H_4^{2-}$ and even benzene, C_6H_6. The E_8^{2+} dications have a bicyclic structure (Figure 6.3); a localized formal positive charge can be written on the "three-coordinate" atoms shared by both five-membered rings. However, the trans-annular distance indicated by the dashed line in Figure 6.3 is appreciably longer (2.83 Å) than the other bonds (2.32 Å), indicative of only very weak bonding. The redox relationships between the selenium cations can be demonstrated by

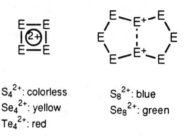

S_4^{2+}: colorless S_8^{2+}: blue
Se_4^{2+}: yellow Se_8^{2+}: green
Te_4^{2+}: red

Figure 6.3. The E_4^{2+} and E_8^{2+} dication structures.

Table 6.2   Redox Reactions of Selenium in Fuming Sulfuric Acid Leading to the Polyatomic Dications Se_4^{2+} and Se_8^{2+}

$$Se(s) \xrightarrow{S_2O_8^{2-}} Se_8^{2+} \underset{N_2H_6^{2+}}{\overset{S_2O_8^{2-}}{\rightleftarrows}} Se_4^{2+} \underset{N_2H_6^{2+}}{\overset{S_2O_8^{2-}}{\rightleftarrows}} SeO_2$$

Oxidation state	0	$+\frac{1}{4}$	$+\frac{1}{2}$	$+4$
Color	Gray	Green	Yellow	Colorless

dissolving selenium in fuming sulfuric acid and using peroxydisulfate, $S_2O_8^{2-}$, as an oxidizing agent and hydrazinium(2+) sulfate, $N_2H_6^{2+}SO_4^{2-}$, as a reducing agent in this system (Table 6.2).

6.3 Compounds with Oxygen–Oxygen Bonds and Related Compounds

6.3.1 Oxygen Fluorides

Oxygen forms fluorides of the general formula O_nF_2 ($n = 1$, 2, and 4). Since fluorine is more electronegative than oxygen, these compounds are oxygen fluorides rather than fluorine oxides.

By far the most stable oxygen fluoride is *oxygen difluoride*, OF_2, which is the one oxygen fluoride that does not have a direct oxygen–oxygen bond. Oxygen difluoride is a pale yellow toxic gas (mp $-224°C$, bp $-145°C$), which can be obtained by reaction of elemental fluorine with 2% aqueous sodium hydroxide or moist potassium fluoride. Oxygen difluoride is relatively unreactive and is stable to 200°C in glass when pure. However, many nonmetals are oxidized or fluorinated by OF_2. Oxygen difluoride explodes upon treatment with steam or the free halogens X_2 ($X = Cl$, Br, I) and is hydrolyzed by base according to the equation

$$OF_2 + 2OH^- \rightarrow O_2\uparrow + 2F^- + H_2O \qquad (6.9)$$

Oxygen fluorides with direct oxygen–oxygen bonds are far less stable than OF_2. *Dioxygen difluoride*, $O_2F_2 = FOOF$, is an unstable yellow-orange solid (mp $-154°C$), which is obtained by passing an electric discharge through mixtures of fluorine and oxygen at 10–20 mm Hg and 77–90 K. Dioxygen difluoride decomposes slowly at $-160°C$ ($\sim 4\%$ per day) by a radical mechanism, and rapidly above $-50°C$, to give O_2 and F_2. Dioxygen difluoride is an extremely potent oxidizing and fluorinating agent. Reaction of O_2F_2 with strong Lewis acids gives salts of the dioxygenyl cation, O_2^+, by the following reaction:

$$2O_2F_2 + 2BF_3 \rightarrow 2O_2^+BF_4^- + F_2\uparrow \qquad (6.10)$$

The structure of O_2F_2 contains an unusually short O—O bond (1.217 Å) compared with H_2O_2 (1.48 Å) and O_2^{2-} (1.49 Å), suggesting that the electronegative fluorine atoms in O_2F_2 decrease the repulsion between lone electron pairs on adjacent oxygen atoms. However, the O—F bonds in O_2F_2 are long (1.575 Å) relative to those in OF_2 (1.409 Å).

Tetraoxygen difluoride, O_4F_2, is a dark red-brown solid (mp − 191°C), which decomposes above its melting point; it is the only known compound containing a chain of four oxygen atoms.

6.3.2 Hydrogen Peroxide

Hydrogen peroxide, H_2O_2, is the most important compound containing a direct oxygen–oxygen bond. It was originally prepared by oxidizing BaO in oxygen to barium peroxide, BaO_2, followed by decomposition of BaO_2 with dilute H_2SO_4 according to the following equations:

$$2BaO + O_2 \rightarrow 2BaO_2 \tag{6.11a}$$

$$BaO_2 + H_2SO_4 \xrightarrow{\ H_2O\ } BaSO_4{\downarrow} + H_2O_2 \tag{6.11b}$$

More practical large-scale preparations of hydrogen peroxide include the following methods.

1. Electrolytic oxidation of sulfate to peroxodisulfate followed by hydrolysis of H_2O_2:

$$2SO_4^{2-} \rightarrow S_2O_8^{2-} + 2e^- \tag{6.12a}$$

$$S_2O_8^{2-} + 2H_2O \rightarrow H_2O_2 + 2HSO_4^- \tag{6.12b}$$

The formation of unwanted O_2 in a side reaction is suppressed by carrying out the electrolytic oxidation at a relatively low temperature, using high current densities and platinum anodes having a high overvoltage toward oxygen. The hydrogen peroxide is isolated by vacuum distillation.

2. Autoxidation of an alkyl 9,10-anthracenediol in an organic solvent:

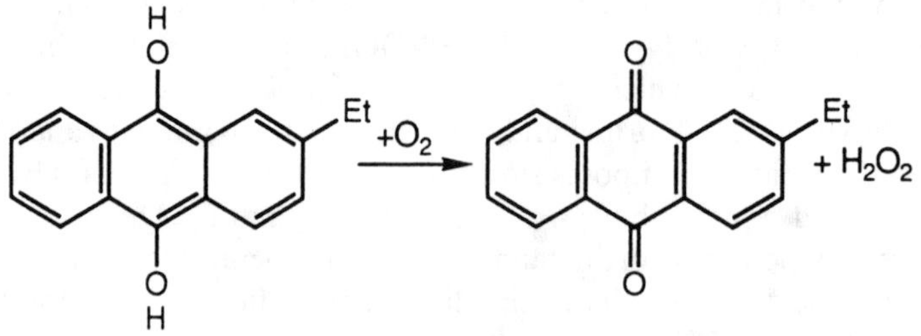

After removal of degradation products from side reactions, the resulting anthraquinone derivative is catalytically reduced by hydrogen gas in the presence of a catalyst, leading to a continuous cyclic process occurring in an organic solvent such as alkylbenzenes. The H_2O_2 is removed from the organic solvent by countercurrent extraction with water, and the resulting 20–40% aqueous H_2O_2 is purified by solvent extraction.

3. Oxidation of isopropanol to acetone and H_2O_2 without catalyst, either in the liquid phase at 90–140°C or in the vapor phase at 350–500°C:

$$(CH_3)_2CHOH + O_2 \rightarrow (CH_3)_2C{=}O + H_2O_2 \qquad (6.13)$$

The hydrogen peroxide is separated by distillation.

Hydrogen peroxide is a colorless liquid (bp 150.2°C, mp −0.43°C), density 1.44 g/mL at 25°C. Because of its high dielectric constant (70 at 20°C) pure hydrogen peroxide is an excellent ionizing solvent. However, its utility as an ionizing solvent is limited by its strong oxidizing properties and its exothermic decomposition in the presence of metals. Hydrogen peroxide is more acidic than water:

$$H_2O_2 \rightleftarrows H^+ + HO_2^- \qquad K_{20°C} = 1.5 \times 10^{-12} \qquad (6.14)$$

However, H_2O_2 is more than 10^6 times *less* basic than H_2O.

Hydrogen peroxide is a strong oxidizing agent in either acidic or basic solution:

$$H_2O_2 + 2H^+ + 2e^- = 2H_2O \text{ (oxidizing agent in acidic solution)}$$
$$E° = 1.77 \text{ V} \qquad (6.15a)$$

$$O_2 + 2H^+ + 2e^- = H_2O_2 \text{ (reducing agent in acidic solution)}$$
$$E° = 0.68 \text{ V} \qquad (6.15b)$$

$$HO_2^- + H_2O + 2e^- = 3OH^- \text{ (oxidizing agent in basic solution)}$$
$$E° = 0.87 \text{ V} \qquad (6.15c)$$

However, H_2O_2 acts as a reducing agent only toward very strong oxidizing agents such as MnO_4^- and Ce^{4+}. Red chemiluminescence is observed when the O_2 generated by oxidation of H_2O_2 is in the excited $^1\Delta_g$ singlet state (see Section 6.2.1.1). In view of the ability of H_2O_2 to act as both an oxidizing and a reducing agent, the decomposition of H_2O_2 to H_2O and O_2 is catalyzed by a number of elements with multiple oxidation states including Fe^{3+}, I_2, and MnO_2. Labeling studies with ^{18}O indicate that the O_2 comes from the H_2O_2 and not the H_2O solvent in the oxidation of H_2O_2 in water to O_2 by Cl_2, MnO_4^-, Ce^{4+}, and so on, and in the catalytic decomposition of H_2O_2 to H_2O and O_2 by Fe^{3+}, I_2, and MnO_2. This indicates that oxidizing agents do not break the O—O bond in H_2O_2 but simply remove electrons. Reduction of H_2O_2 with Fe^{2+} generates ·OH radicals in a mixture known as *Fenton's reagent*, which is useful for the oxidation or dehydrogenation of organic compounds.

The hydrogen atoms in H_2O_2 are capable of hydrogen-bonding to appropriate substrates. For this reason, a number of compounds are known containing hydrogen peroxide of crystallization in which the H_2O_2 is hydrogen bonded to the anion.[15] The carbonate derivatives $M_2CO_3 \cdot nH_2O_2$ (M = NH_4, $n = 1$; M = Na, $n = 1.5$; M = K, Rb, Cs, $n = 3$) were once incorrectly formulated as peroxocarbonates. The fluoride $KF \cdot H_2O_2$ has a strong hydrogen bond of the F—H—O type.

6.3.3 Ionic Peroxides

Ionic peroxides are salts containing discrete O_2^{2-} ions (Table 6.1). They are formed by the alkali metals and the alkaline earth metals (Ca, Sr, Ba) by treating the metal or a lower oxide with elemental oxgyen, for example,

$$4Na + O_2 \rightarrow 2Na_2O \tag{6.16a}$$

$$2Na_2O + O_2 \rightarrow 2Na_2O_2 \tag{6.16b}$$

The EPR spectrum of this reaction product indicates the presence of $\sim 10\%$ of the more highly oxidized superoxide NaO_2 (see Section 6.3.5). Barium peroxide (see equations 6.11) is useful as a source of H_2O_2 upon treatment with sulfuric acid.

Ionic peroxides liberate H_2O_2 upon treatment with water or dilute acids. They also liberate O_2 upon treatment with CO_2:

$$2Na_2O_2 + 2CO_2 \rightarrow 2Na_2CO_3 + O_2 \tag{6.17}$$

This reaction using CO_2 exhaled by human beings is useful for regenerating the oxygen atmosphere in manned spacecraft. Ionic peroxides are powerful oxidizing agents and can convert organic matter readily to carbonate. A number of ionic peroxides form well-crystallized hydrates such as $Na_2O_2 \cdot 8H_2O$ and $M^{II}O_2 \cdot 8H_2O$ (M = Ca, Sr, Ba) with lattices containing water molecules hydrogen-bonded to the peroxide ion.

6.3.4 Covalently Bonded Peroxo Compounds

A number of derivatives of inorganic oxoacids are known in which an —O— unit is replaced by an —O—O— unit. These include the peroxo-monocarbonate ion $[O_2C—O—OH]^-$ in $NaHCO_4 \cdot H_2O$ and the peroxodi-carbonate ion $[O_2C—O—O—CO_2]^{2-}$ in $M_2C_2O_6$ (M = alkali metal) as well as the peroxomonosulfate ion $[O_3S—O—O]^{2-}$ and the peroxodisulfate ion $[O_3S—O—O—SO_3]^{2-}$ mentioned earlier (equations 6.12) as a source of hydrogen peroxide. In addition, peroxocarboxylic acids of the general formula R—C(=O)—O—OH containing a variety of alkyl and aryl groups R are useful oxidants in organic chemistry (e.g., for the conversion of olefins to the corresponding epoxides or sulfides to the corresponding sulfoxides or sulfones). Peroxoacetic acid (colloquially called "peracetic acid") is obtained either by the acid-catalyzed (H_2SO_4) reaction of 50% hydrogen peroxide with acetic acid or by the air oxidation of acetaldehyde according to the equation

$$CH_3C(=O)-H + O_2 \rightarrow CH_3C(=O)-O-OH \qquad (6.18)$$

The autoxidation of a number of organic compounds such as ethers and unsaturated hydrocarbons gives organic peroxo compounds by free radical chain reactions:

$$R-H + X^· \rightarrow R^· + HX \qquad (6.19a)$$

$$R^· + O_2 \rightarrow RO-O^· \qquad (6.19b)$$

$$RO-O^· + R-H \rightarrow RO-OH \qquad (6.19c)$$

Many peroxides obtained by autoxidation of ethers and unsaturated compounds are dangerous explosives. Thus samples of such organic compounds that have been stored for long times should be tested for peroxides before distillation. Peroxides can be removed by washing with acidified ferrous sulfate or passing through a column of activated alumina.

Organic peroxides containing fluorinated groups such as $(FSO_2)O-O(SO_2F)$, $F_5S-O-O-SF_5$, and $F_3C-O-O-CF_3$ are more stable than those containing hydrocarbon groups, since the electron-withdrawing fluorinated groups reduce the repulsion between the oxygen lone electron pairs.

6.3.5 Superoxides

Superoxides contain the paramagnetic O_2^- ion (Table 6.1).[16] They can be obtained by oxidizing the heavier alkali metals with elemental oxygen:

$$M + O_2 \rightarrow M^+O_2^- \qquad (6.20)$$

These superoxides form orange crystals having a structure similar to calcium carbide, $Ca^{2+}C_2^{2-}$. Potassium superoxide is the most common derivative, but it is normally obtained only $\sim 96\%$ pure, with K_2O_2 and KOH impurities. The O_2^- anion in KO_2 has an O—O distance of 1.28 Å (Table 6.1). The chemistry of the superoxide ion has little radical character because the unpaired electron in O_2^- occupies an antibonding π^* orbital (Table 6.1).

The superoxide ion reacts with water to liberate O_2 by the following sequence of reactions:

$$2O_2^- + H_2O \rightarrow O_2\uparrow + HO_2^- + OH^- \quad \text{(fast)} \qquad (6.21a)$$

$$2HO_2^- \rightarrow 2OH^- + O_2\uparrow \quad \text{(slow)} \qquad (6.21b)$$

In addition, superoxides liberate O_2 upon reaction with CO_2:

$$4MO_2(s) + 2CO_2(g) \rightarrow 2M_2CO_3(s) + 3O_2(g) \qquad (6.22)$$

This reaction, like the reaction of sodium peroxide with CO_2 (equation 6.17), is useful for regenerating O_2 from CO_2 in closed systems such as manned spacecraft.

Solutions of the superoxide ion can be obtained in aprotic solvents. Thus KO_2 can be solubilized in dimethyl sulfoxide by addition of 18-crown-6. Metathesis of such solutions of KO_2 with tetramethylammonium carbonate gives tetramethylammonium superoxide, $[Me_4N]^+O_2^-$. In aprotic solvents, O_2^- reacts with alkyl halides as a nucleophile and with many organic compounds as a one-electron oxidant.

6.3.6 Ozonides

Reactions of alkali metal superoxides with ozone in liquid ammonia give the corresponding metal ozonides:

$$MO_2 + O_3 \rightarrow MO_3 + O_2 \qquad M = K, Rb \qquad (6.23)$$

The deep red ozonide ion, O_3^-, has a bent structure (O–O–O = 113.7°; O—O distance = 1.34 Å in RbO_3) and the expected paramagnetism (1.67 Bohr magnetons) for a single unpaired electron. Ozonides have been obtained for all the alkali metals and the alkaline earths calcium, strontium, and barium; their thermal stabilities decrease in the sequences Cs > Rb > K > Na > Li and Ba > Sr > Ca.

6.3.7 The Dioxygenyl Cation

The paramagnetic O_2^+ cation (Table 6.1) is isoelectronic with neutral NO; it is stabilized in salts by large nonoxidizable anions. Preparative methods for the O_2^+ cation include the reaction of O_2F_2 with appropriate Lewis acids (equation 6.10). However, the less inconvenient oxidation of neutral dioxygen in the presence of very strong Lewis acids works in some cases, for example,

$$2O_2 + 2BF_3 + F_2 \xrightarrow[-78°C]{hv} 2O_2^+BF_4^- \qquad (6.24a)$$

$$O_2 + PtF_6 \rightarrow O_2^+PtF_6^- \qquad (6.24b)$$

The PtF_6 in the latter reaction serves both as extremely strong oxidizing agent and as Lewis acid. In the dioxygenyl cation, O_2^+, the O—O distance is 1.12 Å and the stretching frequency $v(O—O)$ is 1905 cm^{-1} (Table 6.1).

6.4 Binary Compounds of Sulfur and Its Heavier Congeners with Hydrogen and Metals

6.4.1 Hydrogen Compounds

The hydrides H_2S (mp $-85.6°C$, bp $-60.7°C$), H_2Se (bp $-41.5°C$), and H_2Te (bp $-2°C$) are very toxic gases with revolting odors; H_2S and H_2Se are even more toxic than HCN. They are very weak acids in aqueous solution, for example,

$$H_2S + H_2O \rightleftarrows H_3O^+ + HS^- \qquad K_{eq} = 1.3 \times 10^{-7} \qquad (6.25a)$$

$$HS^- + H_2O \rightleftarrows H_3O^+ + S^{2-} \qquad K_{eq} = 7.1 \times 10^{-15} \qquad (6.25b)$$

In addition, sulfur is known to form a series of hydrogen compounds of the general type H—(S_n)—H containing sulfur chains; these are called *sulfanes* by analogy to the alkanes. The sulfanes H_2S_n ($2 \leq n \leq 8$) have been isolated in the pure state as reactive yellow liquids with viscosities increasing with the sulfur chain length. A mixture of crude sulfanes separates as an oily liquid upon acidification of an aqueous sodium polysulfide solution at low temperatures:

$$Na_2S_n(aq) + 2HCl(aq) \xrightarrow{-10°C} 2NaCl(aq) + H_2S_n(l) \qquad (6.26)$$

The sodium polysulfide required for the preparation of sulfanes can be obtained by fusion of $Na_2S \cdot 9H_2O$ with various amounts of elemental sulfur, depending on the desired composition of the H_2S_n product. The individual components of the H_2S_n mixture can be distinguished up to $n \approx 10$ by ^1H NMR spectroscopy. Pure sulfanes up to H_2S_4 can be isolated from the crude H_2S_n mixture by fractional distillation at low pressures. The chain lengths of sulfanes can be increased by reactions such as the following:

$$S_nCl_2 + 2H_2S \rightarrow 2HCl{\uparrow} + H_2S_{n+2} \qquad (6.27a)$$

$$S_nCl_2 + 2H_2S_2 \rightarrow 2HCl{\uparrow} + H_2S_{n+4} \qquad (6.27b)$$

6.4.2 Metal Chalcogenides

Almost all metals form binary compounds with sulfur and its heavy congeners; these are frequently made by direct combination of the elements at a sufficiently high temperature. The chemistry of metal sulfides is the best known and is discussed next.

6.4.2.1 Ionic Sulfides

Only the alkali and alkaline earth metals form ionic sulfides; such ionic sulfides are the only sulfides that dissolve in water. Ionic sulfides are extensively hydrolyzed in aqueous solution to give the SH^- anion because the acid strengths of H_2O and SH^- are very similar:

$$S^{2-} + H_2O \rightarrow SH^- + OH^- \qquad K_{eq} \approx 1 \qquad (6.28)$$

Aqueous solutions of ionic sulfides dissolve elemental sulfur to form polysulfide ions S_n^{2-}. The trisulfide (S_3^{2-}) and tetrasulfide (S_4^{2-}) ions are favored in aqueous solution, but crystalline polysulfides containing the S_n^{2-} anions ($3 \leq n \leq 6$)

have been isolated by using large countercations such as Cs^+, NH_4^+, and $H_3N^+CH_2CH_2NH_3^+$. Molten alkali–metal polysulfides are useful solvents (fluxes) for the synthesis at relatively low temperatures (200–450°C) of transition and main group metal sulfides with unusual structures.[17] For stoichiometric compositions of K_2S_x ($x \geq 3$), the melting point is well below 400°C and reaches 160°C for $x = 4$.

Alkali metal sulfides are found in lithium–sulfur and sodium–sulfur batteries, which are useful as high density electrical storage batteries. Thus the sodium–sulfur battery can store five times as much energy as the conventional lead–acid battery.[18] Such batteries have been shown to contain sulfides and polysulfides of the types M_2S, M_2S_2, M_2S_4, and M_2S_5.

Some highly colored sulfide radical anions are known. Thus deep blue solutions of the S_3^- ion are obtained by dissolving alkali metal polysulfides in polar aprotic organic solvents such as acetone, dimethyl sulfoxide, and dimethylformamide. In addition, the blue S_3^- ion and green S_2^- ion are found in the minerals lapis lazuli and ultramarine.

6.4.2.2 Transition Metal Sulfides

Transition metals form a variety of sulfides, many of which are nonstoichiometric.[19] Transition metal sulfides are all insoluble in water and can be made by direct combination of the elements at elevated temperatures or by precipitation from aqueous solution. Transition metal sulfides are more covalent than the corresponding oxides; some have metallic properties. Many transition metal sulfides are found in nature as minerals.

One important type of transition metal sulfide consists of the monosulfides (FeS, CoS, NiS, etc.) with the "nickel arsenide" structure, in which the transition metal atom is octahedrally surrounded by six sulfur atoms and two close metal neighbors at bonding distances (2.60–2.68 Å). The presence of such a metal–metal bonded network is consistent with the alloylike or semimetallic character of this class of transition metal sulfide.

Transition metal sulfides of the type MS_2 contain direct sulfur–sulfur bonds in discrete S_2 units. An important transition metal sulfide of this type is the mineral *pyrite*, FeS_2, which has a distorted NaCl structure with Fe atoms in the Na positions and the centers of discrete S_2 units in the Cl positions. Pyrite may be regarded formally as $Fe^{2+}S_2^{2-}$; since the iron atom is octahedrally coordinated by six sulfur atoms of different S_2 groups, it has the favored 18-electron rare gas electronic configuration.

6.5 Sulfur–Nitrogen Compounds

A number of binary sulfur–nitrogen compounds are known (Figure 6.4).[20–23] Most of these derivatives are "inorganic" heterocyclic compounds containing rings composed exclusively of sulfur and nitrogen atoms in configurations that

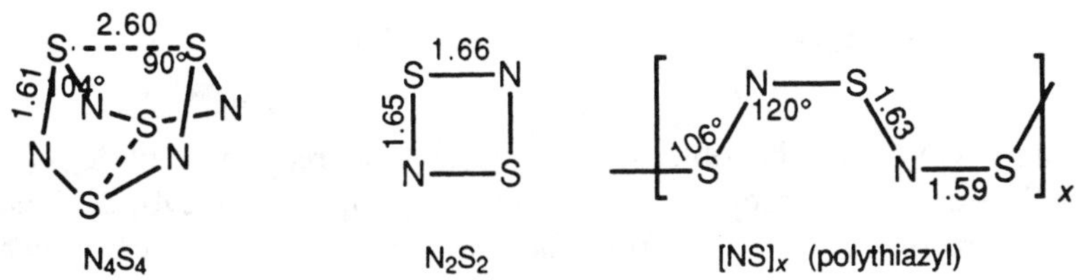

Figure 6.4. Structures of some binary sulfur–nitrogen compounds; bond distances are given in angstrom units.

have no direct nitrogen–nitrogen bonds and maximize the number of sulfur–nitrogen bonds (e.g., N_4S_4 and N_2S_2). However, sulfur–nitrogen chains, such as the polymer $[NS]_x$, are also known. Many sulfur–nitrogen compounds can explode upon shock and must be handled with care.

The most readily available binary sulfur–nitrogen compound is *tetrasulfur tetranitride*, S_4N_4, which was first isolated in 1835 from the reaction of S_2Cl_2 with ammonia. Tetrasulfur tetranitride is a thermochromic air-stable solid (mp 185°C), which is almost colorless at $-190°C$, orange-yellow at 25°C, and red above 100°C. It is an endothermic compound ($\Delta H_f^\circ = 460 \pm 8$ kJ/mol) and can explode on friction or rapid heating. The structure of S_4N_4 (Figure 6.4) consists of an eight-membered ring of alternating sulfur and nitrogen atoms with equal S—N distances of 1.61 Å, suggesting resonance structures similar to the Kekulé forms of benzene. This S_4N_4 ring is folded into a cage with a square set of nitrogen atoms and a bisphenoid set of sulfur atoms. Weak sulfur–sulfur interactions in this structure are indicated by an $S \cdots S$ distance of 2.60 Å, which is appreciably longer than the S—S single bond distance of 2.08 Å but less than the sum of the van der Waals radii.

Reactions of S_4N_4 are of two types: those that preserve the eight-membered S_4N_4 ring, including reduction to $S_4N_4H_4$ with tin(II) chloride in ethanol, fluorination to $S_4F_4N_4$ with AgF in carbon tetrachloride, and oxidation to $[S_4F_4]^{2+}[SbCl_6^-]_2$ with antimony pentachloride; and those that cleave the S_4N_4 ring with reorganization to form other S—N rings, such as the following:

$$S_4N_4 \xrightarrow{Cl_2} S_3N_3Cl_3 \tag{6.29a}$$

$$S_4N_4 \xrightarrow[100°C]{S_8} S_4N_2 \tag{6.29b}$$

$$S_4N_4 \xrightarrow{AlCl_3} S_3N_3^+ AlCl_4^- \tag{6.29c}$$

$$S_4N_4 \xrightarrow{AsF_5} S_3N_2^+ AsF_6^- \tag{6.29d}$$

$$S_4N_4 \xrightarrow{HCl} S_4N_3^+ Cl^- \tag{6.29e}$$

$$S_4N_4 \xrightarrow{N_3^-} S_4N_5^- \qquad (6.29f)$$

The anion $S_3N_3^-$ can be obtained by electrolytic reduction of S_4N_4. The six-membered ring heterocyclic compound S_4N_2 (equation 6.29b) can also be prepared directly in low yield from the reaction of disulfur dichloride with aqueous ammonia.

Pyrolysis of S_4N_4 vapor over silver wool or gauze at 250–300°C and 0.1–1.0 mm Hg gives *disulfur dinitride*, S_2N_2, as a colorless, crystalline, shock-sensitive solid, which can decompose explosively above 30°C. The S_2N_2 ring in disulfur dinitride is nearly a perfect square (Figure 6.4). Disulfur dinitride dimerizes back to S_4N_4 in the presence of trace amounts of nucleophiles or reducing agents. However, upon standing at room temperature in the absence of a catalyst S_2N_2 spontaneously forms the polymer $(SN)_x$, known as *polythiazyl*; this polymer can also be obtained from $S_3N_3Cl_3$ and Me_3SiN_3, thereby avoiding the need to handle the explosive S_2N_2.

Polythiazyl forms a golden-bronze solid, which exhibits metallic-type electrical conductance; it is even superconducting below 0.26 K.[24] Doping polythiazyl with bromine to give $[SnBr_{0.4}]_x$ increases the room temperature conductivity of polythiazyl to 2×10^4 $(\Omega \cdot cm)^{-1}$. The crystal structure of polythiazyl (Figure 6.4) consists of kinked, nearly parallel chains. The ability of π electrons to be extensively delocalized along the sulfur–nitrogen chains accounts for the high conductivity of polythiazyl.

A variety of cyclic $S_n(NH)_{8-n}$ ($n \leq 4$) compounds are known which contain eight-membered rings consisting of sulfur atoms and NH units with no pairs of adjacent NH units and thus no direct nitrogen–nitrogen bonds. The structures of these $S_n(NH)_{8-n}$ compounds are closely related to that of the most stable form of elemental sulfur, S_8 (Section 6.2.2). These $S_n(NH)_{8-n}$ compounds may be separated as colorless crystalline solids by column chromatography of the complicated mixtures obtained by treatment of sulfur halides with liquid ammonia. A single S_7NH derivative, three isomeric $S_6(NH)_2$ derivatives (the 1,3-, 1,4-, and 1,5-isomers), two isomeric $S_5(NH)_3$ derivatives (the 1,3,5- and 1,3,6-isomers), and one $S_4(NH)_4$ derivative (the 1,3,5,7-isomer) can be separated from these reaction mixtures.

Sulfur–nitrogen fluorides are also known.[25] The initial product obtained by the fluorination of S_4N_4 (e.g., by AgF_2 in carbon tetrachloride under relatively mild conditions) is $N_4S_4F_4$, in which the eight-membered ring of S_4N_4 is preserved and the fluorine atoms are bonded to sulfur. Fluorination under more vigorous conditions (e.g., HgF_2 in carbon tetrachloride at higher temperatures) leads to the mononuclear thiazyl fluoride NSF (mp -89°C, bp 0.4°C), a colorless pungent gas, which trimerizes at room temperature to $N_3S_3F_3$. The end product of the fluorination of S_4N_4 is thiotriazyl trifluoride, $N{\equiv}SF_3$ (mp -72.6°C, bp -27.1°C), which is kinetically very stable.

Metal thionitrosyls (i.e., metal complexes containing MNS units) are very rare. However, in a few cases when additional ligands are present, metal thionitrosyls can be prepared from the sulfur–nitrogen derivatives mentioned earlier, for example,

$$3NaCr(CO)_3C_5H_5 + N_3S_3Cl_3 \rightarrow 3C_5H_5Cr(CO)_2NS$$
$$+ 3CO + 3NaCl \quad (6.30)$$

The neutral NS ligand in $C_5H_5Cr(CO)_2NS$ is a three-electron donor ligand like the neutral NO ligand in the closely related $C_5H_5Cr(CO)_2NO$.[26]

6.6 Halides of Sulfur and Its Congeners

6.6.1 Sulfur Fluorides

Sulfur forms several binary fluorides with F/S ratios ranging from 1 in S_2F_2 and SSF_2 to 6 in SF_6 (Figure 6.5). "Lower" sulfur fluorides with F/S ratios less than 4 are relatively unstable compounds.[27] Fluorination of elemental sulfur with silver fluoride at 125°C gives $trans\text{-}S_2F_2$, which readily isomerizes into SSF_2; the latter sulfur fluoride can also be obtained by fluorinating S_2Cl_2 (see equation 6.31b) with potassium fluoride or potassium fluorosulfite in liquid sulfur dioxide. Treatment of SSF_2 with acid catalysts such as HF or BF_3 leads to rapid disproportionation into S_8 and SF_4. Sulfur difluoride, SF_2, unlike H_2S, is very unstable and is only known as a dilute gas.

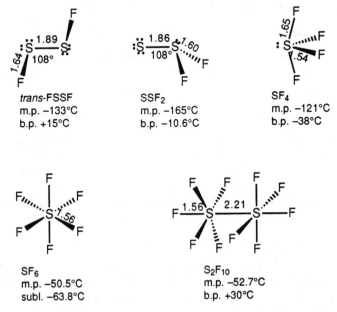

Figure 6.5. Binary sulfur fluorides and their structures; distances are given in angstrom units.

Sulfur tetrafluoride, SF_4, is considerably more stable than the lower sulfur fluorides; it is synthesized by the fluorination of sulfur chlorides, for example,

$$16SCl_2 + 32HF \xrightarrow{\text{pyridine}} S_8 + 8SF_4 + 32HCl \tag{6.31a}$$

$$16S_2Cl_2 + 32NaF \xrightarrow[\text{CH}_3\text{CN}]{\Delta} 8SF_4 + 3S_8 + 32NaCl \tag{6.31b}$$

The sulfur(I) or sulfur(II) chloride starting material in these reactions undergoes disproportionation into sulfur(0) and sulfur (IV), so that only a portion of the sulfur is available ffor formation of SF_4.

Sulfur tetrafluoride is a reactive toxic gas, which is readily hydrolyzed to SO_2 and HF with $O{=}SF_2$ as an intermediate. Sulfur tetrafluoride is amphoteric. Thus SF_4 can function as a Lewis acid to form the SF_5^- anion with fluoride in $CsSF_5$ and $[Me_4N][SF_5]$ and the adduct $C_5H_5N{\rightarrow}SF_4$ with pyridine. Alternatively, SF_4 can function as a Lewis base to form the salt $SF_3^+BF_4^-$ with the Lewis acid boron trifluoride. Sulfur tetrafluoride is a useful fluorinating agent, both for the conversion of metal oxides to metal fluorides and for the selective fluorination of carbon–oxygen and phosphorus–oxygen bonds in organic chemistry in which an oxygen atom is replaced by two fluorine atoms with formation of $O{=}SF_2$, for example,[28]

$$R_2C{=}O + SF_4 \rightarrow R_2CF_2 + O{=}SF_2 \tag{6.32a}$$

$$R_3P{=}O + SF_4 \rightarrow R_3PF_2 + O{=}SF_2 \tag{6.32b}$$

$$RC({=}O)OH + 2SF_4 \rightarrow RCF_3 + 2O{=}SF_2 + HF \tag{6.32c}$$

$$R_2P(O)OH + 2SF_4 \rightarrow R_2PF_3 + 2O{=}SF_2 + HF \tag{6.32d}$$

Organic derivatives of SF_4, namely arylsulfur trifluorides, RSF_3, can be prepared from the corresponding disulfides by treatment with a strong fluorinating agent, namely AgF_2:

$$R{-}S{-}S{-}R + 6AgF_2 \rightarrow 2RSF_3 + 6AgF \quad (R = \text{phenyl, etc.}) \tag{6.33}$$

Arylsulfur trifluorides are fluorinating agents similar to SF_4 and have the advantage of being more readily handled, since they are liquids. The structures of SF_4 (Figure 6.5) and RSF_3 derivatives are both ψ-trigonal bipyramids with the stereochemically active lone pair in the equatorial position.

Sulfur hexafluoride, SF_6, is obtained by the exothermic combustion of sulfur in fluorine:

$$S_8 + 24F_2 \rightarrow 8SF_6 \qquad \Delta H^\circ = -1210 \text{ kJ/mol} \tag{6.34}$$

It is a chemically very *unreactive* nontoxic gas, reacting only with red hot metals and sodium in liquid ammonia. Sulfur hexafluoride is unaffected by molten potassium hydroxide or by steam at 500°C. The low chemical reactivity of SF_6 is related to its high S—F bond strength, coordinate saturation, and lack of polarity. Sulfur hexafluoride is used as an insulating gas in high voltage electrical equipment because of its chemical inertness, low toxicity, and very low electrical conductivity.

Disulfur decafluoride, F_5S—SF_5, is made by the following photochemical reaction:

$$2SF_5Cl + H_2 \xrightarrow{h\nu} S_2F_{10} + 2HCl \qquad (6.35)$$

It is highly toxic, with a physiological action similar to that of phosgene. The structure of F_5S—SF_5 consists of two SF_5 units joined by a relatively long sulfur–sulfur bond in a staggered conformation (Figure 6.5). Despite its long S—S bond, S_2F_{10} is relatively unreactive; it is thus insoluble and unreactive toward water or aqueous alkali. Heating S_2F_{10} to 150°C results in disproportionation into SF_4 and SF_6.

6.6.2 Pentafluorosulfur Derivatives

The stability and low chemical reactivity of sulfur hexafluoride, SF_6, indicates the chemical stability of octahedral sulfur(VI) largely substituted by fluorine. In this connection a variety of very stable derivatives of the pentafluorosulfur(VI) group, —SF_5, are known; the pentafluorosulfur group is analogous to an organic alkyl or aryl group, particularly perfluoroalkyl groups such as CF_3.

A useful starting material for the preparation of pentafluorosulfur derivatives is *pentafluorosulfur chloride,* SF_5Cl, a reactive gas (bp $-15.1°C$, mp $-64°C$), which can be obtained from either of the following reactions:

$$S_2F_{10} + Cl_2 \xrightarrow{200-250°C} 2SF_5Cl \qquad (6.36a)$$

$$SF_4 + ClF \xrightarrow[25°C/1\ h]{CsF} SF_5Cl \qquad (6.36b)$$

The latter reaction proceeds through a $CsSF_5$ intermediate. Pentafluorosulfur bromide, SF_5Br (bp 3°C), has similar properties to SF_5Cl. However, SF_5I does not appear to have been observed and is probably unstable with respect to liberation of I_2 and reduction of the sulfur to a lower oxidation state.

Pentafluorosulfur chloride is much more reactive than SF_6 and reacts readily with OH^- and other nucleophiles. In addition, the SF_5 group in SF_5Cl is more electronegative than chlorine so that the charge distribution is SF_5^-—Cl^+ (i.e., SF_5Cl is a source of "positive chlorine" and functions as a pseudohalogen). As such, SF_5Cl adds to a variety of multiple bonds to give a variety of organic and inorganic SF_5 derivatives, for example,

$$SF_5Cl + RCH\!=\!CH_2 \rightarrow RCHClCH_2SF_5 \qquad (6.37a)$$

$$SF_5Cl + R_2C\!=\!CF_2 \rightarrow R_2CClCF_2SF_5 \qquad (6.37b)$$

$$SF_5Cl + CH_2\!=\!C\!=\!O \rightarrow SF_5CH_2C(O)Cl \qquad (6.37c)$$

$$SF_5Cl + HC\equiv CH \rightarrow ClCH=CHSF_5 \tag{6.37d}$$

$$SF_5Cl + RC\equiv N \xrightarrow{h\nu} RCCl=NSF_5 \tag{6.37e}$$

$$2SF_5Cl + O_2 \xrightarrow{h\nu} SF_5O—OSF_5 + Cl_2 \tag{6.37f}$$

The acid chloride $SF_5CH_2C(O)Cl$ can, in turn, be converted to a variety of other organic SF_5 derivatives and to $SF_4=CH_2$ by the following sequence of reactions:

$$SF_5CH_2C(O)Cl + H_2O \rightarrow SF_5CH_2CO_2H + HCl \tag{6.38a}$$

$$2SF_5CH_2CO_2H + Ag_2O \rightarrow 2SF_5CH_2CO_2Ag + H_2O \tag{6.38b}$$

$$SF_5CH_2CO_2Ag + Br_2 \rightarrow AgBr + CO_2\uparrow + SF_5CH_2Br \tag{6.38c}$$

$$2SF_5CH_2Br + 2Zn + 2HCl \rightarrow ZnCl_2 + ZnBr_2 + 2SF_5CH_3 \tag{6.38d}$$

$$SF_5CH_2Br + Bu^nLi \rightarrow Bu^nBr + LiF + SF_4=CH_2 \tag{6.38e}$$

The presence of a $C=S$ double bond in $SF_4=CH_2$ is indicated by a relatively short $C=S$ distance of 1.554 Å.[29] The SF_5 group, like the CF_3 group, forms a stable hypofluorite, SF_5OF, which can be prepared by the following reaction:

$$OSF_2 + 2F_2 \xrightarrow[25°C]{CsF} SF_5OF \tag{6.39}$$

An intermediate in this reaction is $O=SF_4$.

6.6.3 Selenium and Tellurium Fluorides[30,31]

The tetrafluorides, EF_4 (E = Se, Te), can be prepared by fluorination of the corresponding dioxides with sulfur tetrafluoride:

$$EO_2 + 2SF_4 \rightarrow EF_4 + 2OSF_2 \tag{6.40}$$

Selenium tetrafluoride resembles SF_4, although it is a liquid (mp $-39°C$, bp $106°C$) and thus easier to handle. Tellurium tetrafluoride (mp $130°C$) is a solid consisting of infinite chains of $TeF_3F_{2/2}$ units with ψ-octahedral tellurium(IV) with a stereochemically active lone pair (Figure 6.6).

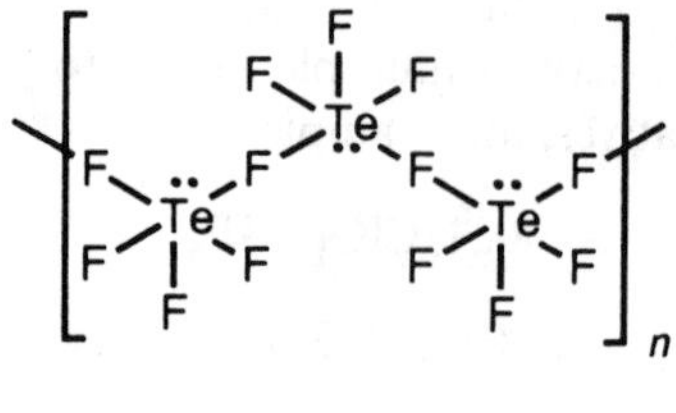

Figure 6.6. The polymeric structure of $TeF_4 = [TeF_3F_{2/2}]_n$.

The limiting products of the fluorination of selenium and tellurium are the corresponding hexafluorides, EF_6 (E = Se, Te). Selenium hexafluoride, unlike SF_6, is toxic. However, SeF_6, like SF_6, is unreactive toward water. Tellurium hexafluoride slowly undergoes hydrolysis to telluric acid, $Te(OH)_6$, through $Te(OH)_nF_{6-n}$ intermediates. The first member of the latter series, $HOTeF_5$ (mp 40°C, bp 60°C), is a strong acid known colloquially as "teflic acid." Teflic acid is more conveniently prepared by the following reaction:

$$5HSO_3F + Te(OH)_6 \rightarrow HOTeF_5 + 5H_2SO_4 \qquad (6.41)$$

Teflic acid is a distillable liquid which forms a variety of derivatives of the very electronegative —$OTeF_5$ group including $AgOTeF_5$, $B(OTeF_5)_3$, and $Xe(OTeF_5)_2$.

6.6.4 Chlorides and Heavier Halides of Sulfur and Its Congeners

6.6.4.1 Sulfur Chlorides

Sulfur forms a variety of binary compounds with chlorine. Chlorination of molten sulfur results in partial cleavage of the S—S bonds to give *sulfur monochloride*, $(SCl)_2 = S_2Cl_2$, as a malodorous orange liquid (mp —76°C, bp 138°C), with a density of 1.677 g/ml at 20°C. It is by far the most stable of the sulfur chlorides, although it is highly reactive with water. The structure of S_2Cl_2 consists of a Cl—S—S—Cl chain with S—Cl and S—S bond distances of 2.05 and 1.95 Å, respectively.

Reaction of sulfur or S_2Cl_2 with excess chlorine at room temperature using a catalyst such as $FeCl_3$ gives dark red liquid *sulfur dichloride*, SCl_2 (mp —122°C, bp 59°C), with a density of 1.621 g/ml at 20°C. Sulfur dichloride has an angular structure (Cl–S–Cl = 102.8°, S—Cl = 2.014 Å) and dissociates readily into S_2Cl_2 unless it is stabilized by addition of PCl_3 or PCl_5. Sulfur dichloride is very reactive and adds across carbon–carbon double bonds. The reaction of SCl_2 with ethylene gives $S(CH_2CH_2Cl)_2$, which is the highly toxic vesicant *mustard gas*.

Reaction of SCl_2 with excess chlorine at —80°C gives white crystals of *sulfur tetrachloride*, SCl_4, which decompose above —30°C to regenerate SCl_2 and Cl_2. Solic SCl_4 has been formulated as the trichlorosulfonium salt, $SCl_3^+Cl^-$; more stable *trichlorosulfonium salts* can be obtained by adding a Lewis acid to a mixture of SCl_2 and Cl_2. Thus reactions of SCl_2 with Cl_2 in the presence of $AlCl_3$ and AdF_5 give $SCl_3^+AlCl_4^-$ and $SCl_3^+AsF_6^-$, respectively.

An important property of sulfur chlorides is their ability to dissolve elemental sulfur to give solutions containing the *dichlorosulfanes* $S_nCl_2 = Cl$—(S_n)—Cl ($n \leq 100$), containing chains of sulfur atoms. Specific dichlorosulfanes can be synthesized by condensation of SCl_2 or S_2Cl_2 with the sulfanes H—(S_n)—H, for example,

$$2S_2Cl_2 + H-(S_n)-H \rightarrow Cl-(S_{n+4})-Cl + 2HCl \qquad (6.42)$$

6.6.4.2 Other Sulfur Halides

Reaction of sulfur with bromine gives *sulfur monobromide*, $(SBr)_2 = S_2Br_2$, a liquid (mp $-46°C$), which decomposes above $90°C$ and has properties similar to those of S_2Cl_2. Elemental sulfur dissolves in S_2Br_2 to give the *dibromosulfanes*, Br—(S_n)—Br. Neutral sulfur iodides appear to be unstable at room temperature. However, a stable cationic sulfur iodide can be obtained by the following reaction in liquid SO_2 or AsF_3 as a solvent[32]:

$$14S_8 + 8I_2 + 24AsF_5 \rightarrow 16S_7I^+AsF_6^- + 8AsF_3 \qquad (6.43)$$

The S_7I^+ cation has a seven-membered S_7 ring with an exocylic iodine atom.

6.6.4.3 Selenium and Tellurium Halides Other than Fluorides

Elemental selenium and tellurium react with chlorine to form the corresponding tetrachlorides, ECl_4 (E = Se, Te). Pale yellow $SeCl_4$ can also be obtained by treatment of SeO_2 with concentrated hydrochloric acid. Selenium tetrachloride is amphoteric, since it is both a donor and an acceptor of chloride ions. Thus reaction of $SeCl_4$ with ammonium chloride gives yellow $(NH_4)_2SeCl_6$; since the lone pair in the $SeCl_6^{2-}$ dianion is *not* stereochemically active, this anion is a regular octahedron. Reaction of $SeCl_4$ with $AlCl_3$ in a suitable nonaqueous solvent (SO_2Cl_2) gives yellow crystalline $SeCl_3^+AlCl_4^-$. Tellurium tetrachloride is a ψ-trigonal bipyramid in the vapor phase and a tetramer $[Cl_3Te(\mu_3\text{-}Cl)]_4$ in the solid state; the tetramer contains a Te_4Cl_4 cube with tellurium and chlorine atoms at alternate vertices.

The lower "selenium monohalides" Se_2X_2 (X = Cl, Br) can be isolated as highly colored heavy liquids from stoichiometric reactions of the elements.[33] However, the selenium dihalides SeX_2 (X = Cl, Br) can be detected only in the vapor phase. Condensation of the selenium dihalides results in disproportionation into Se, Se_2X_2, and SeX_4. Reactions of tellurium tetrahalides with elemental tellurium give a variety of polymeric "tellurium subhalides" with stoichiometries such as Te_3Cl_2, Te_2Br, and TeI.[34]

6.6.4.4 Sulfur and Selenium Oxohalides

Sulfur forms *thionyl halides*, $O{=}SX_2$, and *sulfuryl halides*, O_2SX_2. The $S{=}O$ bonds in these sulfur oxohalides are relatively short (~ 1.45 Å), indicative of multiple $d\pi$–$p\pi$ bonding.

Thionyl chloride, $O{=}SCl_2$ (mp $-101°C$, bp $76°C$), is prepared by the reaction of sulfur dioxide with phosphorus pentachloride:

$$SO_2 + PCl_5 \rightarrow O{=}SCl_2 + O{=}PCl_3 \qquad (6.44)$$

Thionyl chloride is a colorless fuming liquid, which is hydrolyzed readily to HCl and SO_2:

$$O{=}SCl_2 + H_2O \rightarrow 2HCl\uparrow + SO_2\uparrow \qquad (6.45)$$

Since thionyl chloride is a distillable liquid and all its hydrolysis products are gases (equation 6.45), thionyl chloride is a useful reagent for the removal of water without leaving a nonvolatile residue. For this reason thionyl chloride is used for the preparation of anhydrous metal chlorides from the corresponding hydrated metal chlorides. In addition, thionyl chloride is useful for the conversion of organic carboxylic acids to the corresponding acid chlorides:

$$RCO_2H + O{=}SCl_2 \rightarrow RC(O)Cl + SO_2\uparrow + HCl\uparrow \qquad (6.46)$$

The remaining thionyl halides, $O{=}SF_2$, $O{=}SClF$, $O{=}SBr_2$, and $O{=}SBrCl$, are made from $O{=}SCl_2$ by halogen exchange reactions using alkali metal halides under suitable conditions. *Thionyl bromide*, $O{=}SBr_2$ (mp $-50°C$, bp 138°C), is an orange liquid with properties similar to thionyl chloride. Fluorination of $O{=}SF_2$ at 150°C with a platinum metal catalyst gives *thionyl tetrafluoride*, $O{=}SF_4$.

The known sulfuryl halides are O_2SF_2, O_2SCl_2, O_2SFCl, and O_2SFBr. *Sulfuryl fluoride*, O_2SF_2, is a gas (bp $-55.4°C$), which can be obtained by the pyrolysis of barium fluorosulfate:

$$Ba(SO_3F)_2 \xrightarrow{500°C} O_2SF_2\uparrow + BaSO_4 \qquad (6.47)$$

The driving force of this reaction is the high lattice energy of barium sulfate. Sulfuryl fluoride is soluble in water without hydrolysis but is decomposed by aqueous base to give fluorosulfate and fluoride ion by the reaction

$$O_2SF_2 + 2OH^- \rightarrow SO_3F^- + F^- + H_2O \qquad (6.48)$$

Sulfuryl chloride, O_2SCl_2, is a liquid (mp $-54.1°C$, bp 69.1°C), which can be made by the chlorination of sulfur dioxide in the presence of an $FeCl_3$ or active charcoal catalyst. Sulfuryl chloride is rapidly hydrolyzed by water to sulfuric acid and hydrochloric acid. In addition, sulfuryl chloride is a useful chlorinating agent. Of particular interest is the use of an $S_2Cl_2/AlCl_3$ mixture in excess O_2SCl_2 for the perchlorination of aromatic hydrocarbons[35]; this reagent can convert benzene to hexachlorobenzene and naphthalene to octachloronapththalene under relatively mild conditions.

The most important selenium oxohalides are the *selenyl halides*, $O{=}SeX_2$. Selenyl chloride, $O{=}SeCl_2$ (mp 10.9°C, bp 177.2°C), can be prepared by the following reactions:

$$SeO_2(s) + O{=}SCl_2(l) \rightarrow O{=}SeCl_2(l) + SO_2(g) \qquad (6.49a)$$

$$SeO_2 + SeCl_4 \xrightarrow{CCl_4} 2O{=}SeCl_2 \qquad (6.49b)$$

Selenyl chloride is of interest as a nonaqueous solvent. It has a dielectric constant of 46.2 at 20°C, a dipole moment of 2.62 D, and a conductance of 2×10^{-5} $(\Omega \cdot cm)^{-1}$, which can be related to the following self-ionization:

$$2O{=}SeCl_2 \rightleftarrows [O{=}SeCl]^+ + [O{=}SeCl_3]^- \qquad (6.50)$$

6.7 Oxides of Sulfur and Its Congeners[36]

6.7.1 Lower Oxides

The monoxides SO, SeO, and TeO are transient species. Sulfur monoxide can be generated as a reactive intermediate by the dethionylation of ethylene episulfoxide[37]:

$$H_2C\!\!-\!\!CH_2 \xrightarrow[\substack{H_2O/ \\ MeOH}]{NaIO_4} H_2C\!\!-\!\!CH_2 \xrightarrow{\sim 100°C} H_2C\!\!=\!\!CH_2 + \{SO\} \qquad (6.51)$$

Polonium monoxide is a stable but easily oxidized black solid.

Disulfur monoxide, S_2O, which occurs as an intermediate in the combustion of sulfur, is obtained in the gas phase by passing thionyl chloride over powdered Ag_2S at 160°C:

$$O\!\!=\!\!SCl_2 + Ag_2S \rightarrow S_2O + 2AgCl \qquad (6.52)$$

The S_2O molecule is angular like ozone (Section 6.2.1.2) with an S–S–O angle of 118°, an S=O distance of 1.465 Å, and an S=S distance of 1.884 Å. Gaseous S_2O polymerizes at pressures above 1 mm or on condensation at 77 K but survives for several days in the gas phase at less than 1 mm pressure. However, the resulting polymer disproportionates in SO_2 and polysulfur oxides, which decompose at 100°C into SO_2 and S_8.

Other lower sulfur oxides of the type S_nO can be obtained by oxidation of the corresponding cyclo-S_n ring ($5 \leq n \leq 10$) with peroxytrifluoroacetic acid at $-30°C$. The sulfur oxide S_8O (mp 78°C, dec), can also be obtained from $O\!\!=\!\!SCl_2$ and the polysulfane H_2S_7 at $-40°C$. The S_nO oxides contain an S_n ring with a single exocyclic oxygen atom; they decompose readily into S_8 and SO_2.

6.7.2 Dioxides

Sulfur dioxide,[38] SO_2, is a colorless, toxic, and corrosive gas (mp $-75°C$, bp $-10°C$), which is obtained by burning elemental sulfur, metal sulfides, or sulfur-containing fuels in air. Formation of SO_2 from such combustion processes leads ultimately to "acid rain," which is a major ecological problem. Sulfur dioxide itself neither burns nor supports combustion. Sulfur dioxide is monomeric in all phases, with an angular structure. In the gas phase, the O–S–O angle is 119.5° and the S—O distance is 1.43 Å. Despite its low dielectric constant (~ 15) and lack of self-ionization (e.g., to SO^{2+} and SO_3^{2-}), liquid sulfur dioxide is a useful nonaqueous solvent, particularly for redox and complexation reactions as well as reactions of superacids (Section 1.4). Despite the lack of self-ionization of liquid SO_2, thionyl halides can function as acids and sulfites as bases in liquid SO_2 leading to "neutralization reactions" of the following type:

$$O=SCl_2 + M_2SO_3 \rightarrow 2MCl + 2SO_2 \tag{6.53}$$

The mechanisms of such reactions can involve the ions $O=SCl^+$ and SO_3^{2-} by the following sequence:

$$O=SCl_2 \rightleftarrows O=SCl^+ + Cl^- \tag{6.54a}$$

$$M_2SO_3 \rightleftarrows 2M^+ + SO_3^{2-} \tag{6.54b}$$

$$SOCl^+ + SO_3^{2-} \rightarrow SO_2Cl^- + SO_2 \tag{6.54c}$$

$$SO_2Cl^- \rightarrow SO_2 + Cl^- \tag{6.54d}$$

Step 6.54c of the reaction sequence involves an oxide transfer reaction.

Sulfur dioxide is amphoteric and can act as either a Lewis base or a Lewis acid. The Lewis basicity of sulfur dioxide arises from the presence of lone pairs; for this reason sulfur dioxide forms numerous transition metal complexes in which it exhibits a variety of bonding modes (Figure 6.7). The Lewis acidity of SO_2 is exhibited in its reactions with amines, with which it forms 1:1 adducts such as $Me_3N \rightarrow SO_2$.

Selenium and tellurium dioxides are obtained either by burning the elements in air or oxygen or by heating the elements with concentrated nitric acid followed by evaporation and heating to 300°C for SeO_2 or 400°C for TeO_2. *Selenium dioxide* forms colorless crystals, soluble in water, benzene, and glacial acetic acid, which sublime at 315°C to give a vapor consisting of monomeric angular SeO_2 molecules having an O–Se–O angle of 125° and Se—O distances

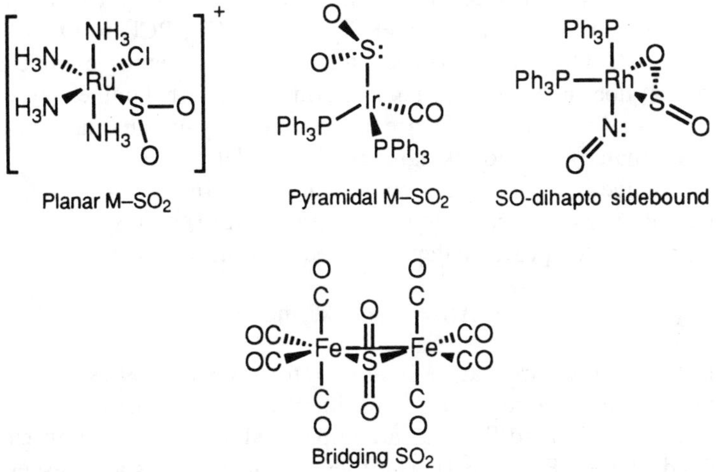

Figure 6.7. Some examples of SO_2 complexes of transition metals exhibiting different bonding modes of the SO_2 ligand.

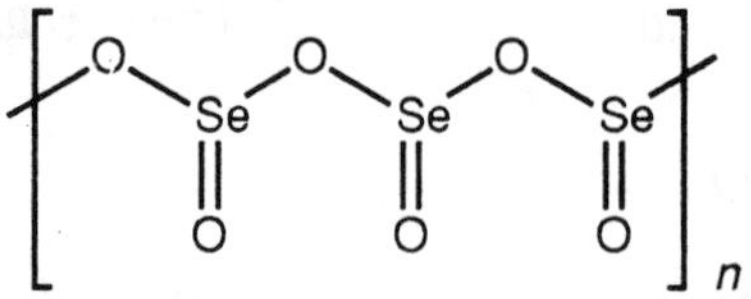

Figure 6.8.   The structure of selenium dioxide.

of 1.61 Å. Solid SeO_2, however, consists of polymeric nonplanar chains of three-coordinate selenium (Figure 6.8) with Se—O distances of 1.73 Å to terminal oxygen atoms and 1.78 Å to bridging oxygen atoms. Selenium dioxide is a useful oxidizing agent in organic chemistry, being reduced to elemental selenium.[39] *Tellurium dioxide* (mp 733°C), has a white (α) and a yellow (β) modification; in both modifications the structure is a three-dimensional polymer with four-coordinate ψ-trigonal bipyramidal tellurium(IV).

Polonium dioxide, PoO_2, can also be obtained by direct combination of the elements and has the fluorite (CaF_2) structure.

6.7.3 Trioxides

Sulfur trioxide is manufactured on a huge scale by the oxidation of SO_2 in air. Although this reaction is favored thermodynamically ($\Delta H° = -95.7$ kJ/mol), it requires a catalyst such as platinum sponge, V_2O_5, or NO to occur at an appreciable rate. Sulfur trioxide vapor can be absorbed by concentrated sulfuric acid to form "fuming sulfuric acid" or *oleum*, from which SO_3 can be isolated by distillation. Sulfur trioxide reacts very exothermically with water to form sulfuric acid. Sulfur trioxide is also a reasonably strong oxidizing agent, oxidizing sulfur to SO_2, SCl_2 to $O{=}SCl_2$ and O_2SCl_2, PCl_3 to $O{=}PCl_3$, and phosphorus to P_4O_{10}. Sulfur trioxide forms donor–acceptor complexes with Lewis bases it does not oxidize, such as pyridine, trimethylamine, and dioxane. These adducts as well as free SO_3 are used in industry for sulfonation reactions to make sulfonated oils and alkylarenesulfonate detergents.

Sulfur trioxide is a very hygroscopic liquid (mp 16.9°C, bp 44.5°C). In the gas phase sulfur trioxide consists of a mixture of SO_3 and S_3O_9 molecules in the pressure- and temperature-dependent equilibrium

$$3SO_3 \rightleftarrows S_3O_9 \qquad \Delta H = -126 \text{ kJ/mol } S_3O_9 \qquad (6.55)$$

Condensation of the vapor at its boiling point gives a colorless liquid, which consists mainly of S_3O_9 molecules; this liquid can be stabilized by addition of traces of boric acid. Solid SO_3 has a complicated structure with at least three well-defined phases (Figure 6.9). The thermodynamically stable, asbestos like α-SO_3 is formed by the condensation of SO_3 on cold surfaces; it consists of cyclic trimers with cyclohexanelike S_2O_3 rings. Polymeric, monoclinic β-SO_3 is

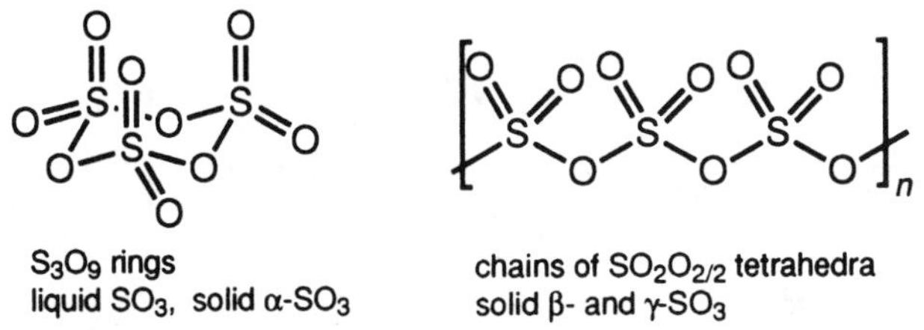

S$_3$O$_9$ rings
liquid SO$_3$, solid α-SO$_3$

chains of SO$_2$O$_{2/2}$ tetrahedra
solid β- and γ-SO$_3$

Figure 6.9.   Structure of sulfur trioxide in various phases.

formed by traces of water below ∼30°C; its structure consists of infinite chains of SO$_2$O$_{2/2}$ tetrahedra. Cooling liquid SO$_3$ below its melting point gives icelike crystals of rhombic γ-SO$_3$, which has an infinite chain structure similar to that of β-SO$_3$ but with cross-linking.

   Selenium trioxide (mp 118°C), is made by the dehydration of selenic acid (see Section 6.8.8) with P$_4$O$_{10}$ at 150–160°C or by the reaction of K$_2$SeO$_4$ with liquid SO$_3$. Selenium trioxide forms colorless hygroscopic crystals (mp 118°C), which are rapidly and exothermically rehydrated by water. In the solid state, selenium trioxide is a tetramer, Se$_4$O$_{12}$, which contains eight-membered Se$_4$O$_4$ rings. In the vapor phase, the tetramer Se$_4$O$_{12}$ is in equilibrium with monomeric SeO$_3$. Selenium trioxide is a stronger oxidizing agent than SO$_3$ and evolves O$_2$ upon heating above 160°C.

   Yellow-orange *tellurium trioxide* is prepared by the dehydration of ortho-telluric acid, Te(OH)$_6$, by heating at 300–360°C. Tellurium trioxide is insoluble in water, in contrast to SO$_3$ and SeO$_3$, but forms tellurates upon treatment with bases. Tellurium trioxide evolves oxygen upon heating above 400°C.

6.8 Oxoacids and Oxoanions of Sulfur, Selenium, and Tellurium

Sulfur, selenium, and tellurium form a variety of oxoacids and oxoanions. However, in most cases the oxoacid corresponding to a stable oxoanion cannot be isolated. Among the sulfur oxoacids, sulfuric acid (H$_2$SO$_4$) is the only free acid indefinitely stable at ambient temperature in the pure state; thiosulfuric acid (H$_2$S$_2$O$_3$) and peroxymonosulfuric acid (H$_2$SO$_5$) have been isolated at low temperatures, and dithionic acid (H$_2$S$_2$O$_6$) can be obtained in concentrated aqueous solutions. The structures of sulfur oxoanions along with their conventional names are depicted in Figure 6.10.

6.8.1 Sulfurous and Disulfurous Acids and Their Anions

Sulfurous acid, H$_2$SO$_3$, and disulfurous acid, H$_2$S$_2$O$_5$, are examples of sulfur oxoacids that do not exist in the free state, although numerous salts of

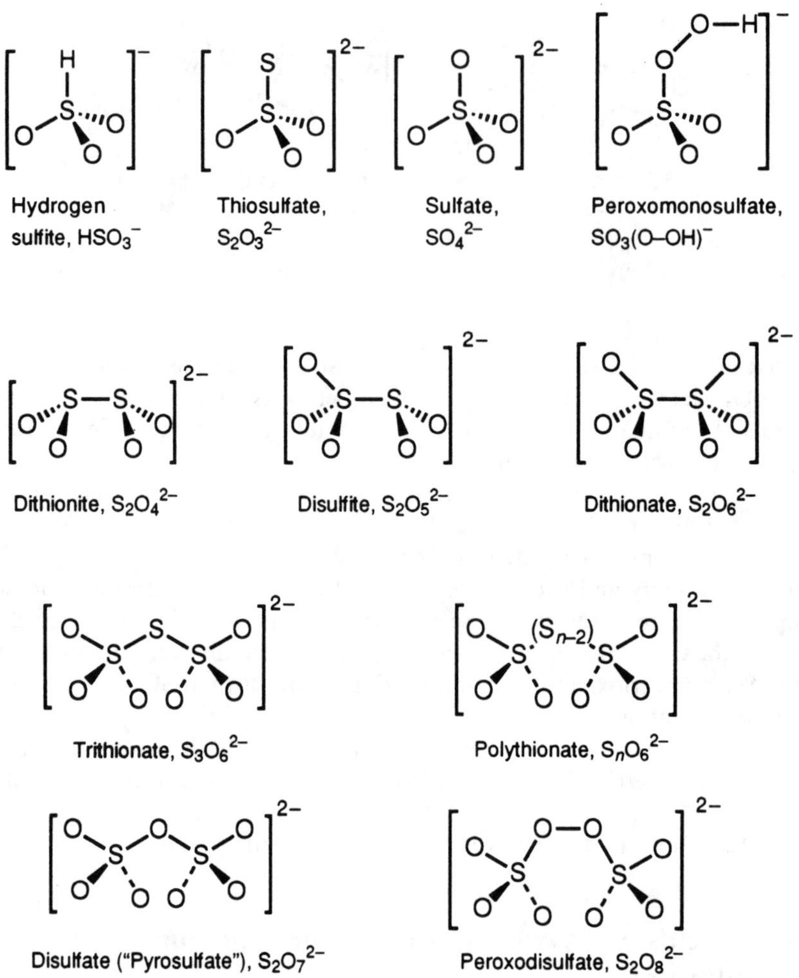

Figure 6.10. Structures of the sulfur oxoanions.

oxoanions derived from them containing the HSO_3^-, SO_3^{2-}, $HS_2O_5^-$, and $S_2O_5^{2-}$ anions are stable solids. Sulfur dioxide is readily soluble in water. However, its aqueous solutions, although acidic, contain negligible quantities of the free acid H_2SO_3. The apparent hydrate $H_2SO_3 \cdot 6H_2O$ is actually the hydrogen-bonded gas hydrate $SO_2 \cdot 7H_2O$. The dissociation constant of H_2SO_3 can be estimated by the following relationship:

$$K_1^{25°C} \text{ of "}H_2SO_3\text{"} = 1.3 \times 10^{-2}$$

$$= \frac{[HSO_3^-][H^+]}{[\text{total dissolved } SO_2] - [HSO_3^-] - [SO_3^{2-}]} \quad (6.56)$$

In this relationship the portion of the SO_2 dissolved but not converted to

HSO_3^- and SO_3^{2-} is treated as if it were the hypothetical, but unknown, H_2SO_3.

The lighter alkali metals Li^+, Na^+, and K^+ give solid salts of the SO_3^{2-} and $S_2O_5^{2-}$ anions. Solid salts of the hydrogen sulfite anion HSO_3^- can be obtained with relatively large counterions such as Rb^+, Cs^+, and R_4N^+; the HSO_3^- anion in these salts has four-coordinate sulfur with an S—H bond. However, the isomeric three-coordinate structure $O_2S(OH)^-$ can be detected in solution by ^{17}O NMR spectroscopy. Salts of the $S_2O_5^{2-}$ and $HS_2O_5^-$ anions can be prepared by the following methods:

$$2MHSO_3 \xrightarrow{\Delta} M_2S_2O_5 + H_2O \tag{6.57a}$$

$$HSO_3^- + SO_2 \xrightarrow{H_2O} HS_2O_5^- \tag{6.57b}$$

Diesters of sulfurous acid exist in the tautomeric forms $(RO)_2S{=}O$, containing three-coordinate sulfur, and $R—S(O)_2(OR)$, containing four-coordinate sulfur.

6.8.2 Sulfuric Acid, Polysulfuric Acids, and Their Salts

Sulfuric acid, H_2SO_4, is the most important commercial sulfur compound. It is produced on a gigantic scale by the oxidation of sulfur to SO_2, followed by catalytic oxidation of the SO_2 to SO_3 and the hydration of SO_3. Pure sulfuric acid is a colorless, oily liquid (mp $10.4°C$), which boils at $290–317°C$, with partial decomposition to water and SO_3. Sulfur trioxide dissolves in H_2SO_4 to give disulfuric acid, $H_2S_2O_7$, and higher polysulfuric acids $H_2S_nO_{3n+1}$. Mixtures with melting point minima containing 20 and 65% of free SO_3 are sold as fuming sulfuric acid or oleum. Concentrated sulfuric acid is reduced by metals to SO_2 and is a strong dehydrating agent; it carbonizes most carbohydrates.

Sulfates and hydrogen sulfates of most electropositive elements can be isolated. Salts of the polysulfuric acids can be obtained by reactions such as the following:

$$2KHSO_4 \xrightarrow{\Delta} K_2S_2O_7 + H_2O \tag{6.58a}$$

$$K_2SO_4 + (n-1)SO_3 \rightarrow K_2S_nO_{3n+1} \tag{6.58b}$$

Since $K_2S_2O_7$ can be prepared by pyrolysis of $KHSO_4$, it was formerly called "potassium pyrosulfate." Polysulfates are hydrolyzed to SO_4^{2-} and undergo thermal decomposition to give sulfate and SO_3.

6.8.3 Thiosulfuric Acid and Thiosulfates

The structure of the thiosulfate anion $S_2O_3^{2-}$ has a central tetrahedral sulfur, which is coordinated to three oxygen atoms and an external sulfur atom (Figure 6.10). In thiosulfate the S—S bond distance of 2.013 Å indicates very little π-bonding, whereas the S—O bond distance of 1.468 Å indicates considerable π-bonding. Thiosulfate ion is a moderate one-electron reducing agent ($E° = 0.169$ V), being oxidized to tetrathionate, $S_4O_6^{2-}$, through formation of a sulfur–sulfur bond. The rapid and quantitative reaction of thiosulfate to tetrathionate upon reaction with iodine is the basis of the iodometric determination of oxidizing agents such as Cu^{2+} by oxidation of iodide to elemental iodine followed by titration with a standard sodium thiosulfate solution with starch as an indicator. Thiosulfate can act as either a one-electron or an eight-electron reducing agent toward Br_2, depending on the conditions; eight-electron oxidation of thiosulfate gives the sulfate ion. The reducing ability of thiosulfate is used to destroy excess chlorine in the bleaching process. Thiosulfate is a powerful complexing agent with thiophilic metals (Cu, Ag, Au, Hg, etc.) through the formation of metal–sulfur bonds. The formation of the silver complex $Ag(S_2O_3)_3^{5-}$ is the basis of the use of solutions of $Na_2S_2O_3 \cdot 5H_2O$ as a "fixer" in photography to dissolve unphotolyzed silver bromide after exposure of the film.

Free thiosulfuric acid cannot be generated in aqueous solution by acidifying thiosulfates; instead, decomposition occurs to give sulfur dioxide and elemental sulfur. However, thiosulfuric acid can be prepared at low temperatures by nonaqueous reactions such as the following:

$$Na_2S_2O_3 + 2HCl + 2Et_2O \xrightarrow[-78°C]{Et_2O} H_2S_2O_3 \cdot 2Et_2O + 2NaCl \qquad (6.59a)$$

$$HSO_3Cl + H_2S \xrightarrow[-10°C]{\text{no solvent}} H_2S_2O_3 + HCl \qquad (6.59b)$$

The product $H_2S_2O_3 \cdot 2Et_2O$ (equation 6.59a) is undoubtedly the diethyl-oxonium salt $[Et_2OH^+]_2[S_2O_3^{2-}]$.

6.8.4 The Dithionite Ion

The dithionite ion is obtained in aqueous solution by the reduction of sulfite ion by zinc dust or by the reduction of $NaHSO_3$ with sodium borohydride. The dithionite ion is not very stable in aqueous solution, disproportionating according to the following equation:

$$2S_2O_4^{2-} + H_2O \rightarrow S_2O_3^{2-} + 2HSO_3^- \qquad (6.60)$$

This decomposition in acid solution is so rapid that the free acid $H_2S_2O_4$ cannot be obtained.

The dithionite ion as the zinc or sodium salt is a very powerful reducing agent in alkaline solution:

$$2SO_3^{2-} + 2H_2O + 2e^- = S_2O_4^{2-} + 4OH^- \qquad E° = -1.12 \text{ V} \qquad (6.61)$$

The initial stage of dithionite reductions appears to involve dissociation of the dithionite ion into the $SO_2^{\overset{\cdot}{-}}$ radical anion, as indicated by the observations of a strong EPR signal and kinetic rate laws that are half-order in dithionite. The ready dissociation of the dithionite ion into the $SO_2^{\cdot-}$ radical anion is consistent with the relative long S—S bond distance of 2.39 Å in the dithionite structure (Figure 6.10).

6.8.5 Dithionic Acid and the Dithionate Ion

The dithionate ion, $S_2O_6^{2-}$, can be prepared by oxidation of sulfur dioxide or sulfite ion with a mild oxidizing agent such as MnO_2:

$$MnO_2 + 2SO_3^{2-} + 4H^+ \rightarrow Mn^{2+} + S_2O_6^{2-} + 2H_2O \tag{6.62}$$

Other sulfur oxoacids, notably sulfate, are produced in this reaction; they can be removed by precipitation with Ba^{2+} and the more soluble $BaS_2O_6 \cdot 2H_2O$ crystallized from the filtrate. Aqueous solutions of dithionates can be boiled without decomposition. Treatment of barium dithionate with sulfuric acid followed by removal of the $BaSO_4$ precipitate gives a solution of free dithionic acid, a relatively stable moderately strong acid that decomposes slowly upon concentration or warming.

6.8.6 Polythionates (Sulfanedisulfonates)

A series of polysulfanedisulfonate anions $S_nO_6^{2-}$ ($n \leq$ at least 22) are known which have long sulfur chains similar to the H_2S_n and S_nCl_2 derivatives; these are called *polythionates*. Individual polythionates can be separated by chromatography, although only the lower members are well characterized.[40] Both the free acids, $H_2S_nO_6$, and the hydrogen polythionate anions, $HS_nO_6^-$, are unstable.

Solutions containing mixtures of the polythionates can be obtained by treatment of thiosulfate with sulfur dioxide in the presence of As_2O_3 or by reaction of H_2S with aqueous SO_2; the latter solution, which has been known for more than a century, is called *Wackenroeder's liquid* after H. W. F. Wackenroeder (1846), who first studied this material. In addition, trithionate and tetrathionate can be made from thiosulfate by the following specific methods:

$$2S_2O_3^{2-} + 4H_2O_2 \rightarrow S_3O_6^{2-} + SO_4^{2-} + 4H_2O \tag{6.63a}$$

$$2S_2O_3^{2-} + I_2 \rightarrow S_4O_6^{2-} + 2I^- \tag{6.63b}$$

6.8.7 Peroxoacids and Salts of Sulfur

Peroxosulfates are known containing either one or two sulfur atoms. The *peroxodisulfate anion*, $S_2O_8^{2-}$, is obtained by the electrolysis of sulfates at low temperatures and high current densities; alkali metal and ammonium peroxodisulfates can be isolated. The peroxodisulfate ion is a very convenient powerful oxidizing agent[41] and can oxidize Mn^{2+} to permanganate and Cr^{3+} to chromate:

$$S_2O_8^{2-} + 2e^- \rightarrow 2SO_4^{2-} \qquad\qquad E^\circ = 2.01 \text{ V} \qquad\qquad (6.64)$$

There is evidence in peroxodisulfate oxidations for the intermediacy of the sulfate radical anions, $SO_4^{\cdot-}$, formed by rupture of the O—O bond in $S_2O_8^{2-}$. Silver frequently catalyzes peroxodisulfate oxidations through a Ag^{2+} intermediate.

Peroxomonosulfuric acid, $H_2SO_5 = H_2[O_3S—O—O]$, also known as *Caro's acid*, can be obtained by hydrolysis of $H_2S_2O_8$ to H_2SO_4 and H_2SO_5, or by reaction of $\sim 100\%$ hydrogen peroxide with sulfuric acid or chlorosulfonic acid. Pure peroxomonosulfuric acid, as obtained from hydrogen peroxide and chlorosulfonic acid, is a colorless, explosive solid (mp 45°C). Salts of the $HSO_5^- = H—O—O—SO_3^-$ anion such as $KHSO_5$ are known. The structure of the HSO_5^- anion in such salts has two different S—O distances, namely three 1.45 Å S—O distances and one much longer (1.62 Å) S—O$_2$ distance.

6.8.8 Oxoacids of Selenium and Tellurium

Selenous acid, H_2SeO_3, can be prepared by dissolving SeO_2 in water or by oxidizing elemental selenium with nitric acid. Solid selenous acid consists of layers of pyramidal SeO_3 groups linked by hydrogen bonds. Selenous acid is a weak acid ($K_1 = 3.5 \times 10^{-3}$; $K_2 = 5 \times 10^{-8}$). Neutralization of selenous acid with alkalies gives hydrogen selenites, M^IHSeO_3, and selenites, $M_2^ISeO_3$. Concentrated hydrogen selenite solutions also contain the diselenite ion, $Se_2O_5^{2-}$ in which both selenium atoms are three-coordinate with an Se—O—Se structural unit, in contrast to the disulfite ion, $S_2O_5^{2-}$ (Figure 6.10), in which one of the sulfur atoms is four-coordinate and there is a direct S—S bond. Selenous acid is a moderately strong oxidizing agent:

$$H_2SeO_3 + 4H^+ + 4e^- = Se + 3H_2O \qquad\qquad E^\circ = 0.74 \text{ V} \qquad\qquad (6.65)$$

Selenous acid is thus reduced by N_2H_4, SO_2, H_2S, or HI to give red selenium.

Tellurous acid arises from hydrolysis of the tellurium tetrahalides but has not been well characterized; it is a weak acid, like selenous acid. Tellurites containing the $HTeO_3^-$ or TeO_3^{2-} ions can be obtained by dissolving TeO_2 in an alkali; polytellurites, $M_2Te_nO_{2n+1}$, are also known.

Selenic acid, H_2SeO_4 (mp 62°C), can be obtained by fusion of elemental selenium with potassium nitrate or by oxidation of SeO_2, selenous acid, or selenites with strong oxidizing agents such as H_2O_2, bromine or chlorine in water, MnO_4^-, F_2, or ozone. Selenic acid forms colorless hygroscopic crystals (mp 58°C), and is as strong an acid as sulfuric acid. Most salts of the $HSeO_4^-$ and SeO_4^{2-} anions are isomorphous to the corresponding salts of the HSO_4^- and SO_4^{2-} anions. Selenic acid is a much stronger oxidizing agent than sulfuric acid:

$$SeO_4^{2-} + 4H^+ + 2e^- = H_2SeO_3 + 2H_2O \qquad\qquad E^\circ = 1.15 \text{ V} \qquad\qquad (6.66)$$

However, many selenic acid oxidations are not kinetically fast. Nevertheless selenic acid has enough oxidizing ability to dissolve metallic gold and palladium (and even platinum in the presence of chloride). Selenic acid evolves oxygen upon heating above 200°C.

Telluric acid (mp 136°C) consists of hydrogen-bonded $Te(OH)_6$ molecules; the central tellurium atom in telluric acid is thus six-coordinate, in contrast to the central sulfur atom in H_2SO_4 and the central selenium atom in H_2SeO_4. Telluric acid can be obtained as colorless crystals by the oxidation of tellurium or tellurium dioxide with H_2O_2, Na_2O_2, CrO_3, $KMnO_4/HNO_3$, $HClO_3$, and so on. In water $Te(OH)_6$ is a weak, dibasic acid ($K_1 \approx 10^{-7}$) from which $MTeO(OH)_5$ and $M_2TeO_2(OH)_4$ salts can be isolated by treatment with base; these salts are not isomorphous with the corresponding sulfates and selenates, since they contain TeO_6 octahedra. Salts of the type M_6TeO_6 can be obtained by precipitation of tellurates with Ag^+ (M = Ag) or by fusion of telluric acid with excess NaOH (M = Na). Telluric acid is a strong, but kinetically slow oxidizing agent ($E° = 1.02$ V).

6.9 Miscellaneous Selenium and Tellurium Compounds

6.9.1 Organoselenium and Organotellurium Derivatives

Organoselenium and organotellurium derivatives are known in the $+2$, $+4$, and $+6$ oxidation states; they have a number of applications in organic synthesis.[42-46] The $+2$ derivatives include compounds of the types R_2E, REH, and REX (R = alkyl or aryl group; E = Se or Te; X = halogen). Compounds of the type R_2E and REH with lower molecular weight alkyl and even aryl groups are among the most malodorous compounds known, but such compounds with higher molecular weight aryl groups are much more readily handled. The $+4$ derivatives include compounds of the types R_2EX_2, REX_3, $R_3E^+X^-$, and R_4Te (R = Me, Ph), which are prepared by reactions such as the following for the tellurium derivatives:

$$R_2Te + X_2 \rightarrow R_2TeX_2 \tag{6.67a}$$

$$R_2Te + RX \rightarrow R_3Te^+X^- \tag{6.67b}$$

$$TeCl_4 + 4RLi \rightarrow R_4Te \; (R = Me) + 4LiCl \tag{6.67c}$$

Organoselenium and organotellurium derivatives in the $+6$ oxidation state are much rarer. However, hexamethyltellurium can be obtained as a volatile white solid starting from tetramethyltellurium by the following sequence of reactions[47]:

$$Me_4Te + XeF_2 \rightarrow Me_4TeF_2 + Xe \tag{6.68a}$$

$$Me_4TeF_2 + Me_2Zn \rightarrow ZnF_2 + Me_6Te \tag{6.68b}$$

Hexamethyltellurium is thermally stable to 140°C, whereas tetramethyltellurium decomposes completely after 4 hours at 120°C.

6.9.2 Miscellaneous Coordination Compounds of Selenium and Tellurium

Selenium and tellurium in the $+2$ and $+4$ oxidation states form a variety of coordination compounds with standard ligands, particularly "soft" ligands with donor halogen (Cl, Br, I) or sulfur atoms. The one lone pair in Se(IV) and Te(IV) is not stereochemically active in six-coordinate derivatives of the type $E^{IV}X_6^{2-}$ (E = Se, Te; X = Cl, Br), which are regular octahedra; they are obtained by treating EO_2 with halide ions and the corresponding hydrogen halide and then adding a large cation to give a crystalline salt. Similarly, the six-coordinate tellurium(IV) complex *trans*-(tmtu)$_2$TeCl$_4$ (tmtu = tetramethylthiourea, S=C(NMe$_2$)$_2$) is obtained by treatment of a hydrochloric acid solution of TeO$_2$ with tetramethylthiourea. This complex is also octahedral, with no evidence for stereochemical activity of the lone pair.

The lone pair(s) in selenium and tellurium complexes in the $+2$ and $+4$ oxidation states and coordination numbers below 6 are often stereochemically active and lead to some unusual coordination polyhedra (Figure 6.11). Thus the tellurium(IV) in the six-coordinate complex Te(S$_2$CNEt$_2$)$_2$(CH$_3$)I has

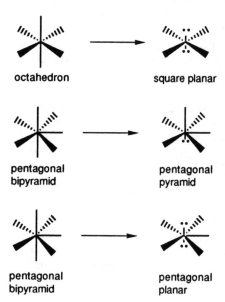

Figure 6.11. Square planar coordination is derived from a ψ-octahedron with two stereochemically active lone pairs in trans positions; pentagonal pyramidal coordination and pentagonal planar coordination are derived from ψ-pentagonal bipyramids with one or two sterochemically active lone pairs, respectively, in axial positions.

pentagonal pyramidal coordination derived from a ψ-pentagonal bipyramid with a sterochemically active lone pair in an axial position. Similarly, the tellurium (II) in the five-coordinate complex $[Et_4N][Te^{II}(S_2COEt)_3]$, in which two of the ethyl xanthate ligands are bidentate and the third ligand monodentate, has planar pentagonal coordination derived from a ψ-pentagonal bipyramid with sterochemically active lone pairs in both axial positions. Square planar tellurium(II) complexes, such as yellow $(tmtu)_2TeCl_2$, obtained by pyrolysis of the trans-$(tmtu)_2TeCl_4$ just discussed, are derived from a ψ-octahedron with two sterochemically active lone pairs in trans positions.

References

1. Rodger, C.; Sheppard, N.; McFarlane, C.; McFarlane, W., in *NMR and the Periodic Table*, R. K. Harris and B. E. Mann, Eds., Academic Press, London, 1978, pp. 383–400.

2. Rodger, C.; Sheppard,, N.; McFarlane, H. C. E.; McFarlane, W., *Selenium-77 and Tellurium-125*, in *NMR and the Periodic Table*, R. K. Harris and B. E. Mann, Eds., Academic Press, London, 1978, pp. 402–419.

3. Biswas, A. K., Ed., *Strategy for the Ozone Layer*, Pergamon Press, Oxford, 1980.

4. Riess, J. G.; Le Blanc, M., Solubility and Transport Phenomena in Perfluorochemicals Relevant to Blood Substitution and Other Biomedical Applications, *Pure Appl. Chem.*, 1982, **54**, 2383–2406.

5. Wasserman, H. H.; Murray, R. W., *Singlet Oxygen*, Academic Press, New York, 1979.

6. Randy, B.; Rabek, J. F., Eds., *Singlet Oxygen: Reactions with Organic Compounds and Polymers*, Wiley, Chichester, 1978.

7. Razumovskii, S. D.; Zaikov, G. E., *Ozone and Its Reactions with Organic Compounds*, Elsevier, Amsterdam, 1984.

8. Bailey, P. S., *Ozonization in Organic Chemistry*, Academic Press, New York: Vol. 1, *Olefinic Compounds*, 1978; Vol. 2, *Nonolefinic Compounds*, 1982.

9. Meyer, B., Elemental Sulfur, *Chem. Rev.*, 1976, **76**, 367–388.

10. Steudel, R., Homocyclic Sulfur Molecules, *Top. Curr. Chem.*, 1982, **102**, 149–197.

11. Jennings, R. H.; Yannopoulos, J. C., *Reccovery and Refining of Selenium*, in *Selenium*, R. A. Zingaro and W. C. Cooper, Eds., Van Nostrand Reinhold, New York, 1974, pp. 31–86.

12. Stuke, J., Optical and Electronic Properties of Selenium, in *Selenium*, R. A. Zingaro and W. C. Cooper, Eds., Van Nostrand Reinhold, New York, 1974, pp. 174–297.

13. Bagnall, K. W., The Chemistry of Polonium, *Q. Rev.*, 1957, **11**, 30–48.

14. Gillespie, R. J., Ring, Cage, and Cluster Compounds of the Main Group Elements. *Chem. Soc. Rev.*, 1979, **8**, 315–352.

15. Jones, D. P.; Griffith, W. P., *J. Chem. Soc. Dalton Trans.*, 1980, 2526–2532.

16. Sawyer, D. T.; Valentine, J. S., How Super Is Superoxide? *Acc. Chem. Res.*, 1981, **14**, 393–400.

17. Kanatzidis, M. G., *Chem. Mater.*, 1990, **2**, 353–363.

18. Sudworth, J. L.; Tilley, A. R., Eds., *The Sodium–Sulfur Battery*, Chapman and Hall, New York, 1985.

19. Hulliger, F., Crystal Chemistry of Transition Metal Chalcogenides, *Struct. Bonding (Berlin)*, 1968, **4**, 83–229.

20. Roesky, H. W., Cyclic Sulfur–Nitrogen Compounds, *Adv. Inorg. Chem. Radiochem.*, 1979, **22**, 239–301.

21. Chivers, T., Electron-Rich Sulfur–Nitrogen Heterocycles, *Acc. Chem. Res.*, 1984, **17**, 166–171.

22. Chivers, T., Synthetic Methods and Structure–Reactivity Relationships in Electron-Rich Sulfur–Nitrogen Rings and Cages, *Chem. Rev.*, 1985, **85**, 341–365.

23. Oakley, R. T., Cyclic and Heterocyclic Thiazenes, *Prog. Inorg. Chem.*, 1988, **36**, 299–391.

24. Labes, M. M.; Love, P.; Nichols, L. F., Poly(sulfur Nitride)—A Metallic Superconducting Polymer, *Chem. Rev.*, 1979, **79**, 1–15.

25. Glemser, O.; Mews, R., Chemistry of Thiazyl Fluoride (NSF) and Trifluoride (NSF_3): A Quarter Century of Sulfur–Nitrogen–Fluorine Chemistry, *Angew. Chem. Int. Ed. Engl.*, 1980, **19**, 883–899.

26. Lichtenberger, D. L.; Hubbard, J. L., *Inorg. Chem.*, 1985, **24**, 3835–3841.

27. Seel, F., Lower Sulfur Fluorides, *Adv. Inorg. Chem. Radiochem.*, 1974, **16**, 297–333.

28. Boswell, G. A., Jr.; Ripka, W. C.; Scribner, R. M.; Tullock, C. W., Fluorination by Sulfur Tetrafluoride, *Org. React.*, 1974, **21**, 1–124.

29. Kleemann, G.; Seppelt, K., *Chem. Ber.*, 1983, **116**, 645–658.

30. Cohen, B.; Peacock, R. D., Fluorine Compounds of Selenium and Tellurium, *Adv. Fluorine Chem.*, 1970, **6**, 343–385.

31. Engelbrecht, E.; Sladky, F., Selenium and Tellurium Fluorides, *Adv. Inorg. Chem. Radiochem.*, 1981, **24**, 189–223.

32. Klapötke, T.; Passmore, J., Sulfur and Selenium Iodine Compounds: From Nonexistence to Significance, *Acc. Chem. Res.*, 1989, **22**, 234–240.

33. Kniep, R.; Korte, L.; Mootz, D., *Z. Naturforsch.*, 1983, **38b**, 1–6.

34. Kniep, R.; Rabenau, A., Subhalides of Tellurium, *Top. Curr. Chem.*, 1983, **111**, 145–192.

35. Ballester, M.; Molinet, C.; Castañer, J., *J. Am. Chem. Soc.*, 1960, **82**, 4254–4258.

36. Schenk, P. W.; Steudel, R., Oxides of Sulphur, in *Inorganic Sulphur Chemistry*, G. Nickless, Ed. Elsevier, Amsterdam, 1968, pp. 367–418.

37. Hartzell, G. E.; Paige, J. N., Ethylene Episulfoxide, *J. Am. Chem. Soc.* 1966, **88**, 2616–2617.

38. Schroeter, L. C., *Sulfur Dioxide*, Pergamon Press, Oxford, 1966.

39. Rabjohn, N., Selenium Dioxide Oxidation, *Org. Reacts.*, 1976, **24**, 261–415.

40. Steudel, R.; Holdt, G.; Göbel, T.; Hazen, W., *Angew. Chem. Int. Ed. Engl.*, 1987, **26**, 151–153.

41. Minisci, F.; Cittterio, A.; Giordano, C., Electron-Transfer Processes: Peroxydisulfate, a Useful Versatile Reagent in Organic Chemistry, *Acc. Chem. Res.*, 1983, **16**, 27–32.

42. Irgolic, K. J.; Kudchadker, M. V., The Organic Chemistry of Selenium, in *Selenium*, R. A. Zingaro and W.C. Cooper, Eds., Van Nostrand Reinhold, New York, 1974.

43. Liotta, D., New Organoselenium Methodology, *Acc. Chem. Res.*, 1980, **17**, 28–35.

44. Paulmeier, C., *Selenium Reagents and Intermediates in Organic Synthesis*, Pergamon Press, Oxford, 1986.

45. Engman, L., Synthetic Applications of Organotellurium, Chemistry, *Acc. Chem. Res.*, 1985, **18**, 274–279.

46. Krief, A.; Hevesi, L., *Organoselenium Chemistry*, Vol. I, *Functional Group Transformations*, Springer, Berlin, 1988.

47. Ahmed, L.; Morrison, J. A., *J. Am. Chem. Soc.*, 1990, **112**, 7411–7413.

CHAPTER
7

Halogens and the Noble Gases

7.1 General Aspects of Halogen and Noble Gas Chemistry

The halogens (fluorine, chlorine, bromine, iodine, and astatine) are one electron short of the noble gas configuration. They therefore readily form the halide monoanions X^- (X = F, Cl, Br, I, At) or a single covalent bond R—X. Furthermore, elemental halogens are the single-bonded dimers $:\ddot{X}—\ddot{X}:$ and are very reactive both toward reduction to halides, X^-, and toward the formation of covalent bonds R—X with suitable substrates. Elemental fluorine, $F_2 = :\ddot{F}—\ddot{F}:$, is unusually reactive because of the low strength of its fluorine–fluorine bond combined with the high strengths of bonds of fluorine to other elements. Halogens are not found in nature in the elemental state because of their high reactivity.

All halogens except fluorine form compounds with formal positive oxidation states; these are hypervalent compounds involving nd orbitals or $(n + 1)p$ orbitals in their bonding. Examples of positive halogen compounds include halogen fluorides such as ClF_3, ClF_5, BrF_3, BrF_5, IF_5, and IF_7; halogen oxides such as Cl_2O, ClO_2, Cl_2O_7, and I_2O_5; halogen oxoanions of the general types XO_n^- (X = Cl, Br, I; n = 1–4); halogen oxofluorides such as F_3BrO and FIO_3; and organoiodine compounds of the types R_2I^+, R_3I, $(RIO)_n$, $(RIO_2)_n$, and so on. In general, compounds with iodine in positive oxidation states are significantly more stable than similar compounds with bromine and chlorine.

The noble gases (helium, neon, argon, krypton, xenon, and radon) already have a very favorable electronic configuration. For this reason their ability to combine with other elements is very limited; in fact until 1962 noble gases were

173

Table 7.1 Spin $\frac{1}{2}$ Isotopes of the Halogens and Noble Gases

Isotope	Natural Abundance (%)	NMR Sensitivity for Equal Numbers of Nuclei*
Fluorine-19	100	0.941
Helium-3	10^{-5}–10^{-7}	0.443
Xenon-129	11.78	0.277

* The sensitivities are given relative to that of the proton, $^1H = 1.000$.

believed to be incapable of forming any compounds at all. However, during the past 30 years the chemistry of xenon has developed considerably. Since the chemistry of xenon in the $+2$, $+4$, $+6$, and $+8$ oxidation states parallels the isoelectronic chemistry of iodine in the $+1$, $+3$, $+5$, and $+7$ oxidation states the chemistry of xenon is conveniently discussed in the same chapter as the chemistry of iodine. A few krypton compounds are also known. However, no compounds of helium, neon, or argon are known which are stable under ambient conditions.

The chemistry of both the halogens and the noble gases is purely nonmetallic; even the heaviest elements, namely astatine and radon, do not exhibit any metallic properties.

The halogens and noble gases having spin $\frac{1}{2}$ isotopes suitable for NMR studies are listed in Table 7.1. Fluorine-19 is the most sensitive NMR nucleus next to ^1H and ^3H (Section 1.2); ^{19}F NMR spectroscopy is thus an important method for the identification and characterization of fluorine compounds. Although ^3He is also one of the most sensitive NMR nuclei, it is of limited value owing to its rarity and the absence of any known helium compounds. NMR spectroscopy based on ^{129}Xe has proven to be important in the characterization of xenon chemistry in solution.

7.2 The Elemental Halogens and Noble Gases

7.2.1 Fluorine

Fluorine is more abundant in the earth's crust than chlorine (0.065% v. 0.055%) and is the thirteenth most abundant element. Concentrated deposits of fluorine occur in fluorite (CaF_2), cryolite (Na_3AlF_6), and fluoroapatite ($Ca_3(PO_4)_2 \cdot Ca(F, Cl)_2$). However, only fluorite is extensively processed commercially for recovery of fluorine and its compounds.

The only stable isotope of fluorine is ^{19}F, which is an excellent NMR nucleus (Table 7.1). The radioactive fluorine isotope, ^{18}F, with a half-life of 109.7 minutes is available and can be used as a tracer, albeit with some difficulty.[1]

Elemental fluorine is a toxic yellow gas (mp $-218.6°C$, bp $-188.1°C$), its structure consists of F_2 molecules with an F—F distance of 1.43 Å. Fluorine is not only the most strongly oxidizing element but also one of the strongest known oxidants of any type:

$$F_2(g) + 2e^- = 2F^- \qquad E° = +2.87 \text{ V} \qquad (7.1)$$

Fluorine can be handled in glass apparatus, provided traces of hydrogen fluoride are removed (unlike fluorine, hydrogen fluoride attacks glass rapidly). Hydrogen fluoride can be removed from fluorine by passing the gas through a solid fluoride MR (M = Na, K), which reacts with any HF impurity to form the corresponding MHF_2 salt.

Elemental fluorine is not liberated from the electrolysis of fluorides in aqueous solution because of its strong oxidizing power; instead, oxygen is liberated. However, elemental fluorine can be prepared by the electrolysis of molten $KF \cdot nHF$ ($n = 2$–3, mp 70–100°C; $n = 1$, mp 150–270°C). Electrolytic cells for the preparation of fluorine are constructed from steel, copper or Monel metal, materials that become coated with an unreactive layer of fluoride. Steel or copper cathodes and ungraphitized carbon anodes are used for the electrolysis.

The chemical synthesis of fluorine from compounds not requiring fluorine for their preparation is very difficult. However, the following reaction[2] provides a good yield chemical synthesis of fluorine:

$$2K_2MnF_6 + 4SbF_5 \rightarrow 4KSbF_6 + 2MnF_3 + F_2\uparrow \qquad (7.2)$$

The necessary reagents can be prepared from hydrogen fluoride using the following reactions:

$$2KMnO_4 + 2KF + 10HF + 3H_2O_2 \xrightarrow{\text{50\% aq HF}}$$
$$2K_2MnF_6\downarrow + 8H_2O + O_2\uparrow \qquad (7.3a)$$

$$SbCl_5 + 5HF \rightarrow SbF_5 + 5HCl \qquad (7.3b)$$

Reaction 7.2 involves displacement of the weaker Lewis acid MnF_4 by the stronger Lewis acid SbF_5; however, free MnF_4 is unstable and decomposes to $MnF_3 + \frac{1}{2}F_2$ when heated.

7.2.2 Chlorine

Chlorine occurs in nature as the chloride ion in ocean water as well as in salt lakes and salt deposits; it is the twelfth most abundant element in crustal rocks.

Elemental chlorine is a greenish-yellow toxic gas (mp -101.0°C, bp -34.9°C), it has a pungent odor and attacks the mucous membranes. The structure of elemental chlorine consists of Cl_2 molecules with a Cl—Cl distance of 1.99 Å. Elemental chlorine is a reasonably strong oxidizing agent:

$$Cl_2 + 2e^- = 2Cl^- \qquad E° = +1.3595 \text{ V} \qquad (7.4)$$

It spontaneously ignites with many metals such as the alkali and alkaline earth metals, copper, iron, arsenic, antimony, and bismuth. Among nonmetals, hydrogen reacts after ignition explosively in a chain reaction. Chlorine dissolves in water by the reaction

$$Cl_2 + H_2O \rightarrow HCl + HOCl \tag{7.5}$$

Passing chlorine into a dilute aqueous solution of $CaCl_2$ at 0°C leads to the precipitation of feathery crystals of *chlorine hydrate*, $Cl_2 \cdot 7.3H_2O$; this hydrate has a clathrate structure in which Cl_2 molecules occupy holes in a hydrogen-bonded water network (Section 1.3).

Elemental chlorine can be prepared by the electrolysis of brine (concentrated aqueous sodium chloride). It can also be obtained by chemical methods such as the oxidation of hydrochloric acid with MnO_2 or $KMnO_4$.

7.2.3 Bromine

Bromine, like chlorine, occurs in ocean water as bromide ion, but it is present in much smaller amounts than chlorine. Elemental bromine is a dense, mobile dark red corrosive liquid (mp -7.3°C, bp $+59.5$°C); the Br—Br distance in its Br_2 molecules is 2.28 Å. Bromine can be liberated from ocean water or other solutions of bromide ion by chlorination at pH ~ 3.5 and removed in a current of air. Bromine is moderately soluble in water (33 g/L at 25°C) and forms a crystalline clathrate hydrate. Bromine is more readily soluble in nonpolar solvents such as CS_2 and CCl_4 than in water.

7.2.4 Iodine and Its Charge-Transfer Complexes

Iodine is found in ocean water as iodide ion and is concentrated in seaweed, which can be burned to give an ash rich in iodide salts. In addition, Chilean nitre ($NaNO_3$) contains iodine as calcium iodate, $Ca(IO_3)_2$. Elemental iodine, I_2, is a black solid (mp 113.6°C, bp 185.2°C), which readily sublimes to give a purple vapor. Iodine is readily liberated from iodide salts by oxidation.

Iodine is only slightly soluble in water (0.33 g/L at 25°C) but dissolves readily in a variety of organic solvents to give a wide range of colors at various wavelengths, λ, including:

1. Bright violet in CCl_4 and saturated hydrocarbons ($\lambda_{max} = 520$–540 nm).
2. Pink-red aromatic hydrocarbons ($\lambda_{max} = 490$–510 nm).
3. Deep brown in alcohols and amines ($\lambda_{max} = 460$–480 nm).

The reasons for these colors can be related to the bonding in the I_2 molecule, which can be represented by $1\sigma_g^2 1\sigma_u^2 \pi_u^4 2\sigma_g^2 \pi_g^4$ with an empty $2\sigma_u$ antibonding orbital. The violet color of gaseous iodine arises from an allowed $\pi_g \rightarrow 2\sigma_u$ transition; complexation with the solvent increases this energy separation with an obvious effect on the color. Some charge-transfer complexes with relatively strong donors can be isolated and their structures determined by X-ray crystallography. Thus the trimethylamine adduct $Me_3N \rightarrow I_2$ has a linear $N \rightarrow I$—I unit with an $N \rightarrow I$ distance of 2.27 Å and an I—I distance of 2.83 Å;

the latter distance is longer than the 2.66 Å I—I distance in I_2, in accord with electron donation from the trimethylamine nitrogen lone pair into the $2\sigma_u$ antibonding orbital. Reactions of I_2 with sufficiently strong Lewis bases can lead to decomposition of the original adduct through disproportionation reactions, for example, (C_5H_5N = pyridine):

$$C_5H_5N + I_2 \rightarrow C_5H_5N{\rightarrow}I_2 \tag{7.6a}$$

$$2C_5H_5N{\rightarrow}I_2 \rightarrow [(C_5H_5N)_2I]^+I_3^- \tag{7.6b}$$

The intense blue complex formed by iodine with the amylose form of starch is related to these charge-transfer complexes; its structure contains a linear array of I_5^- ($=I_2\cdots I^-\cdots I_2$) repeating units inside the amylose helix.[3]

7.2.5 Astatine[4]

Astatine has no stable isotopes; the longest lived isotopes are ^{210}At and ^{211}At, with half-lives of 8.3 and 7.2 hours respectively. The isotope ^{211}At is obtained by bombarding ^{209}Bi with α-particles (helium-4 nuclei) in a large cyclotron. Since the short half-life of astatine presents the accumulation of macroscopic samples, astatine chemistry can be investigated only by tracer methods, typically in the concentration range 10^{-11}–$10^{-15}M$. However, a few astatine compounds, such as HAt, CH_3At, AtCl, AtBr, and AtI, have been detected by mass spectrometry.

7.2.6 Noble Gases

The noble gases occur as minor constituents of the atmosphere: $\sim 0.0005\%$ for helium, $\sim 0.002\%$ for neon, $\sim 0.9\%$ for argon, $\sim 0.001\%$ for krypton, and $\sim 0.000\,009\%$ for xenon. In addition, helium is found up to $\sim 7\%$ in certain US natural gases as a result of α-particles obtained in the radioactive decay of the heaviest elements, notably uranium and thorium. All isotopes of radon are radioactive; sometimes different radon isotopes are given specific names reflecting their origin from different radioactive decay series (e.g., "actinon" and "thoron," indicating formation from actinium and thorium, respectively). Radon is normally collected over radium salt solutions as a decay product.

7.3 Halides

Except for the noble gases helium, neon, and argon, elements of the periodic table form some *halides* (i.e., binary compounds with the halogens). Specific halides of the elements discussed in this book are treated in the chapters on the individual elements. Some general features of halide chemistry are discussed in Sections 7.3.1–7.3.6. The following methods are useful for the preparation of anhydrous halides.

7.3.1 Halogenation of the Elements

Anhydrous transition metal chlorides, bromides, and iodides can be made by dry reactions of the free metals with Cl_2, Br_2, and I_2, respectively, at elevated temperatures. Ether solvates of many transition metal halides can be made by halogenating the metal in an ether solvent such as diethyl ether, tetrahydrofuran, or 1,2-dimethoxyethane; halogenations in such ethers often occur at ambient conditions. The oxidation state of the metal in the halide obtained from halogenation of the metal increases in the series $I_2 < Br_2 < Cl_2 < F_2$ in accord with the oxidizing power of the free halogen. Fluorination of transition metals can often give unusually high oxidation states, particularly in the case of the later transition metals where platinum(VI) is known in PtF_6, gold(V) in AuF_5, and so on.

7.3.2 Halogen Exchange Reactions

Halogen exchange reactions often are used to obtain metal fluorides from the corresponding chlorides using reagents such as SbF_3, SbF_3Cl_2, ZnF_2, AsF_3, alkali metal fluorides, and KSO_2F. Excess alkali metal halides in an appropriately inert polar solvent are often useful for converting chlorides to the corresponding bromides or iodides when the system does not complex with excess halide ion. Boron trichloride is occasionally useful for the conversion of fluorides to the corresponding chlorides in systems resistant to the high Lewis acidity of BCl_3.

7.3.3 Halogenation of Oxides with Halogen Compounds

Many oxides can be converted to the corresponding halides by treatment with appropriate halogen compounds, often at elevated temperatures, for example,

$$TiO_2 + 2CCl_4 \xrightarrow{\Delta} TiCl_4 + 2O{=}CCl_2 \tag{7.7a}$$

$$UO_2 + 2CCl_2{=}CClCCl_3 \xrightarrow{\Delta} UCl_4 + 2CCl_2{=}CClC(O)Cl \tag{7.7b}$$

$$La_2O_3 + 6NH_4Cl \xrightarrow{\Delta} 2LaCl_3 + 6NH_3 + 3H_2O \tag{7.7c}$$

7.3.4 Dehydration of Hydrated Halides

Many halides can be prepared as hydrates and then dehydrated by treatment with reagents that react with water, leaving no residue that is difficult to separate. Two reagents useful for this purpose are thionyl chloride (Section 6.6.4.4) and 2,2-dimethoxypropane, which react with water in hydrated halides as follows:

$$O{=}SCl_2 + H_2O \rightarrow 2HCl{\uparrow} + SO_2{\uparrow} \tag{7.8a}$$

$$CH_3C(OCH_3)_2CH_3 + H_2O \rightarrow (CH_3)_2C{=}O + 2CH_3OH \tag{7.8b}$$

The volatile by-products from these reactions, namely, HCl, SO_2, acetone, and methanol, are relatively easy to remove in most cases. Thionyl chloride is the more powerful of the two dehydrating agents, but obviously can be used only for chlorides, 2,2-Dimethoxypropane cannot be used to dehydrate halides that are solvolyzed by methanol.

7.3.5 Oxidative Fluorinations

Many of the more aggressive fluorinating agents use fluorine derivatives in relatively high oxidation states; this can lead to an increase in the oxidation state of the derivative undergoing fluorination. The oxidative fluorinating agents can be classified into the following types:

7.3.5.1 Halogen Fluorides

The fluorinating power (and oxidizing ability) of the commonly available halogen fluorides increases in the sequence $IF_5 < BrF_3 < ClF < IF_7 < BrF_5 < ClF_3$.

7.3.5.2 High Oxidation State Metal Fluorides

High oxidation state fluorides that have been used as oxidizing fluorinating agents include AgF_2, CoF_3, MnF_3, CeF_4, PbF_4, BiF_5, and UF_6; elemental fluorine is required for the efficient preparation of most of these reagents.

7.3.5.3 Xenon Fluorides

Xenon difluoride, XeF_2 (Section 7.7.1.1), is a strong fluorinating agent; it has the major advantage that it liberates gaseous elemental xenon upon release of its fluorine, leaving no solid or liquid by-products, thereby simplifying the separation of pure fluorination products.

7.3.6 Physical Characteristics of Halides

Halides in which the lattice consists of discrete ions rather than molecular units are considered to be ionic halides; by this criterion most metal halides are predominantly ionic. The ionic radius of the fluoride ion, F^-, is 1.19 Å, which is very close to the 1.26 Å radius of the oxide ion, O^{2-}. For this reason, monovalent metal fluorides and bivalent metal oxides have similar formulas and crystal structures.[5] Most metal halides are readily soluble in water *except* for the following.

1. Fluorides of the small highly electropositive ions lithium, alkaline earths, lanthanides, and actinides.
2. Chlorides, bromides, and iodides of the highly polarizable metals Cu^I, Ag^I, Au^I, Hg^I, Pb^{II}, and Tl^I in their lowest oxidation states.

Many metals exhibit their highest oxidation states in their hexafluorides MF_6 (M = U, Np, Pu, Mo, W, Re, Os, Ir, Pt). Such metal hexafluorides are gases or volatile liquids exhibiting covalent structures with octahedral metal coordination because of the unfavorable energetics of forming ionic structures with M^{6+} ions.

7.4 Halogen Oxides

Fluorine forms the binary compounds OF_2, O_2F_2, and O_4F_2 with oxygen. However, since fluorine is more electronegative than oxygen, these compounds are best regarded as *oxygen fluorides* rather than *fluorine oxides* and are therefore discussed in the oxygen chapter (Section 6.3.1).

7.4.1 Chlorine Oxides[6]

All chlorine oxides are highly endothermic, explosive compounds and therefore cannot be formed by direct reaction between Cl_2 and O_2. There is a crude analogy between the chlorine oxides ClO_n and Cl_2O_m and the nitrogen oxides NO_{n-1} and N_2O_{m-2}, respectively, except that the chlorine oxides are far less stable. Thus ClO_2, $ClO_2 \rightleftarrows Cl_2O_6$, and Cl_2O_7 are analogues of NO, $NO_2 \rightleftarrows N_2O_4$, and N_2O_5, respectively.

7.4.1.1 Dichlorine Monoxide, Cl_2O

Dichlorine monoxide, Cl_2O, is a yellow-brown gas (mp $-116°C$, bp $2°C$); it decomposes readily explosively to $2Cl_2 + O_2$ upon mild heating or sparking but can be handled safely in 1 M carbon tetrachloride solution.[7] Dichlorine monoxide can be prepared by the reaction of chlorine with yellow mercuric oxide:

$$2Cl_2 + 2HgO \rightarrow HgO \cdot HgCl_2 + Cl_2O \qquad (7.9)$$

The structure of Cl_2O is bent with a Cl–O–Cl angle of $111°$ and a Cl–O bond length of 1.71 Å. Dichlorine monoxide is readily soluble in water, in which it forms hypochlorous acid (Section 7.5.1) according to the following equilibrium:

$$Cl_2 + H_2O \rightleftarrows 2HOCl \qquad (7.10)$$

A solution of Cl_2O in carbon tetrachloride is a powerful chlorinating agent for either the side chain or nuclear chlorination of benzenoid compounds; its reactivity toward nuclear chlorination of benzenoids can be enhanced by the addition of a strong acid such as CF_3CO_2H.[8]

7.4.1.2 Chlorine Dioxide, ClO_2

Chlorine dioxide, ClO_2 (mp $-59°C$, np $11°C$), is a highly reactive and explosive yellow gas, which can be made safe to handle by dilution with air to partial pressures of ClO_2 < 50 mm Hg.[9] Chlorine dioxide is best synthesized from a chlorate salt by treatment with a mildly reducing acid such as SO_2/H_2SO_4 or oxalic acid, for example,

$$2KClO_3 + 2H_2C_2O_4 \rightarrow 2ClO_2\uparrow + 2CO_2\uparrow + K_2C_2O_4 + 2H_2O \qquad (7.11)$$

The use of oxalic acid has the advantage that the hazardous ClO_2 is diluted with the safe and relatively unreactive CO_2 by-product; however, SO_2/H_2SO_4 is a less expensive reagent for large-scale preparations. Chlorine dioxide has a bent structure, with a bond angle O–Cl–O of 118°C and a bond length Cl—O of 1.47 Å. Although ClO_2 has un unpaired electron, it has no tendency to dimerize; in this respect it resembles NO more than NO_2 (Section 4.5). Chlorine dioxide disproportionates in aqueous alkali to give the chlorite (ClO_2^-) and chlorate (ClO_3^-) ions:

$$2ClO_2 + 2OH^- \rightarrow ClO_2^- + ClO_3^- + H_2O \qquad (7.12)$$

Chlorine dioxide is manufactured on a massive scale for the bleaching of wood pulp for water treatment.[10]

7.4.1.3 Dichlorine Hexoxide, Cl_2O_6

Dichlorine hexoxide, Cl_2O_6, is an unstable red liquid (mp 3.5°C). which is obtained by reaction of ClO_2 with ozone. In the solid state it appears to be chloryl perchlorate, $ClO_2^+ClO_4^-$; it reacts with water to form a mixture of chlorate and perchlorate ions:

$$Cl_2O_6 + 3H_2O \rightarrow 2H_3O^+ + ClO_3^- + ClO_4^- \qquad (7.13)$$

7.4.1.4 Dichlorine Heptoxide, Cl_2O_7

Dichlorine heptoxide, Cl_2O_7 (mp -91.5°C, bp 82°C) is the most stable chlorine oxide, but nevertheless is explosive. It is the anhydride of perchloric acid, $HClO_4$, and thus can be made by the dehydration of perchloric acid with P_4O_{10} at -10°C followed by careful distillation at -35°C and 1 mm Hg. Reaction of Cl_2O_7 with water regenerates perchloric acid. Reactions of Cl_2O_7 with alcohols, ROH, give the dangerously explosive alkyl perchlorates, $ROClO_3$.

7.4.2 Bromine Oxides

Bromine oxides are much less stable than even chlorine oxides, and until relatively recently it was believed that they could not exist. No bromine oxides are stable at room temperature. The oxide BrO_2, a yellow solid at low temperatures, can be obtained by ozonation of bromine in $CFCl_3$ at -50°C; it is thermally unstable above -40°C and decomposes violently at ~ 0°C. The oxide Br_2O, a volatile brown solid (mp -17.5°C), can be obtained by the thermal decomposition of BrO_2 in vacuum at low temperatures or by the reaction of Br_2 with yellow HgO analogous to the preparation of Cl_2O (equation 7.9); it decomposes above $-60°C$ and appears to have a symmetrical bent structure analogous to that of Cl_2O.

7.4.3 Iodine Oxides

Iodine oxides are *much* more stable and have formulas and structures totally different from the oxides of chlorine and bromine. The much greater stability of iodine oxides can be at least partially attributed to the lower electronegativity of iodine relative to chlorine and bromine, leading to strong I—O bonds because of the greater electronegativity difference.

The most important iodine oxide is "iodine pentoxide," I_2O_5,[11] a white crystalline solid, obtained by heating iodic acid, HIO_3 (see equation 7.14) to 240°C. In contrast to oxides of chlorine and bromine iodine pentoxide is stable to 300°C. Iodine pentoxide functions as the anhydride of iodic acid and regenerates iodic acid upon treatment with water, in which it is very soluble. Iodine pentoxide has a polymeric structure containing O_2I—O—IO_2 units built into a three-dimensional network through strong I$\cdots$O interactions. The most important reaction of I_2O_5 is its ability to oxidize H_2S, HCl, and CO under mild conditions. The reaction of I_2O_5 with CO proceeds rapidly and quantitatively at room temperature according to the equation:

$$5CO + I_2O_5 \rightarrow I_2 + 5CO_2 \tag{7.14}$$

This reaction provides a basis for the iodometric determination of CO. Combined with the pyrolysis of oxygen-containing organic compounds at high temperatures to give CO quantitatively, reaction 7.14 provides a basis for the Unterzaucher direct determination of oxygen in organic compounds.[12]

Other far less important iodine oxides include yellow $I_2O_4 = IO^-IO_3^-$, obtained by partial hydrolysis of $(IO)_2SO_4$, and yellow $I_4O_9 = I(IO_3)_3$, obtained by ozonation of elemental iodine. These iodine oxides have polymeric structures in which IO_3 groups cross-link polymeric I—O—I—O— chains in I_2O_4 and iodine atoms in $I(IO_3)_3$.

7.5 Halogen Oxoacids and Anions

Halogens form oxoacids of the following general types:

Hypophalous acids, HOX, forming the *hypohalite* anions, XO^-, in which the halogen has the +1 formal oxidation state.

Halous acids, HXO_2, forming the *halite* anions, XO_2^-, in which the halogen has the +3 formal oxidation state.

Halic acids, HXO_3, forming the *halate* anions, XO_3^-, in which the halogen has the +5 formal oxidation state.

Perhalic acids, HXO_4, forming the *perhalate* anions, XO_4^-, in which the halogen has the +7 formal oxidation state.

The halogen oxoacids are more or less powerful oxidizing agents depending on the conditions in view of the necessarily positive halogen formal oxidation states; their ultimate reduction products are the corresponding halide ions.

The ultimate source of essentially all the halogen oxoacids and their salts is the disproportionation of the free halogen in base, which initially forms hypohalites according to the following general equation:

$$X_2 + 2OH^- \rightarrow X^- + XO^- + H_2O \tag{7.15}$$

Halates, XO_3^-, are obtained by further disproportionation of the hypohalites (see Section 7.5.1). In general, the redox chemistry of halogen oxoacids, their anions, and related halogen oxides is rather complicated. Chemical systems exhibiting unusual time-dependent behavior such as oscillations, chaos, and traveling waves frequently depend on the complicated redox behavior of halogen oxoacids and their salts.[13]

7.5.1 Hypohalous Acids and Hypohalites

Hypofluorous acid, HOF (mp $-117°C$), is a gas at ambient temperature; it is obtained by the reaction of fluorine with water at $-40°C$. Hypofluorous acid decomposes spontaneously at room temperature to HF and O_2 with a half-life of 30 minutes at $25°C$. The H–O–F angle in HOF is $97°C$, which is the smallest known angle around oxygen for an unconstrained oxygen atom. Salts of hypofluorous acid are unknown. However, hypofluorous acid is the parent of a small series of covalent compounds containing —OF groups (CF_3OF, SF_5OF, O_3ClOF, FSO_2OF, etc.).

The other hypohalous acids are generated in aqueous solution by treatment of a suspension of mercuric oxide with the free halogen:

$$2X_2 + 2HgO + H_2O \rightarrow HgO \cdot HgX_2 + 2HOX \tag{7.16}$$

Hypohalous acids are weak acids with pK values of 7.5, 8.7, and 11.0 for HOCl, HOBr, and HOI, respectively. Hypohalites are generated by reactions of halogens with cold base (equation (7.15); such reactions are rapid, with favorable equilibrium constants. The most stable hypochlorites are those of lithium, calcium, strontium, and barium. Hypochlorites are used widely as bleaches and water disinfectants (e.g., swimming pools).

The most important reaction of hypohalites is their further disproportionation into the corresponding halates:

$$3XO^- \rightarrow XO_3^- + 2X^- \tag{7.17}$$

The hypochlorite ion disproportionates very slowly at or below room temperature, with the result that treatment of Cl_2 with aqueous alkali in the cold gives reasonably pure solutions of $Cl^- + ClO^-$. However, rapid disproportionation occurs at $\sim 75°C$ to give good yields of ClO_3^- (equation 7.17: X = Cl), although only one-sixth of the original Cl_2 is converted to ClO_3^-. The hypobromite ion disproportionates moderately fast at room temperature to bromate and bromide (equation 7.17: $X = Br$), which means that solutions of hypobromite can be made only at $0°C$. The hypoiodite ion, which disproportionates very rapidly at all temperatures, is unknown in solution. Thus the conversion of iodine to iodide and iodate is quantitative according to the following equation:

$$3I_2 + 6OH^- \rightarrow 5I^- + 3H_2O + IO_3^- \tag{7.18}$$

7.5.2 Halous Acids and Halites

The halous acids, HXO_2, do not arise directly from hydrolytic disproportionation of the halogens; thus the following reaction has an unfavorable equilibrium constant:

$$2HOCl \rightleftharpoons Cl^- + H^+ + HOClO \qquad K \approx 10^{-5} \tag{7.19}$$

Chlorites can be made by disproportionation of ClO_2 in alkali to give equal amounts of ClO_3^- and ClO_2^- (equation 7.12), directly from chlorates by reduction with oxalic acid, or by reduction of ClO_2 with hydrogen peroxide:

$$2ClO_2 + H_2O_2 \rightarrow 2HClO_2 + O_2\uparrow \tag{7.20}$$

Sodium chlorite, $NaClO_2$, is prepared by the ton for bleaching and stripping of textiles. Aqueous solutions of $NaClO_2$ are strongly oxidizing and dry $NaClO_2$ forms explosive mixtures with oxidizable materials. Heavy metal chlorites (e.g., those of Ag^+ Hg_2^{2+}, Tl^+, Pb^{2+}) are explosive solids. The chlorite ion is stable to prolonged boiling in alkaline solution but rapidly decomposes to ClO_2 in acid solution. Solutions of chlorous acid can be obtained by treatment of $Ba(ClO_2)_2$ with sulfuric acid:

$$Ba(ClO_2)_2 + H_2SO_4 \rightarrow BaSO_4\downarrow + 2HClO_2 \tag{7.21}$$

However, attempts to isolate chlorous acid by concentration leads to disproportionation according to the following equation:

$$5HClO_2 \rightarrow 4ClO_2 + Cl^- + H_3O^+ + H_2O \tag{7.22}$$

Chlorous acid is the least stable of the four chlorine oxoacids $HClO_n$ ($1 \le n \le 4$) and is a moderately strong acid ($K^a = 1.1 \times 10^{-2}$ at $25°C$).

Bromites are known and have properties similar to those of the chlorites. Thus crystalline barium bromite, $Ba(BrO_2)_2 \cdot H_2O$, can be made by treatment of barium hypobromite with Br_2 and pH 11.2 and $0°C$. Iodites are unknown.

7.5.3 Halic Acids and Halates: Ingredients for Chemical Oscillators

Halates, XO_3^- ($X = Cl, Br, I$) can be made by reactions of the free halogens with base as already noted (see, e.g., equation 7.18). Such methods have the disadvantage that only one-sixth of the halogen at most is converted to halate. This difficulty is avoided by using the electrolytic oxidation of the corresponding halides for the syntheses of halates. For example, $NaClO_3$ is manufactured on a large scale by the electrolysis of NaCl in a diaphragmless cell, which promotes efficient mixing of the Cl_2 produced at the anode with the OH^- produced at the cathode. The halates are good oxidizing agents: $KClO_3$ is used for the manufacture of matches and explosives.

Iodates and bromates participate in reactions with complicated time-dependent behavior. For example, chemical oscillations can be obtained from the Belousov–Zhabotinskii reaction,[14] which uses a mixture of potassium bromate, an organic compound preferably with enolizable hydrogens such as malonic acid ($CH_2(CO_2H)_2$), and a strong one-electron oxidizing agent such as Ce^{4+} or $[Fe^{III}(2,2'-bipyridyl)_3]^{3+}$. If the iron(III) complex is used as the one-electron oxidizing agent, the color of the Belousov–Zhabotinskii mixture using appropriate concentrations of reactants may oscillate between the dark red of $[Fe^{II}(2,2'-bipyridyl)_3]^{2+}$ and the lighter blue of $[Fe^{III}(2,2'-bipyridyl)_3]^{3+}$ for more than an hour, with periods that can be varied from a few seconds to a few minutes. Concurrent oscillations in other solution parameters such as bromide concentration and pH can be monitored by the use of suitable electrodes. Determination of the mechanisms of such oscillating chemical reactions has proven to be a very complicated problem, the discusssion of which is beyond the scope of this book.

The only halic acid that can be obtained in the pure state is iodic acid, HIO_3, a white solid obtained by the oxidation of iodine with concentrated nitric acid. The structure of iodic acid consists of pyramidal $HOIO_2$ molecules connected by hydrogen bonds.

7.5.4 Perhalic Acids and Perhalates

Perchlorate salts containing the ClO_4^- anion are the most stable oxo compounds of chlorine. Perchlorates are generally obtained by electrolytic oxidation of the corresponding chlorates. Perchlorates of most metals are known.[15] All perchlorates are soluble in water except for a few salts with large monovalent cations of low charge density, $M^+ClO_4^-$, including perchlorates of the heavier alkali metals. The perchlorate ion is thermodynamically a good oxidant in aqueous solution:

$$ClO_4^- + 2H^+ + 2e^- = ClO_3^- + H_2O \qquad E° = +1.23 \text{ V} \qquad (7.23)$$

However, oxidations by perchlorate are often kinetically unfavorable. Nevertheless, Ru(II) will reduce perchlorate to chlorate, and V(II), V(III), Mo(III), and Ti(III) will reduce perchlorate to chloride in aqueous solution. Strong reducing agents that are kinetically incapable of reducing perchlorate include Eu(II) and Cr(II), possibly because these compounds are incapable of forming the M=O derivatives required as intermediates.[16]

Perbromates are very difficult to prepare, since it is *much* more difficult to oxidize bromate to perbromate than chlorate to perchlorate: compare equation 7.23 to the following equation:

$$BrO_4^- + 2H^+ + 2e^- = BrO_3^- + H_2O \qquad E° = +1.76 \text{ V} \qquad (7.24)$$

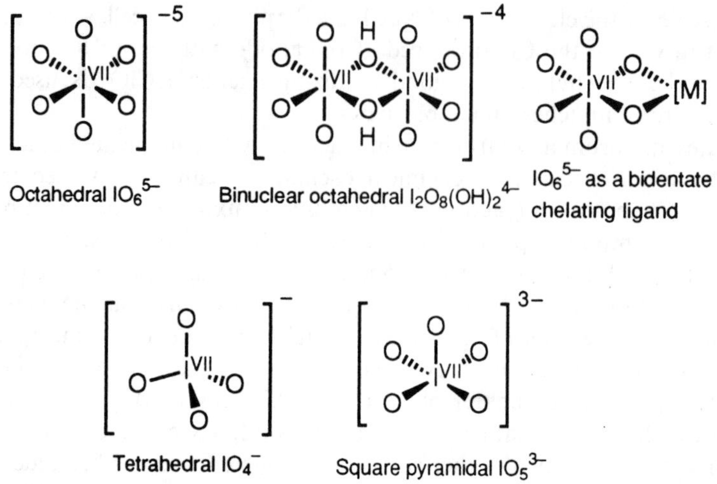

Figure 7.1. The periodate ions and the complexation of IO_6^{5-} to metals [M] as a bidentate chelating agent.

This is another consequence of the relative instability of the highest oxidation states of the elements of the As, Se, Br row of the periodic table relative to those in the rows above (P, S, Cl) or below (Sb, Te, I)—compare Section 5.52 discussing $AsCl_5$. For many years after the discovery of all the other perhalic acids, the perbromate ion was believed to be incapable of existing as a stable species.[17] Finally, it was discovered that the bromate ion could be oxidized to the perbromate ion by the very strongest known oxidizing agents; the best choice of such an oxidizing agent is elemental fluorine in 5 M NaOH solution, but XeF_2 and electrolytic methods also work.[18] Perbromic acid, $HBrO_4$, is stable indefinitely to 100°C in 6 M aqueous solution and is a very strong oxidant, although dilute perbromic acid is a very sluggish oxidant at 25°C.[19]

The chemistry of the *periodates* is more complicated, since ions with either tetrahedral (IO_4^-), square pyramidal (IO_5^{3-}), or octahedral (IO_6^{5-} and $(HO)_2I_2O_8^{4-}$) iodine coordination are known, whereas in the other perhalates only tetrahedral halogen coordination is known (Figure 7.1). Periodates are obtained by the oxidation of iodate, iodine, or iodide either electrolytically or with chlorine in strongly basic solution. Stable free periodic acids include H_5IO_6 ("orthoperiodic acid"), which undergoes dehydration upon heating at 100°C in vacuum to give HIO_4 ("metaperiodic acid"). Further attempted dehydration of HIO_4 to I_2O_7 results in oxygen evolution to give I_2O_5. Treatment of periodic acid with concentrated sulfuric acid gives colorless crystalline $[I(OH)_6][HSO_4]$ containing the iodine(VII) *cation* $I(OH)_6^+$, which is isoelectronic and isostructural with the species $Te(OH)_6$ (Section 6.8.8) and $Sb(OH)_6^-$ (section 5.9.2); the salts $I(OH)_6^+HSO_4^-$ and $[I(OH)_6^+]_2SO_4^{2-}$ can be isolated from these reaction mixtures.[20] The IO_6^{5-} is an excellent bidentate chelating ligand

(Figure 7.1) for stabilizing unusually high metal oxidation states such as d^8 Cu(III) and Ag(III) in square planar complexes such as $Na_3K[H_3Cu(IO_6)_2] \cdot 14H_2O$ and Mn(IV) and Ni(IV) in octahedral complexes such as $Na_7[H_4Mn(IO_6)_3] \cdot 17H_2O$. A factor in the ability of IO_6^{5-} to stabilize high metal oxidation states undoubtedly is its high (-5) negative charge. Periodates are useful as strong and generally rapid oxidizing agents. They thus can oxidize Mn(II) to permanganate, as well as many organic compounds, rapidly and quantitatively. Metaperiodates, $M^+IO_4^-$ (M = Na, K), are useful oxidizing agents for converting sulfides, R_2S, to the corresponding sulfoxides, $R_2S{=}O$, without overoxidation to the corresponding sulfones, R_2SO_2.

7.5 Interhalogen and Polyhalogen Compounds, Cations, and Anions

Interhalogen compounds contain two or more different halogens and no other elements. Polyhalogen cations and anions contain only a single type of halogen and have the general formulas X_n^+ and X_n^-, respectively. These species can be classified into the following types:

1. *Neutral molecules* are binary compounds of the type XX'_{2n+1}; they are closed-shell volatile diamagnetic species. The diatomic molecules ClF, BrF, BrCl, IBr, and ICl have properties intermediate between those of the constituent halogens.
2. *Cations* include I_2^+, Cl_3^+, $I_3Cl_2^+$, and $IBrCl^+$; they may be either diamagnetic closed-shell ions or paramagnetic ions with an unpaired electron.
3. *Anions* include the triiodide anion, I_3^-, which is a well-known constituent of solutions of iodine in aqueous iodide. Other binary and ternary combinations include $IBrCl^-$ and $IBrCl_3^-$. The anions are generally obtained as crystalline salts with large cations.

7.5.1 Halogen Fluoride Molecules: Fluoridic Nonaqueous Solvents

Halogen fluorides are obtained by the fluorination of heavier halogens or their compounds under suitable conditions, as illustrated by the following reactions for the preparations and interconversion of chlorine fluorides:

$$Cl_2 + F_2 \xrightarrow{220-250°C} ClF \qquad \text{(mp } -156°C, \text{ bp } -100°C) \qquad (7.25a)$$

$$Cl_2 + 3F_2 \rightarrow 2ClF_3 \qquad \text{(mp } -76°C, \text{ bp } +11.8°C) \qquad (7.25b)$$

$$Cl_2 + ClF_3 \xrightarrow{250-350°C} 3ClF \qquad (7.25c)$$

$$ClF_3 + F_2 \rightarrow ClF_5 \qquad \text{(bp } -14°C) \qquad (7.25d)$$

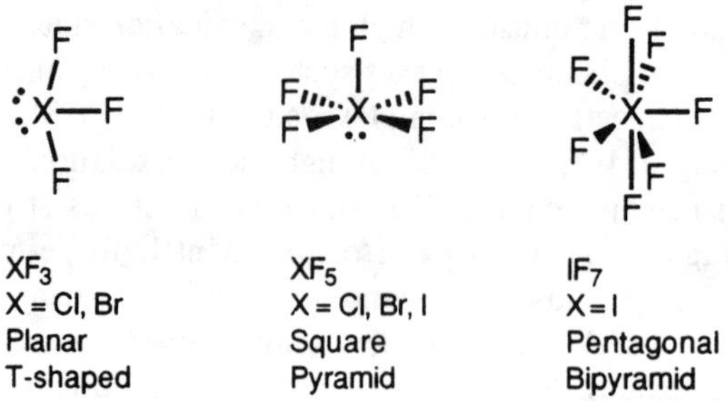

XF$_3$
X = Cl, Br
Planar
T-shaped

XF$_5$
X = Cl, Br, I
Square
Pyramid

IF$_7$
X = I
Pentagonal
Bipyramid

Figure 7.2. Structures of the halogen fluorides XF$_3$, XF$_5$, and XF$_7$ showing the stereochemically active lone pairs in XF$_3$ and XF$_5$.

$$KCl + 3F_2 \xrightarrow[\text{bomb}]{200°C} KF + ClF_5 \tag{7.25e}$$

$$ClF_5 \xrightarrow{>165°C} ClF_3 + F_2 \tag{7.25f}$$

The bromine fluorides BrF (mp $-33°C$, bp $20°C$), BrF$_3$ (mp $9°C$, bp $126°C$), and BrF$_5$ (mp $-60°C$, bp $+41°C$), as well as the iodine fluorides IF$_5$ (mp $10°C$, bp $101°C$) and IF$_7$ (subl $5°C/1$ atm) are prepared analogously by the fluorination of Br$_2$, I$_2$, or KI.

The F/X ratios in the stable halogen fluorides increase as the atomic weight of X increases. Thus ClF$_7$ and BrF$_7$ cannot be prepared, where IF$_7$ is stable. Similarly IF$_3$, although it can be obtained as a yellow powder by the fluorination of I$_2$ in Freon at $-78°C$, is unstable above $-35°C$ with respect to disproportionation into IF$_5$ and I$_2$, whereas ClF$_3$ and BrF$_3$ are stable. The structures of the halogen fluorides (Figure 7.2) of the types XF$_3$ and XF$_5$ exhibit a stereochemically active lone pair.

Halogen fluorides are very reactive toward water or organic substances. For example, ClF$_3$ spontaneously ignites asbestos, wood, and other building materials and was used for incendiary bombs during the Second World War. Halogen fluorides are also very powerful fluorinating agents, with fluorinating ability decreasing in the sequence ClF$_3$ > BrF$_3$ > BrF$_5$ > IF$_7$ > ClF > IF$_5$ > BrF. Thus the very reactive ClF$_3$ converts most chlorides to fluorides and reacts even with refractory oxides such as MgO, CaO, Al$_2$O$_3$, MnO$_2$, Ta$_2$O$_5$, and MoO$_3$ to form the corresponding fluorides.

Some of the lower halogen fluorides, namely ClF, BrF$_3$, and IF$_5$, have unusually high entropies of vaporization, which can be attributed to association through fluorine bridging. In addition, BrF$_3$ has an unusually high conductivity for a molecular compound: $>8 \times 10^{-3}$ $(\Omega \cdot cm)^{-1}$ at $25°C$ in contrast to 3.9×10^{-9} $(\Omega \cdot cm)^{-1}$ and 9.1×10^{-8} for ClF$_3$ and BrF$_5$, respectively. This relatively high conductivity for BrF$_3$ has been attributed to self-ionization by fluoride transfer according to the following general scheme:

$$2BrF_3 \rightleftarrows BrF_2^+ + BrF_4^- \tag{7.26}$$

Bromine trifluoride is thus an ionizing solvent undergoing fluoride transfer analogous to ionizing solvents such as water or liquid ammonia undergoing proton transfer. The term *fluoridic solvent* can be used to describe BrF_3 and other solvents ionizing by fluoride transfer just as the term *protic solvent* is used to describe solvents ionizing by proton transfer (Section 1.4). Continuing this analogy, sources of BrF_2^+, such $BrF_2^+SbF_6^-$ obtained by reaction of BrF_3 with the Lewis acid SbF_5, function as "acids" in liquid BrF_3, and sources of BrF_4^-, such as $K^+BrF_4^-$ obtained by reaction of BrF_3 with KF, function as "bases" in liquid BrF_3.

The ability of BrF_3 to form both the adduct $BrF_3 \cdot SbF_5 = BrF_2^+SbF_6^-$ with the Lewis acid SbF_5 and the adduct $KF \cdot BrF_3 = K^+BrF_4^-$ with the Lewis base F^- indicates the amphoteric nature of BrF_3. Several other halogen fluorides such as ClF and ClF_3 are amphoteric, as indicated by the following reactions:

$$2ClF + AsF_5 \rightarrow Cl_2F^+AsF_6^- \tag{7.27a}$$

$$ClF + CsF \rightarrow Cs^+ClF_2^- \tag{7.27b}$$

$$ClF_3 + AsF_5 \rightarrow ClF_2^+AsF_6^- \tag{7.28a}$$

$$ClF_3 + KF \rightarrow K^+ClF_4^- \tag{7.28b}$$

7.5.2 The X_n^+ and Related Interhalogen Cations

7.5.2.1 The I_2^+ and Related X_2^+ Cations

It has long been known that elemental iodine dissolves in fuming sulfuric acid and other oxidizing strong acids to give bright blue paramagnetic solutions; the source of the blue color has been shown to be the I_2^+ cation. A solid derivative of the I_2^+ cation can be obtained by oxidation of I_2 with SbF_5 according to the following equation[21]:

$$2I_2 + 5SbF_5 \rightarrow 2I_2^+Sb_2F_{11}^- + SbF_3 \tag{7.29}$$

The I—I distance of 2.67 Å in neutral I_2 decreases to 2.56 Å in I_2^+ (Figure 7.3), in accord with the removal of the antibonding electron upon oxidation corresponding to a bond order of 1.5 in I_2^+. Solutions of the I_2^+ cation in HSO_3F dimerize upon cooling below $-60°C$ to give the brown diamagnetic I_4^{2+} dication with a rectangular structure (Fig. 7.3). A related Br_2^+ cation is known as a $Br_2^+Sb_3F_{16}^-$ salt.

7.5.2.2 Other X_n^+ Cations

Several *diamagnetic* cations of the type X_n^+ ($n = odd$ number) are known, such as X_3^+ (Cl, Br, I), Br_5^+, and I_5^+; these are obtained by oxidizing the free halogen with acidic oxidizing agents such as $ClF_2^+AsF_6^-$, AsF_5, and $BrSO_3F$. The structures of these cations are depicted in Figure 7.3.

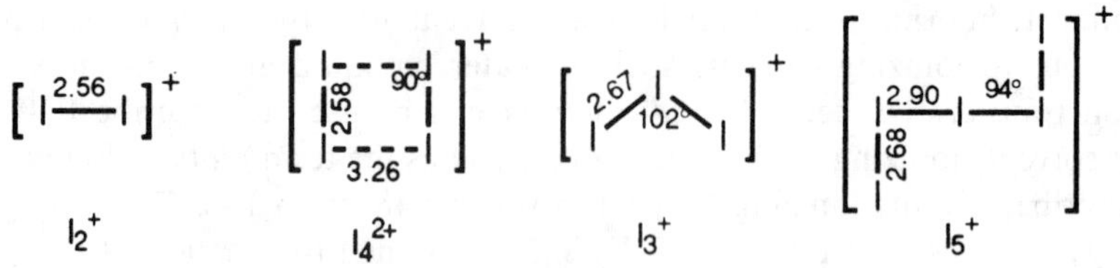

Figure 7.3. Structures of some homoatomic iodine cations; distances are given in angstrom units.

7.5.2.3 Interhalogen Cations

Some interhalogen cations can be obtained from an interhalogen molecule and a halogen acceptor, as illustrated by the preparation of $ClF_2^+AsF_6^-$ in equation 7.28a. Since ClF_7 and BrF_7 are not available, a strongly fluorinating Lewis acid is required to prepare the octahedral cations ClF_6^+ and BrF_6^+, for example,

$$2ClF_5 + 2PtF_6 \rightarrow ClF_6^-PtF_6^- + ClF_4^+PtF_6^- \tag{7.30a}$$

$$BrF_5 + KrF^+AsF_6^- \rightarrow BrF_6^+AsF_6^+ + Kr\uparrow \tag{7.30b}$$

In all known structures of these interhalogen cation salts, the halogenated anions make very close contacts with the interhalogen cations through halogen bridges.[22]

7.5.3 Polyhalide Anions

7.5.3.1 Polyiodide Anions

The triodide anion, I_3^-, is the only polyiodide anion of importance in aqueous chemistry. Formation of I_3^- by the following equilibrium accounts for the greater solubility of elemental iodine in aqueous potassium iodide than in pure water:

$$I_2 + I^- \rightleftarrows I_3^- \qquad K_{eq} = \sim 700 \text{ at } 20°C$$

The black solid salt $KI_3 \cdot H_2O$ can be isolated from such solutions. The triiodide ion has a linear structure, which is symmetrical in solution but may be unsymmetrical in crystalline solids when not required by the crystal symmetry. A few larger polyiodide anions are known, including I_5^-, I_9^-, I_4^{2-}, and I_8^{2-}; these can be isolated as salts with large cations.

7.5.3.2 Other Polyhalide Anions

A variety of polyhalide anions of the general type $X_kY_mZ_n^-$ are known, where $k + m + n$ is an odd number, these anions are made by reactions of halides with an appropriate free halogen or interhalogen. The central atom is always the halogen with the highest atomic number. The triatomic ions are linear and the pentatomic ions are square planar.

7.6 Other Halogen Compounds

7.6.1 Oxohalogen Fluorides Including Perchloryl Fluoride

Oxahalogen fluorides contain fluorine, oxygen, and a heavier halogen (X) with both X—F and X—O bonds; examples are octahedral $O{=}IF_5$ and ψ-trigonal bipyramidal $O{=}BrF_3$ with a stereochemically active lone pair in an equatorial position. The most important oxohalogen fluoride is *perchloryl fluoride*, $FClO_3$ (mp $-147.8°C$, bp $-46.7°C$), which is a toxic gas synthesized by action of F_2 or FSO_3H on $KClO_4$ or by the following reaction:

$$KClO_4 + 2HF + SbF_5 \xrightarrow{40-50°C} FClO_3{\uparrow} + KSbF_6 + H_2O \tag{7.31}$$

Perchloryl fluoride, which may be regarded as the acid fluoride of perchloric acid, is thermally stable to 500°C and resists hydrolysis, like SF_6 (Section 6.6.1). Perchloryl fluoride is a powerful oxidizing agent and a selective fluorinating agent. It can be used to introduce the *perchloryl group* ($—ClO_3$) into organic compounds. Thus reaction of perchloryl fluoride with benzene in the presence of a Lewis acid ("Friedel–Crafts type") catalyst such as BF_3 gives perchloryl-benzene, $C_6H_5ClO_3$, with a direct bond from a phenyl carbon to the perchloryl chlorine; such perchlorylarenes are liquids stable under ambient conditions.

7.6.2 Miscellaneous Compounds with Oxygen–Halogen Bonds

7.6.2.1 Halogen Perchlorates

Halogen perchlorates, $XOClO_3$, are known for all four halogens (X = F, Cl, Br, I), although iodine perchlorate is not well characterized and $BrOClO_3$ and $ClOClO_3$ are unstable and shock-sensitive. Pure fluorine perchlorate, $FOClO_3$ (mp $-167°C$, bp 16°C), can be made by the pyrolysis of tetrafluoroammonium perchlorate (Section 4.7.1)[23]:

$$NF_4^+ClO_4^- \xrightarrow{\Delta} NF_3 + FOClO_3 \tag{7.32}$$

Fluorine perchlorate is useful for the synthesis of anhydrous metal perchlorates. Also, addition of fluorine perchlorate across the $C{=}C$ double bond of fluoroolefins provides a route to perfluoroalkyl perchlorates, for example,

$$CF_2{=}CF_2 + FOClO_3 \rightarrow CF_3CF_2OClO_3 \tag{7.33}$$

7.6.2.2 Halogen Fluorosulfates

Halogen fluorosulfates, $XOSO_2F$, like halogen perchlorates, are known for all four halogens ($X = F, Cl, Br, I$); they can be obtained by reaction of the elemental halogen, X_2, with $FSO_2O\!-\!OSO_2F$. Fluorine fluorosulfate can be made more easily by direct addition of fluorine to sulfur trioxide.

7.6.3 Trivalent Iodine Derivatives of Oxoacids

A number of iodine(III) derivatives of oxoanions with covalent I—O bonds can be synthesized from elemental iodine by reactions such as the following:

$$2I_2 + 3AgClO_4 \xrightarrow{\text{Et}_2\text{O}} 3AgI\downarrow + I(OClO_3)_3 \tag{7.34a}$$

$$I_2 + HNO_3 + (CH_3CO)_2O \rightarrow I(O_2CCH_3)_3 + \cdots \tag{7.34b}$$

$$I_2 + HNO_3 + H_3PO_4 \rightarrow IPO_4 + \cdots \tag{7.34c}$$

These derivatives are sensitive toward moisture and decompose only a little above room temperature. They hydrolyze with disproportionation of the iodine(III), for example,

$$5IPO_4 + 9H_2O \rightarrow I_2 + 3HIO_3 + 5H_3PO_4 \tag{7.35}$$

7.6.4 Organic Derivatives of Polyvalent Iodine

A number of organic derivatives of polyvalent iodine are known.[24] In general, aryl derivatives of polyvalent iodine are considerably more stable than corresponding alkyl derivatives. Organic derivatives of polyvalent iodine may be classified into the following pages:

1. Iodine(III) compounds with *one* carbon–iodine bond, including *aryliodine dihalides*, RIX_2, and *iodosylarenes*, RIO.
2. Iodine(III) compounds with two carbon–iodine bonds to the same iodine atom, such as *diaryliodonium salts*, $R_2I^+X^-$.
3. Iodine(III) compounds with three carbon–iodine bonds to the same iodine atom, such as *triphenyliodine*, Ph_3I.
4. Iodine(V) compounds with one carbon–iodine bond, such as *iodoxyarenes*, RIO_2, and *iodosarene difluorides*, $RI(O)F_2$.
5. Iodine(V) compounds with two carbon–iodine bonds to the same iodine atom, such as *diphenyliodyl acetate*, $Ph_2I(O)OCOMe$.

Chlorination of aryl iodides results in formation of aryliodine dichlorides, $RICl_2$. Phenyliodine dichloride, $PhICl_2$, has been shown to have a T-shaped three-coordinate iodine atom based on a ψ-trigonal bipyramid with lone pairs in the two equatorial positions similar to the halogen trifluoride structures (Figure 7.2). Hydrolysis of aryliodine dichlorides in the presence of a mild base such as sodium carbonate leads to the corresponding iodosylarenes, RIO.

Treatment of arenes with iodosylarenes in the presence of H_2SO_4 gives diaryliodonium salts, $R_2I^+X^-$ in an electrophilic aromatic substitution reaction with the protonated iodosylarene, $RIOH^+$, as the electrophile:

$$RH + RIO + H_2SO_4 \rightarrow R_2I^+HSO_4^- + H_2O \tag{7.36}$$

Unsymmetrical diaryliodonium salts, $RR'I^+X^-$, can also be made by a procedure of this type. Arylation of diaryliodonium salts with aryllithium derivatives at low temperatures gives yellow triaryliodine derivatives:

$$R_2I^+X^- + RLi \rightarrow R_3I + LiX \tag{7.37}$$

These triaryliodine derivatives have limited thermal stability, undergoing ready decomposition to biaryl and the corresponding aryl iodide:

$$R_3I \rightarrow R{-}R + RI \tag{7.38}$$

Iodoxyarenes, RIO_2, are obtained by the disproportionation or oxidation of iodosylarenes, for example,

$$2RIO \rightarrow RI + RIO_2 \tag{7.39a}$$

$$RIO + OCl^- \rightarrow RIO_2 + Cl^- \tag{7.39b}$$

The most important of these organic derivatives of polyvalent iodine is *iodosylbenzene*, $[C_6H_5IO]_n$, which has a polymeric structure with an I—O—I—O— backbone. Because of its polymeric nature, iodosylbenzene is generally insoluble in solvents with which it does not react. Iodosylbenzene reacts with alcohols to give $C_6H_5I(OR)_2$ derivatives.[25] Iodosylbenzene is a useful oxygen transfer agent.[26]

7.7 Noble Gas Compounds

Until 1962 the noble gases were believed to be incapable of forming any stable chemical compounds; they were often called the "inert gases" because of their low chemical reactivity and apparent inability to form compounds. This myth was shattered in 1962 when Bartlett[27] found that elemental xenon reacted with the strongly oxidizing fluoride PtF_6 to give a stable red solid, which Bartlett formulated as "$Xe^+PtF_6^-$." Although Bartlett's formulation of this solid was subsequently shown to be incorrect, his discovery is nevertheless highly significant in demonstrating for the first time the ability of noble gases to form compounds. Shortly after Barlett's discovery of "$Xe^+PtF_6^-$," elemental xenon was shown to combine directly with elemental fluorine to give the binary fluorides XeF_2, XeF_4, or XeF_6, depending on the reaction conditions.

Noble gas compounds correspond to halogen compounds in positive oxidation states. Even-numbered oxidation states, $+2n$, of noble gases are isoelectronic with the immediately preceding odd-numbered oxidation states, $2n - 1$, of the immediately preceding halogen atom. Thus the extensive chemistry of iodine is positive oxidation states serves as a model for xenon chemistry with iodine in the $+1$, $+3$, $+5$, and $+7$ oxidation states corresponding to xenon in the $+2$, $+4$, $+6$, and $+8$ oxidation states, all of which are known. The much lower stability of compounds with bromine in positive oxidation states (e.g., compare the stability of bromine and iodine oxides in Sections 7.4.2 and 7.4.3) translates into much more limited krypton chemistry. The lightest noble gases, helium, neon, and argon, still are not known to form any stable compounds. The chemistry of radon, like the chemistry of its halogen neighbor astatine, is limited by the radioactivity of the element.

7.7.1 Xenon Compounds

7.7.1.1 Xenon Fluorides

The three xenon fluorides, XeF_2 (mp 129°C), XeF_4 (mp 117°C), and XeF_6 (mp 49.5°C) are all made by direct combination of xenon and fluorine under various reaction conditions, which must be carefully controlled to give pure products. All three xenon fluorides are volatile solids that can be readily sublimed; they can be stored indefinitely in nickel or Monel containers. The higher fluorides XeF_4 and XeF_6 are susceptible to hydrolysis, so moisture must be excluded when handling them. Xenon difluoride is more stable in water. An aqueous solution of XeF_2 has a half-life of ~ 7 hours at 0°C and is a very powerful oxidizing agent, converting $2Cl^-$ to Cl_2, Ce^{III} to Ce^{IV}, Cr^{III} to Cr^{VI}, Ag^I to Ag^{II}, and even BrO_3^- to BrO_4^-.

Xenon difluoride is a convenient mild fluorinating agent, since the gaseous elemental xenon by-product is readily removed from the reaction mixture. Reactions of olefins with XeF_2 result in addition of F_2 to the C=C double bond.

Xenon fluorides, like most halogen fluorides, can function as fluoride donors to give fluoroxenon cations,[28] as exemplified by the following reactions:

$$XeF_2 + AsF_5 \rightarrow XeF^+ AsF_6^- \tag{7.40a}$$

$$2XeF_2 + SbF_5 \rightarrow Xe_2F_3^+ SbF_6^- \tag{7.40b}$$

$$XeF_4 + BiF_5 \rightarrow XeF_3^+ + BiF_6^- \tag{7.40c}$$

$$2XeF_6 + 2RuF_6 \rightarrow 2XeF_5^+ + 2RuF_6^- + F_2 \tag{7.40d}$$

$$2XeF_6 + AuF_3 + F_2 \rightarrow Xe_2F_{11}^+ + AuF_6^- \tag{7.40e}$$

In the binuclear cations such as $Xe_2F_3^+$ and $Xe_2F_{11}^+$ the two xenon atoms are bridged by a single fluorine atom. In addition, many of the "anions" in the fluoroxenon salts just listed are joined to the fluoroxenon cation by fluorine bridges. Based on the foregoing fluoroxenon chemistry, the original reaction of xenon with PtF_6 appears to be more complicated than Bartlett thought; it is now interpreted as follows:

$$Xe + 2PtF_6 \xrightarrow{25°C} XeF^+PtF_6^- + PtF_5 \qquad (7.41a)$$

$$XeF^+PtF_6^- + PtF_5 \xrightarrow{60°C} XeF^+Pt_2F_{11}^- \qquad (7.41b)$$

$$Xe + PtF_6 \rightarrow XeF^+PtF_5^- \qquad (7.41c)$$

The product obtained at 25°C appears to contain both the platinum(V) derivative $XeF^+PtF_6^-$ (equation 7.41a) and the platinum(IV) derivative $XeF^+PtF_5^-$ (equation 7.41c). Formation of the XeF^+ cation by reaction of elemental xenon with a strong oxidizing agent such as PtF_6 (equation 7.41a) or $O_2^+SbF_6^-$ (Section 6.3.7) is accompanied by the production of a green paramagnetic Xe_2^+ intermediate, which can be detected by EPR spectroscopy because of its unpaired electron.[29]

Xenon hexafluoride is amphoteric: besides functioning as a fluoride donor to form the XeF_5^+ and $Xe_2F_{11}^+$ cations, it can function as a fluoride acceptor to give the XeF_7^- anion, for example,

$$XeF_6 + MF \rightarrow M^+XeF_7^- \qquad (M = Rb, Cs) \qquad (7.42)$$

Pyrolysis of these $M^+XeF_7^-$ salts at 20°C (M = Rb) to 50°C (M = Cs) gives the corresponding $(M^+)_2XeF_8^{2-}$ salts:

$$2M^+XeF_7^- \xrightarrow{\Delta} XeF_6\uparrow + (M^+)_2XeF_8^{2-} \qquad (7.43)$$

The salts Rb_2XeF_8 and Cs_2XeF_8 are the most thermally stable xenon compounds known, they decompose only above 400°C. The XeF_8^{2-} anion can also be made directly from XeF_6 and nitrosyl fluoride (Section 4.7.3) as its nitrosonium salt:

$$2ONF + XeF_6 \rightarrow (NO^+)_2XeF_8^{2-} \qquad (7.44)$$

7.7.1.2 Xenon–Oxygen Compounds

Xenon, like the halogens, does not combine directly with oxygen to form oxides; however, xenon–oxygen compounds can be made by indirect methods based on the hydrolysis of xenon–fluorine compounds. Thus xenon trioxide, XeO_3, is obtained by the aqueous hydrolysis of XeF_4 or XeF_6. It is stable in aqueous solutions up to 11 M but is deliquescent and dangerously explosive in the solid states. The potential formation of explosive XeO_3 upon hydrolysis is a distinct hazard in studying xenon chemistry. The structure of XeO_3 consists of trigonal pyramidal XeO_3 units linked by bridged oxygen atoms in a molecular lattice.[30] An aqueous solution of XeO_3 is an extremely strong ($E° = 2.10$ V) but kinetically slow oxidizing agent.

Xenon trioxide is acidic and forms xenate salts of the hypothetical xenic acid H_2XeO_4 ($pK_1 = 2.8$), with bases (e.g. $NaHXeO_4 \cdot 1.5H_2O$ and $CsHXeO_4$).

However, reaction of XeO_3 with excess base results is disproportionation of the xenon(VI) into xenon(VIII) and elemental xenon:

$$2XeO_3 + 4OH^- \rightarrow XeO_6^{4-} + Xe\uparrow + O_2\uparrow + 2H_2O \qquad (7.45)$$

Salts of the XeO_6^{4-} anion are called *perxenates* and resemble periodates in that the xenon is octahedrally coordinated to six oxygen atoms (compare Figure 7.1). Perxenates can also be made by oxidation of xenon(VI) with ozone. Stable perxenates include the sodium salts $Na_4XeO_6 \cdot 8H_2O$ and $Na_4XeO_6 \cdot 6H_2O$, in which the presence of XeO_6^{4-} octahedra has been confirmed by X-ray diffraction. Solutions of alkali metal perxenates are alkaline by hydrolysis:

$$H_2XeO_6^{2-} + OH^- \rightleftarrows HXeO_6^{3-} + H_2O \qquad K \approx 4 \times 10^3 \qquad (7.46a)$$

$$HXeO_6^{3-} + OH^- \rightleftarrows XeO_6^{4-} + H_2O \qquad K < 3 \qquad (7.46b)$$

Thus the main perxenate ion, present even at pH 11–13, is $HXeO_6^{3-}$ rather than XeO_6^{4-}, and H_4XeO_6 is an anomalously weak acid compared with the octahedral highest oxidation state oxoacids of the neighboring elements, namely H_5IO_6 and H_6TeO_6.

The other binary xenon oxide is xenon tetroxide, XeO_4, which is a highly unstable explosive yellow gas formed by treating barium perxenate, Ba_2XeO_6, with cold concentrated sulfuric acid; it decomposes into its elements above 0°C.

7.7.1.3 Xenon Oxofluorides

The xenon(VI) oxofluorides $XeOF_4$ and XeO_2F_2 can be obtained by the following reactions:

$$XeF_6 + H_2O \rightarrow XeOF_4 + 2HF \qquad (7.47a)$$

$$XeO_3 + XeOF_4 \rightleftarrows 2XeO_2F_2 \qquad (7.47b)$$

$$2XeO_3 + XeF_6 \rightleftarrows 3XeO_2F_2 \qquad (7.47c)$$

$$XeO_3 + 2XeF_6 \rightleftarrows 3XeOF_4 \qquad (7.47d)$$

In the hydrolysis of XeF_6 to form $XeOF_4$ (equation 7.47a), the amount of water must be limited to prevent formation of XeO_3. Both $XeOF_4$ and XeO_2F_2 have stereochemically active lone pairs. The structure of $XeOF_4$ is thus a ψ-octahedron, with an $XeOF_4$ square pyramid having the oxygen atom in the apical position. The structure of XeO_2F_2 is a ψ-trigonal bipyramid having the lone pair in an equatorial position, analogous to the structure of SF_4 (Figure 6.5). The oxofluoride $XeOF_4$ is a Lewis acid that reacts with fluoride to give the $XeOF_5^-$ anion:

$$XeOF_4 + CsF \rightarrow Cs^+XeOF_5^- \qquad (7.48)$$

The $XeOF_5^-$ anion has a distorted octahedral structure with a stereochemically active lone pair.

7.7.1.4 *Other Xenon Compounds*

Xenon(IV) forms colorless $OXe(OTeF_5)_4$ and dark red $Xe(OTeF_5)_6$ with the highly electronegative —$OTeF_5$ group (Section 6.6.2); the latter compound is obtained from XeF_6 and $B(OTeF_5)_3$. Compounds containing a direct xenon–nitrogen bond are rare, but $FXeN(SO_2F)_2$, a white solid decomposing at 70°C, can be made by the following reaction[31]:

$$XeF_2 + HN(SO_2F)_2 \rightleftharpoons HF + FXeN(SO_2F)_2 \qquad (7.49)$$

The key to the stability of these compounds appears to be the high electronegativities of both the —$OTeF_5$ and —$N(SO_2F)_2$ groups.

7.7.2 Krypton and Radon Chemistry

Krypton chemistry is very limited compared with xenon chemistry. Krypton difluoride, KrF_2, can be obtained by passing an electric discharge through a Kr/F_2 mixture, but higher krypton fluorides are unknown. Krypton difluoride is a white solid that decomposes into its elements at room temperature; it is a highly reactive fluorinating agent and can, for example, fluorinate NF_3 to NF_4^+ (Section 4.7.1). Krypton difluoride is an endothermic compound ($\Delta H_f = +63$ kJ/mol), whereas xenon difluoride is an exothermic compound ($\Delta H_F = -105$ kJ/mol). The cations KrF^+ and $Kr_2F_3^+$ are formed by reactions of KrF_2 with strong fluoride acceptors such as AsF_5 and SbF_5.

Radon chemistry, like that of astatine, is limited to tracer studies because of the short half-life of the element. There is some evidence for species such as RnF_2 and $RnF^+TaF_6^-$.

References

1. Winfield, J. M., Preparation and Use of 18-Fluorine Labelled Inorganic Compounds, *J. Fluorine Chem.*, 1980, **16**, 1–17.

2. Christe, K. O., *Inorg. Chem*, 1986, **25**, 3721–3722.

3. Teitelbaum, R. C.; Ruby, S. L.; Marks, T. J., *J. Am. Chem. Soc.*, 1980, **102**, 3322–3328.

4. Brown, I., Astatine: Its Organonuclear Chemistry and Biomedical Applications, *Adv. Inorg. Chem.*, 1987, **31**, 43–88.

5. Portier, J., Solid State Chemistry of Ionic Fluorides, *Angew. Chem. Int. Ed. Engl*, 1976, **15**, 475–486.

6. Masschelein, W. J., *Chlorine Dioxide: Chemistry and Environmental Impact of Oxychlorine Compounds*, Ann Arbor Science Publishers, Ann Arbor, MI, 1979, Chapters 3–9.

7. Renard, J. J.; Bolker, H. I., The Chemistry of Chlorine Monoxide (Dichlorine Monoxide), *Chem. Rev.*, 1976, **76**, 487–505.

8. Marsh, F. D.; Farnham, W. B.; Sam, D. J.; Smart, B. E., *J. Am. Chem. Soc.*, 1982, **104**, 4680–4682.

9. Gordon, G.; Kieffer, R. G.; Rosenblatt, D. H., The Chemistry of Chlorine Dioxide, *Prog. Inorg. Chem.*, 1972, **15**, 201–286.

10. Masschelein, W. J., *Chlorine Dioxide: Chemistry and Environmental Impact of Oxychlorine Compounds*, Ann Arbor Science Publishers, Ann Arbor, MI, 1979, Chapters 10–14.

11. Selte, K.; Kjekshus, A., The Crystal Structure of I_2O_5, *Acta Chem. Scand.*, 1970, **24**, 1912.

12. Unterzaucher, J., *Chem. Ber.*, 1940, **73**, 391–404.

13. Field, R. J.; Burger, M., eds., *Oscillations and Travelling Waves in Chemical Systems*, Wiley, New York, 1985.

14. Ruoff, P.; Varga, M.; Kőrős, E., How Bromate Oscillators Are Controlled, *Acc. Chem. Res.*, 1988, **21**, 326–332.

15. Schmuacher, J. C., Ed., *Perchlorates: Their Properties, Manufacture, and Uses*, Reinhold, New York, 1960.

16. Hills, E. F.; Sharp, C.; Sykes, A. G., *Inorg. Chem.*, 1986, **25**, 2566–2569.

17. Appelman, E. H., Non-existent Compounds: Two Case Histories, *Acc. Chem. Res.*, 1973, **6**, 113–117.

18. Appelman, E. H., *Inorg. Chem.*, 1969, **2**, 223–227.

19. Appelman, E. H., Kläning, U. K.; Thompson, R. C., *J. Am. Chem. Soc.*, 1979, **101**, 929–934.

20. Siebert, H., Worerner, U., *Z. anorg. allgem. Chem.*, 1973, **398**, 193–197.

21. Wilson, W. W.; Thompson, R. C.; Aubke, F., *Inorg. Chem.*, 1980, **19**, 1489–1493.

22. Birchall, T.; Myers, R. D. *Inorg. Chem.*, 1981, **20**, 2207–2211.

23. Christie, K. O.; Curtis, E. C., *Inorg. Chem.*, 1982, **21**, 2938–2945.

24. Banks, D. F., Organic Polyvalent Iodine Compounds, *Chem. Rev.*, 1966, **66**, 243–266.

25. Schardt, B. C.; Hill, C. L., *Inorg. Chem.*, 1983, **22**, 1563–1565.

26. Moriarty, R. M.; Prakash, O., Hypervalent Iodine in Organic Synthesis, *Acc. Chem. Res.*, 1986, **19**, 244–250.

27. Bartlett, N., *Proc. Chem. Soc.*, 1962, 218.

28. Selig, H.; Holloway, J. H., Cationic and Anionic Complexes of the Noble Gases, *Top. Curr. Chem.*, 1984, **124**, 33–90.

29. Brown, D. R.; Clegg, M. J.; Downs, A. J.; Fowler, R. C.; Minihan, A. R.; Norris, M. R.; Stein, L., *Inorg. Chem.*, 1992, **31**, 5041–5052.

30. Templeton, D. H; Zalkin, A.; Forrester, J. D.; Williamson, S. M., Crystal and Molecular Structure of Xenon Trioxide, *J. Am. Chem. Soc.*, 1963, **85**, 817.

31. DesMarteau, D. D., *J. Am. Chem. Soc.*, 1978, **100**, 6270–6271.

CHAPTER

8

Boron

8.1 General Aspects of Boron Chemistry

Boron is always trivalent and never monovalent despite its $2s^2 2p$ electronic structure. In addition, the presence of only three valence orbitals makes boron the simplest example of a *hypoelectronic element*, that is, an element with fewer valence electrons than orbitals. The ionization potentials of boron are so high (e.g., 8.296 eV for the first ionization potential) that the total energy required to produce B^{3+} cations cannot be balanced by energy gained through solid lattice formation or solvation. For this reason, simple electron loss to form B^{3+} cations does not occur in boron chemistry either in the solid state or in solution. Instead, covalent bond formation plays a major role in boron chemistry as in the chemistry of other nonmetals.

Many simple compounds of boron of the type BX_3 (X = F, Cl, Br, I, alkyl, aryl, etc.) contain a three-coordinate boron atom with a valence sextet rather than the favored octet. In such compounds the boron atom uses three trigonal planar sp^2 hybrids to form three bonds, leaving an empty p orbital perpendicular to the plane (Figure 8.1). Such trigonal boron compounds are usually Lewis acids, since they can receive an electron pair from a Lewis base, such as a tertiary amine, $R_3N:$, to complete the electron octet of the boron atom. The boron atom in such adducts (e.g., $R_3N \rightarrow BX_3$) has tetrahedral sp^3 hybridization similar to the ubiquitous sp^3 hybridization in carbon chemistry (Chapter 2). Tetrahedral sp^3 boron hybridization is also found in anionic boron derivatives such as BH_4^- and $B(C_6H_5)_4^-$, which are isoelectronic with the carbon compounds CH_4 and $C(C_6H_5)_4$. A few cationic tetrahedral boron derivatives such as $(2,2'-\text{bipyridyl})B(C_6H_5)_2^+$ are also known.[1]

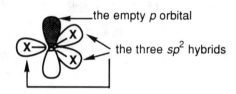

Figure 8.1. The trigonal boron hybridization in BX₃ derivatives.

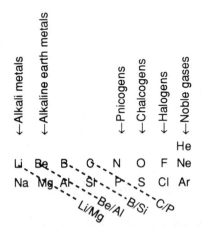

Figure 8.2. A section of the periodic table showing the diagonal relationships Li/Mg, Be/Al, B/Si, and C/P.

In the chemistry of some of the lightest elements, there is a *diagonal relationship* because of the similar electronegativities and sizes of diagonal pairs of elements such as Li/Mg, Be/Al, B/Si, and C/P (Figure 8.2). The chemistry of boron thus resembles that of silicon (Chapter 3) more than that of its heavier congeners, namely aluminum, gallium, indium, and thallium (Chapter 9). Respects in which boron resembles silicon more than its heavier congener, aluminum, include the following:

1. Elemental boron is a semiconductor like elemental silicon, whereas aluminum is unambiguously metallic.
2. Boron and silicon hydrides are both volatile and highly flammable, whereas the only binary aluminum hydride, $(AlH_3)_x$, is a nonvolatile polymer.
3. Boron and silicon halides are readily hydrolyzed, whereas aluminum halides form the stable hydrated $Al(H_2O)_6^{3+}$ ion in aqueous solution.
4. The oxides B_2O_3 and SiO_2 form glasses and dissolve metal oxides on fusion in contrast to the refractory Al_2O_3.
5. Boric acid, $B(OH)_3$, and silicic acid, $SiO_2 \cdot xH_2O$, are weak acids with no basic properties, whereas $Al(OH)_3$ is amphoteric (i.e., having both acidic and basic properties).

The diagonal relationship, of course, does not extend to the stoichiometries of the analogous series of compounds, since the primary valences of the diagonal pairs Li/Mg, Be/Al, B/Si, and C/P differ by one unit, (i.e., $n/n + 1$). This is particularly evident in a comparison of the boron and silicon hydrides. Thus although boron and silicon hydrides are both volatile and highly flammable, the structures of silicon hydrides (Section 3.4.1) consist solely of two-electron, two-center bonds, whereas the structures of boron hydrides (Section 8.3) contain many two-electron, three-center bonds, since boron, but not silicon, is hypoelectronic.

8.2 Elemental Boron and Metal Borides

Boron is not an abundant element. Thus its concentration in the earth's crust is only 9 ppm compared with 18 ppm for lithium, 13 ppm for lead, and 8.1 ppm for thorium. Boron is readily available, however; it is highly concentrated in mineral deposits in the United States, Russia, and Turkey. Important boron minerals include *borax*, $Na_2[B_4O_5(OH)_4] \cdot 8H_2O$, *kernite*, $Na_2[B_4O_5(OH)_4] \cdot 2H_2O$, and *tourmaline*, an aluminosilicate (Section 3.5.3.5) containing $\sim 10\%$ of boron.

Boron has two nonradioactive isotopes: ^{10}B, with a spin of 3 and 19.1–20.3% natural abundance, and ^{11}B, with a spin of $\frac{3}{2}$ and 79.7–80.9% natural abundance. The more abundant boron isotope, ^{11}B, is a useful NMR nucleus despite its quadrupole moment $(0.0355/10^{-28} \text{ m}^2)$.

Elemental boron is a black solid with a metallic luster and poor electrical conductivity.[2] It is difficult to prepare in a pure state because of its high melting point (2297°C), the corrosiveness of liquid boron, and the tendency of the elemental boron structures to incorporate small but significant amounts of impurities in their lattices. For these reasons elemental boron was not isolated until 1892, by Moissan. The following reactions can be used to prepare elemental boron in various states of purity:

1. High temperature reduction of B_2O_3 with an electropositive metal, for example,

$$B_2O_3 + 3Mg \rightarrow 2B + 3MgO \tag{8.1}$$

   This method was first used by Moissan to prepare elemental boron in 90–98% purity.

2. Electrolysis of KBF_4 in a KCl/KF fused salt mixture at 800°C. This method gives $\sim 95\%$ pure elemental boron.

3. Reduction of boron tribromide with elemental hydrogen at 800–1100°C on a heated tantalum wire.[3] This method is useful for preparing pure crystalline boron.

4. Thermal decomposition of BI_3 or various boron hydrides.

Table 8.1 Well-Characterized Allotropes of Elemental Boron

Allotrope	Atoms per Unit Cell	Structural Units in the Unit Cell
α-Rhombohedral boron	12	One B_{12} icosahedron
β-Rhombohedral boron	105	One B_{84} unit, two B_{10} groups, and one B atom
α-Tetragonal boron	50	Four B_{12} icosahedra and two B atoms
β-Tetragonal boron	188	$B_{21} \cdot 2B_{12} \cdot B_{2.5}$

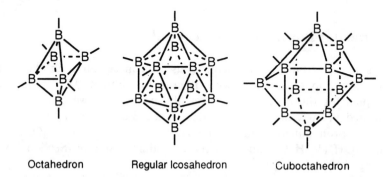

| Octahedron | Regular Icosahedron | Cuboctahedron |

Figure 8.3. Boron polyhedra found in elemental boron and boron-rich borides.

Elemental boron has four well-characterized allotropes, whose structures are summarized in Table 8.1. These structures are based on B_{12} icosahedra (Figure 8.3) as fundamental building blocks and are rather complicated. In such B_{12} icosahedra three of the four valence orbitals of the boron are used for the skeletal bonding of the icosahedron, leaving one outwardly directed orbital for bonding to neighboring B_{12} icosahedra or isolated boron atoms between boron icosahedra. The simplest elemental boron structure is exhibited by the α-rhombohedral allotrope (Table 8.1), in which the nearly regular B_{12} icosahedral building blocks are linked by two-center and three-center bonds in a deformed cubic close packing.[4] The thermodynamically most stable allotrope of elemental boron is β-rhombohedral boron, which with 105 boron atoms in the unit cell, has a considerably more complicated structure[5] (Table 8.1) than that of α-rhombohedral boron.

Borides are defined as compounds containing boron and a less electronegative element, typically a metal. Metal borides may be synthesized by the high temperature reactions of a metal with boron or a mixture of B_2O_3 and carbon. The structures of binary metal borides are relatively complicated, exhibiting the following features[6,7] as the boron/metal ratio is increased:

1. Boron/metal ratios of 0.25–0.6: isolated boron atoms as in Mn_4B, Re_7B_3, Ni_3B, and Be_2B.

2. Boron/metal ratios of 0.6–0.8: bonded pairs of boron atoms as in V_3B_2 and Cr_5B_3.
3. Boron/metal ratios of 0.8 to <2.0: zigzag chains of boron atoms as in MB (M = Ti, Hf; V, Nb, Ta; Cr, Mo, W; Mn, Fe, Co, Ni) or double chains of boron atoms as in M_3B_4 (M = Ti, V, Nb, Ta, Cr, Mn, Ni).
4. Boron/metal ratios of 2.0 to <4.0: two-dimensional hexagonal nets of boron atoms MB_2 (M = Mg; Al, Sc, Y, Gd; Ti, Zr, Hf; V, Nb, Ta; Cr, Mo, W; Mn, Tc, Re; Ru, Os, U, Pu) and M_2B_5 (M = Ti, Mo, W).
5. Boron/metal ratios of 4.0 or more: three-dimensional boron networks containing B_6 octahedra in MB_4 and MB_6 (M = La), B_{12} cuboctahedra in YB_{12}, B_{12} icosahedra in NaB_{15} or C_3B_{12}, or $B_{12}(B_{12})_{12}$ "icosahedra of icosahedra" in YB_{66} (Figure 8.3).[8,9]

Many metal borides have unusual physical and chemical properties. For example:

1. Lanthanide borides of the general formula LnB_6 are among the best thermionic emitters known.
2. The borides TiB_2 and ZrB_2 exhibit electrical conductivities about 10 times as great as those of metals.
3. Ternary boride superconductors with critical temperatures up to 11.5 i include $LnRh_4B_4$, $Ba_{0.67}Pt_3B_2$, $LnRhB_2$, and $LnOs_3B_2$ (Ln = lanthanide element).

8.3 Boranes and Related Compounds

Boron forms a large number of both neutral and anionic binary compounds with the hydrogen; these are called *boranes*. The systematics of boranes are totally different from those of hydrocarbons, largely because of the prevalence of three-center B—H—B and B—B—B bonding in boranes.

An important method for the characterization of boranes in solution is ^{11}B NMR spectroscopy.[10] Even though ^{11}B has a quadrupole moment, its field gradient in boranes is low enough to permit reasonably sharp ^{11}B NMR spectra to be obtained. The number of *terminal* hydrogens bonded to a given boron atom can be determined from the multiplicity of its resonance with BH and BH_2 groups, leading to doublets and triplets, respectively (in the *absence* of proton decoupling). Proton NMR spectra are rarely useful for boranes because signals are broadened excessively by the quadrupolar boron nucleus. In some borane derivatives such as $B_3H_8^-$ (Section 8.3.2.2), the NMR spectra indicate higher apparent symmetry than is actually known from structural determinations by X-ray and neutron diffraction, indicating *fluxional* properties (i.e., structural rearrangements on a time scale that is fast relative to the NMR time scale).

8.3.1 Neutral Boranes

Known *boranes* fall into two homologous series:

B_nH_{n+4}: B_2H_6, B_5H_9, B_6H_{10}, B_8H_{12}, $B_{10}H_{14}$, $B_{16}H_{20}$, $B_{18}H_{22}$ (two isomers)

B_nH_{n+6}: B_4H_{10}, B_5H_{11}, B_6H_{12}, B_9H_{15} (two isomers), $B_{10}H_{16}$

The structures of some of the more important boranes are depicted in Figure 8.4.

The presence of the three-center, two-electron B—H—B and B—B—B bonds in these boranes arises from the "electron deficiency" in these compounds, which relates to the hypoelectronicity of boron: that is, there are not enough electrons to form two-center, two-electron bonds between all adjacent pairs of atoms. A semitopological approach to describe the balance between the types of bonds in these borane structures has been developed by Lipscomb[11–13]; this approach is colloquially called the "*styx* nomenclature" because of letters chosen to represent the variables. In the *styx* nomenclature the letters s, t, y, and x stand for the numbers of three-center B—H—B bonds, three-center B—B—B bonds, two-center B—B bonds, and BH_2 groups, respectively. The structure of a given borane can then be described by a four-digit *styx number* corresponding to the values for s, t, y, and x, respectively. Thus the *styx* numbers for the structures for B_2H_6, B_4H_{10}, B_5H_9, B_5H_{11}, B_6H_{10}, and $B_{10}H_{14}$ are 2002, 4012, 4120, 3203, 4220, and 4450, respectively (Figure 8.4). Since each boron atom in a stable borane uses all four orbitals of its sp^3 manifold and is bonded to at least one terminal hydrogen atom, the *styx* numbers for a given borane must satisfy the following *equations of balance*, where b is the number of boron atoms:

$$2s + 3t + 2y + x = 3b \qquad \text{(orbital balance)} \qquad (8.2a)$$

$$s + 2t + 2y + x = 2b \qquad \text{(electron balance)} \qquad (8.2b)$$

Note that the presence of a single terminal B—H bond in a borane is regarded as "normal" and therefore is not directly indicated by a digit in the *styx* number.

The simplest neutral borane is diborane(6) or simply *diborane*,[14] B_2H_6, which is a toxic gas (mp $-164.9°C$, bp $-92.6°C$). The structure of diborane (Figure 8.4) has four terminal hydrogen and two bridging hydrogen atoms. In diborane the B—H distances to the bridging hydrogen atoms are longer (1.329 Å) than those to the terminal hydrogen atoms (1.192 Å), suggesting that the three-center B—H—B bonds are weaker than the two-center B—H bonds owing to the presence of fewer bonding electrons relative to the number of orbitals. Diborane can be obtained by treatment of a tetrahydroborate (see Section 8.3.2.1) with a Lewis acid or a protonic acid, for example,

$$3NaBH_4 + 4BF_3 \xrightarrow[\text{or Et}_2\text{O}]{\text{diglyme}} 3NaBF_4 + 2B_2H_6\uparrow \qquad (8.3a)$$

$$2NaBH_4 + H_2SO_4 \rightarrow Na_2SO_4 + B_2H_6\uparrow + 2H_2\uparrow \qquad (8.3b)$$

The need for previously prepared $NaBH_4$ for a source of diborane can be

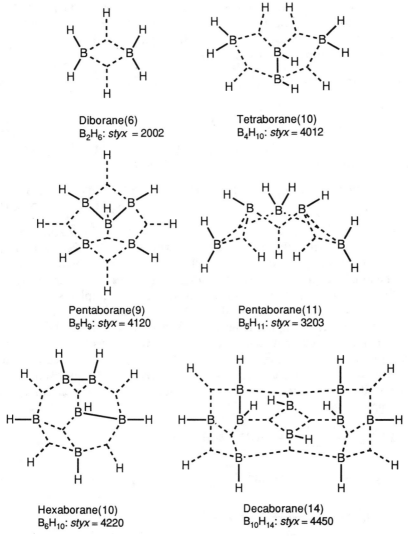

Figure 8.4. Structures of some neutral binary boranes and their *styx* numbers (see text). Two-center, two-electron bonds are depicted by solid lines; three-center, two-electron bonds are depicted by broken lines.

circumvented by treatment of solid sodium hydride (Section 1.5) directly with BF_3 at 180°C, but the diborane produced in this reaction must be trapped as it is formed, to prevent subsequent thermal decomposition. Diborane burns very exothermically with a green flame:

$$B_2H_6 + 3O_2 \rightarrow B_2O_3 + 3H_2O \qquad \Delta H = -2165 \text{ kJ/mol} \qquad (8.4)$$

Diborane has the highest heat of combustion per unit weight except for H_2,

BeH_2, and $Be(BH_4)_2$ and thus has been considered for use in rocket fuels. However, undesirable properties of its B_2O_3 combustion product have limited this application. Diborane can undergo either symmetrical or unsymmetrical cleavage upon treatment with Lewis bases:

$$\text{symmetrical cleavage: } B_2H_6 + 2L \rightarrow 2L \rightarrow 2BH_3 \qquad (8.5a)$$

$$\text{unsymmetrical cleavage: } B_2H_6 + 2L \rightarrow [L_2BH_2]^+[BH_4]^- \qquad (8.5b)$$

Symmetrical cleavage of B_2H_6 with hydride ion regenerates the tetra-hydroborate anion BH_4^-. The stability of the adducts obtained by the symmetrical cleavage of diborane increases in the following sequence:

$$PF_3 < CO < Et_2O < Me_2O < C_4H_8O < C_4H_8S < Et_2S < Me_2S$$
$$< C_5H_5N < Me_3N < H^-$$

The strong π-bonding ligands CO and PF_3 form the unstable adducts $H_3B \leftarrow CO$ and $H_3B \leftarrow PF_3$ upon treatment with CO and PF_3, respectively; these compounds appear to have a small amount of back-bonding in which B—H bonding electrons are partially shared with the same π-acceptor orbitals of CO and PF_3 that are used for their transition metal complexes.

Boranes were initially identified by Alfred Stock in the volatile acid hydrolysis products of impure magnesium boride, MgB_2, using glass vacuum line techniques to fractionate the mixture of reactive air-sensitive products; this original work led to the characterization of the six boranes depicted in Figure 8.4.[15] The more stable higher boranes, such as B_4H_{10}, B_5H_9, and $B_{10}H_{14}$, can be obtained by the pyrolysis of diborane under suitably chosen conditions, for example,

$$2B_2H_6 \xrightarrow[120°C]{25°C/10 \text{ days or}} B_4H_{10} + H_2 \qquad (8.6a)$$

$$5B_2H_6 \xrightarrow[3 \text{ s residence}]{250°C} 2B_5H_9 + 6H_2 \qquad (8.6b)$$

$$5B_2H_6 \xrightarrow[Me_2O]{150°C} B_{10}H_{14} + 8H_2 \qquad (8.6c)$$

Tetraborane(10), B_4H_{10} (mp $-120°C$, bp $18°C$), and *pentaborane(9)* (mp $-46.8°C$, bp $60°C$), are toxic flammable liquids at room temperature. Tetra-borane(10) decomposes fairly rapidly at $25°C$. Pentaborane(9) is stable at $25°C$ but decomposes slowly at $150°C$ and is spontaneously flammable in air. *Decaborane(14)*, $B_{10}H_{14}$, is a toxic volatile solid (mp $99.5°C$, bp $213°C$), which is air-stable under ambient conditions. Pentaborane(9) and decaborane(14) have been made on a large scale for potential use as rocket fuel.

Under nonoxidizing conditions, pentaborane(9) and decaborane(14) are stable enough to undergo electrophilic substitution without destruction of the B_5 and B_{10} cages, respectively. For example, halogenation of $B_{10}H_{14}$ with iodine gives mixtures of $B_{10}H_{13}I$ and $B_{10}H_{12}I_2$ isomers. Decaborane(14) and pentaborane(9)[16] undergo electrophilic alkylation upon treatment with alkyl halides in the presence of aluminum chloride.

All the neutral boranes have structures with some bridging hydrogen atoms (e.g., Figure 8.4). Except for B_2H_6, the bridging hydrogen atoms are acidic enough to be removed by a strong base (e.g., potassium hydride: Section 1.5), to give the corresponding monoanion. This process results in the replacement of a three-center B—H—B bond by a two-center B—B bond. Treatment of the monoanions generated by this process with diborane leads to expansion of the boron cage, for example,

$$B_4H_{10} + KH \rightarrow K^+B_4H_9^- + H_2\uparrow \tag{8.7a}$$

$$2K^+B_4H_9^- + B_2H_6 \rightarrow 2K^+B_5H_{12}^- \tag{8.7b}$$

$$K^+B_5H_{12}^- + HCl \rightarrow H_2\uparrow + B_5H_{11} + KCl \tag{8.7c}$$

$$B_5H_9 + KH \rightarrow K^+B_5H_8^- + H_2\uparrow \tag{8.7a'}$$

$$2K^+B_5H_8^- + B_2H_6 \rightarrow 2K^+B_6H_{11}^- \tag{8.7b'}$$

$$K^+B_6H_{11}^- + HCl \rightarrow \ + B_6H_{12} + KCl \tag{8.7c'}$$

Such reaction sequences can be used to prepare relatively unstable boron hydrides of the type B_nH_{n+6}, which are not accessible from the pyrolysis of diborane. The acidity of the bridging hydrogen atoms in boron hydrides increases with the sizes of the boron framework, for example,

$$B_5H_9 < B_6H_{10} < B_{10}H_{14} < B_{16}H_{20} < n\text{-}B_{18}H_{22}$$

for the boron hydrides of the type B_nH_{n+4}. Thus $B_{10}H_{14}$ can be titrated in aqueous alcoholic media as a monobasic acid, $pK_a = 2.70$.

8.3.2 Borane Anions

8.3.2.1 Tetrahydroborate ("Borohydride") and Other Mononuclear Hydroborates

The tetrahydroborate anion, BH_4^-, colloquially called the "borohydride" ion, is isoelectronic with methane and is formally a hydride adduct of BH_3. Thus $LiBH_4$ can be obtained by treatment of lithium hydride with diborane in diethyl ether:

$$2LiH + B_2H_6 \rightarrow 2LiBH_4 \tag{8.8}$$

The most commonly used tetrahydroborate is sodium borohydride,[17] which can be obtained by the reaction of sodium hydride with excess trimethyl borate at elevated temperatures:

$$4NaH + (MeO)_3B \xrightarrow{250-270°C} NaBH_4 + 3NaOMe \tag{8.9}$$

After hydrolysis with water, the $NaBH_4$ can be extracted from the reaction mixture with isopropylamine. On an industrial scale $NaBH_4$ can be obtained from borax, silica, sodium, and hydrogen at 450–500°C.

The BH_4^- ion and its substitution products such as BH_3CN^-, $BH(OBu^t)_3^-$, and $BHEt_3^-$, are useful as reducing agents and hydride sources.[18] Thus sodium borohydride and related compounds are useful for reduction of aldehydes and ketones to the corresponding alcohols:

$$RR'C{=}O \xrightarrow{\ NaBH_4\ } RR'CHOH \tag{8.10}$$

In contrast to simple saline hydrides such as M^+H^- (M = Li, Na, K, etc.), sodium and potassium borohydrides hydrolyze only slowly in water and ethanol but rapidly in methanol. Aqueous borohydride solutions can be stabilized by addition of base. Studies of the hydrolysis of BH_4^- in D_2O and BD_4^- in H_2O suggest that the hydrolysis of $NaBH_4$ proceeds through a transient BH_5 intermediate with a lifetime of $\sim 10^{-10}$ second. Since boron has only four valence orbitals, this BH_5 intermediate, **8.1**, must have an H—B—H three-center, two-electron bond (compare equation 1.28 in Section 1.4):

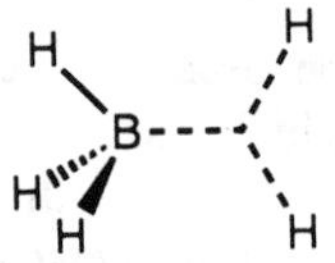

8.1

The cyanotrihydroborate ("cyanoborohydride") ion, BH_3CN^-, is more stable than $NaBH_4$ toward hydrolysis because of the electron-withdrawing cyano group. Its sodium salt, $NaBH_3CN$, can be used in moderately acid solutions. The lithium derivative $LiBHEt_3$ is an exceptionally powerful S_N2 nucleophile toward alkyl halides.

Tetrahydroborate and related hydroborates can function as ligands in transition metal chemistry through the formation of M—H—B three-center bonds. Thus the tetrahydroborate ion can function as a monodentate, bidentate, or tridentate ligand depending on whether one, two, or three M—H—B three-center bonds are formed (Figure 8.5).

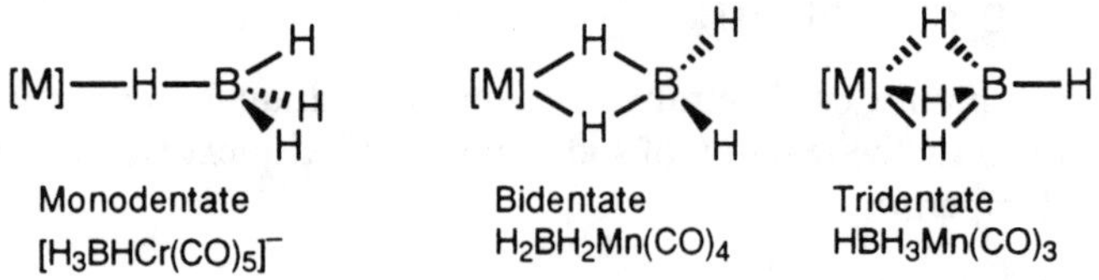

Monodentate
$[H_3BHCr(CO)_5]^-$

Bidentate
$H_2BH_2Mn(CO)_4$

Tridentate
$HBH_3Mn(CO)_3$

Figure 8.5. The tetrahydroborate ion as a monodentate, bidentate, and tridentate ligand through M—H—B three-center bond formation. Hydrogens participating in such three-center M—H—B bonds are drawn as simple bridges for clarity.

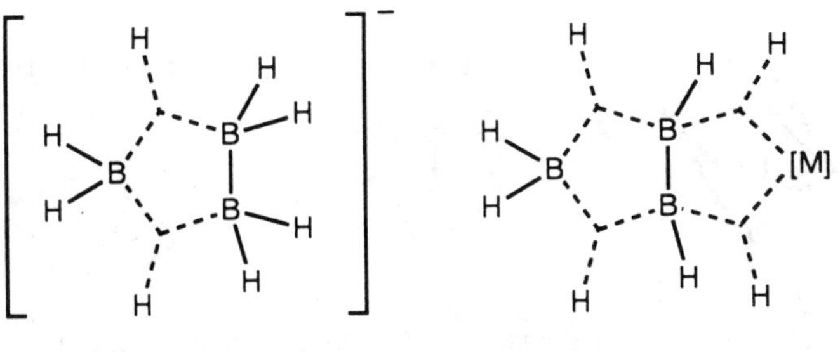

The B$_3$H$_8^-$ anion   The B$_3$H$_8^-$ anion as a
                                          bidentate ligand

Figure 8.6. The octahydrotriborate anion as a bidentate ligand in transition metal chemistry.

8.3.2.2 The Octahydrotriborate Anion

Reaction of BH$_4^-$ with diborane in a suitable solvent results in formation of the octahydroborate anion, B$_3$H$_8^-$, for example,

$$NaBH_4 + B_2H_6 \xrightarrow[100°C]{diglyme} NaB_3H_8 + 2H_2\uparrow \qquad (8.11)$$

The use of previously isolated diborane can be avoided by generating it in situ from NaBH$_4$ and Et$_2$O·BF$_3$, leading to the following more convenient preparation of NaB$_3$H$_8$:

$$5NaBH_4 + 4Et_2O·BF_3 \rightarrow 2NaB_3H_8 + 3NaBF_4 + 2H_2\uparrow + 4Et_2O \quad (8.12)$$

The B$_3$H$_8^-$ anion is of interest in being a simple example of a *fluxional* borane anion. Thus its static structure (Figure 8.6) has two bridging and six terminal hydrogens with boron atoms of two different types but shows single ^{11}B and ^{1}H NMR resonances owing to rapid cycling of the bridging hydrogens on the NMR time scale. The octahydrotriborate ion is a good bidentate ligand in transition metal chemistry (Figure 8.6) in complexes such as [B$_3$H$_8$Cr(CO)$_4$]$^-$ through the formation of M—H—B bonds (compare Figure 8.5).

8.3.2.3 Deltahedral Borane Anions[19]

The most stable borane anions have the general formula B$_n$H$_n^{2-}$ ($6 \leq n \leq 12$) and structures based on *deltahedra* (polyhedra in which all faces are triangles), with the maximum number of vertices of degrees 4 and 5, where the *degree* of a polyhedral vertex is the number of edges meeting at that vertex. Deltahedra with 6, 7, 8, 9, 10, and 12 vertices are possible in which all vertices have degrees 4 or 5, whereas at least one vertex of degree 6 is required for an 11-vertex deltahedron.[20] The deltahedra found in deltahedral borane anions are depicted in Figure 8.7.

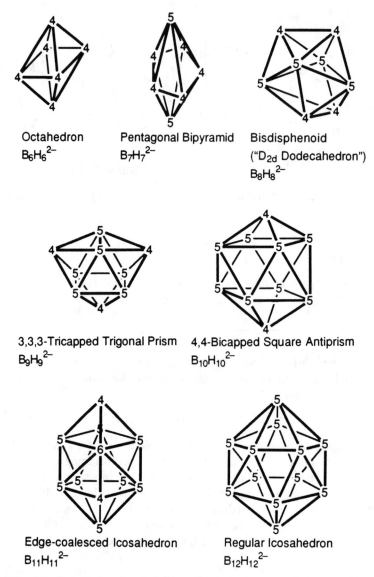

Figure 8.7. The deltahedra found in the borane anions $B_nH_n^{2-}$ ($6 \leq n \leq 12$). The vertices are labeled by their degrees.

The octahydrotriborate anion is a useful precursor for syntheses of many of the $B_nH_n^{2-}$ anions. Thus pyrolysis of NaB_3H_8 at 200–230°C leads to the formation of $B_{10}H_{10}^{2-}$ and $B_{12}H_{12}^{2-}$. Pyrolysis of CsB_3H_8 leads not only to the formation of $B_{10}H_{10}^{2-}$ and $B_{12}H_{12}^{2-}$ but aslo $B_9H_9^{2-}$. Oxidation of $B_9H_9^{2-}$ leads to formation of $B_8H_8^{2-}$ and $B_7H_7^{2-}$. The $B_{10}H_{10}^{2-}$ anion can also be made by heating $B_{10}H_{14}$ with tertiary amines:

$$B_{10}H_{14} + 2R_3N \xrightarrow{130°C} [R_3NH]_2[B_{10}H_{10}] + H_2\uparrow \qquad (8.13)$$

The $B_{12}H_{12}^{2-}$ anion can be made directly from $NaBH_4$ and diborane under sufficiently vigorous conditions, for example,

$$5B_2H_6 + 2NaBH_4 \xrightarrow[180°C]{Et_3N} Na_2B_{12}H_{12} + 13H_2\uparrow \qquad (8.14)$$

The deltahedral borane anions are distinguished from other borane derivatives by unusual kinetic stability. They thus may be regarded as "three-dimensional aromatic systems" (Section 8.3.4), analogous to two-dimensional cyclic hydrocarbon aromatic systems such as cyclopentadienide, benzene, or tropylium (see Section 8.6). The hydrolytic stability of the deltahedral boranes decreases in the approximate sequence:

$$B_{12}H_{12}^{2-} > B_{10}H_{10}^{2-} > B_{11}H_{11}^{2-} > B_9H_9^{2-} \sim B_8H_8^{2-} \sim B_6H_6^{2-} > B_7H_7^{2-}$$

Thus the water-soluble $M_2B_{12}H_{12}$ salts (M = alkali metal) are stable to 800°C and resist alkali or 3 N hydrochloric acid at 100°C despite the thermodynamically favored hydrolysis to boric acid and hydrogen. The free acids of $B_{10}H_{10}^{2-}$ and $B_{12}H_{12}^{2-}$ can be isolated as stable oxonium salts $(H_3O^+)_2B_{10}H_{10}^{2-}$ and $(H_3O^+)_2$-$B_{12}H_{12}^{2-}$ by passing solutions of the anions through an acidic ion exchange resin.

One or more hydrogens in the anions $B_{10}H_{10}^{2-}$ and $B_{12}H_{12}^{2-}$ can be replaced by other groups using electrophiles such as RCO^+, PhN_2^+, and X^+ (X = Cl, Br, I). The $B_nH_n^{2-}$ anions are much more reactive toward such electrophilic substitution than benzene. In $B_{10}H_{10}^{2-}$ the apical hydrogens at the two vertices of degree 4 are replaced preferentially by electrophiles over the equatorial hydrogens at the eight vertices of degree 5. Halogenation of $B_{10}H_{10}^{2-}$ and $B_{12}H_{12}^{2-}$ can lead to complete substitution of hydrogen with halogen without rupture of the boron deltahedron. The resulting perhalo derivatives $B_{10}X_{10}^{2-}$ and $B_{12}X_{12}^{2-}$ (X = Cl, Br, I) are very thermally stable and resistant toward hydrolysis in contrast to simple boron halides.

The icosahedral anion $B_{12}H_{12}^{2-}$ is very stable toward oxidation, even by strong oxidants. However, the $B_{10}H_{10}^{2-}$ anion undergoes oxidative coupling upon reaction with a strong one-electron oxidant such as Fe^{3+} or Ce^{4+}. In the resulting coupled product $B_{20}H_{18}^{2-}$ two B_{10} bicapped square antiprisms (Figure 8.7) are linked together through a B—B—B three-center bond. Reduction of $B_{20}H_{18}^{2-}$ with sodium in liquid ammonia gives the tetraanion $B_{20}H_{18}^{4-}$ in which the two B_{10} bicapped square antiprisms are linked through only a simple B—B two-center bond.

8.3.3 Carboranes

Replacement of anionic BH^- vertices in the deltahedral $B_nH_n^{2-}$ anions by the isoelectronic and isolobal CH vertices leads to the so-called *carboranes*,[21-23]

$CB_{n-1}H_n^-$ and particularly $C_2B_{n-2}H_n$; the latter are of particular interest because they are neutral compounds. The terminal hydrogen atoms in deltahedral carboranes such as those of the type $C_2B_{n-2}H_n$ can be replaced by other monovalent groups, leading to extensive derivative chemistry.

Neutral deltahedral carboranes $C_2B_{n-2}H_n$ are generally obtained by reactions of alkynes with boranes. Use of substituted alkynes $RC\equiv CR'$ leads to deltahedral carboranes $RR'C_2B_{n-2}H_{n-2}$ in which the substituents R and R' are bonded to the carbon atoms. Thus icosahedral carboranes of the general type $1,2\text{-}RR'C_2B_{10}H_{10}$ can be prepared on a large scale from decaborane by the following sequence of reactions:

$$B_{10}H_{14} + 2Et_2S \rightarrow B_{10}H_{12}(SEt_2)_2 + H_2\uparrow \qquad (8.15a)$$

$$B_{10}H_{12}(SEt_2)_2 + RC\equiv CR' \rightarrow RR'C_2B_{10}H_{10} + H_2\uparrow + 2Et_2S \qquad (8.15b)$$

Reaction of B_5H_9 with acetylene leads to the smaller deltahedral carboranes $1,5\text{-}C_2B_3H_5$, $1,6\text{-}C_2B_4H_6$, and $2,4\text{-}C_2B_5H_7$, containing a C_2B_3 trigonal bipyramid, a C_2B_4 octahedron, and a C_2B_5 pentagonal bipyramid, respectively.

Isomeric $C_2B_{n-2}H_n$ carboranes are possible, depending on which two vertices of the deltahedron are occupied by the two carbon atoms. The thermodynamic stability of carboranes is maximized if the carbon atoms occupy the vertices of the lowest degree (i.e., lowest "coordination number") and keep as far apart as possible from other carbon atoms.[24] The preparation of carboranes from acetylenes and boranes introduces the carbon atoms into the boron cage as a C_2 unit so that the initially produced carborane isomer has two adjacent carbon atoms. Pyrolysis of such isomers leads to rearrangements producing isomers in which the carbons have moved further apart from each other. The "ortho-, meta-, para- nomenclature" for benzenoid organic compounds has been adapted to designate the three possible isomers of the icosahedral carborane skeleton $C_2B_{10}H_{12}$ (Figure 8.8). Thus, reaction of

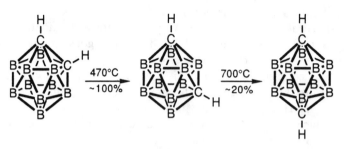

1,2-dicarbadecaborane	1,7-dicarbadecaborane	1,12-dicarbadecaborane
"ortho-carborane"	"meta-carborane"	"para-carborane"
m.p. 320°C	m.p. 265°C	m.p. 261°C

Figure 8.8. Rearrangements of the icosahedral carboranes $C_2B_{10}H_{12}$. The terminal hydrogens attached to each of the boron atoms are omitted for clarity.

acetylene with a suitable decaborane derivative (e.g., equations 8.15) leads initially to 1,2-dicarbadecaborane ("orthocarborane", mp 320°C). Pyrolysis of orthocarborane at 470°C leads to a good yield of 1,7-dicarbadecaborane ("metacarborane", mp 265°C). Further pyrolysis of metacarborane at 700°C with a contact time of a few seconds leads to further isomerization, giving a $\sim 20\%$ yield of 1,12-dicarbadecaborane ("paracarborane"). The high pyrolysis temperatures required for these isomerizations are an excellent indication of the high stability of the icosahedral C_2B_{10} skeleton.

The C—H hydrogen atoms in carboranes are similar in acidity to the C—H hydrogens in acetylenes, from which they can be prepared. For this reason the C—H hydrogens in carboranes can be metallated with organolithium compounds to give the corresponding lithium derivatives, for example,

$$1,2\text{-}C_2B_{10}H_{12} + 2Bu^nLi \rightarrow 1,2\text{-}Li_2C_2B_{10}H_{10} + 2Bu^nH \qquad (8.16)$$

The lithiated carborane has a chemical reactivity similar to that of other organolithium compounds (Section 10.3.3). It thus reacts with CO_2 to form the corresponding carboxylic acid and with aldehydes or ketones to form the corresponding hydroxy compounds ("carboranyl alcohols").

8.3.4 Electron Counting, Structure, and Bonding in Polyhedral Boranes and Carboranes: Three-Dimensional Aromaticity

Structures consisting of two-center B—B and B—H bonds and/or three-center B—H—B and B—B—B bonds can be drawn for the deltahedral boranes. However, such structures do not reflect the special stability of deltahedral boranes relative to boron hydrides with more open structures (e.g., those in Figure 8.4) any more than the olefinic Kekulé structures of benzene reflect its special stability. In addition, structures consisting of only two-center and three-center bonds for the most symmetrical deltahedral boranes such as octahedral $B_6H_6^{2-}$ and icosahedral $B_{12}H_{12}^{2-}$ do not depict the high effective symmetries of these compounds.

Electron counting in polyhedral boranes and carboranes is based on the number of *skeletal electrons* associated with the bonding within the polyhedron. Electrons associated with external bonds (e.g., to hydrogen atoms in $B_nH_n^{2-}$ or $C_2B_{n-2}H_n$) are not included in the count of skeletal electrons. Thus B—H vertices (or their equivalent such as B—R or B—X vertices) are donors of *two* skeletal electrons, since one of the three boron valence electrons is needed for the external B—H bond. Similarly C—H vertices are donors of *three* skeletal electrons, since one of the four carbon valence electrons is needed for the external C—H bond. "Extra" hydrogen atoms beyond one terminal hydrogen (or equivalent) per vertex are donors of one skeletal electron each, whether terminal or bridging. Similarly, each additional negative charge leads to an additional skeletal electron.

These general rules for skeletal electron counting lead to *Wade's rules*,[25] and subsequently to the *polyhedral skeletal electron pair approach*[26] for the favored number of skeletal electrons for various types of boron polyhedra, which may be stated as follows.

$B_nH_n^{x-}$: The preferred structure is an n-vertex deltahedron with the maximum number of degree 4 and 5 vertices and a preferred charge of -2 (i.e., $x = 2$). This structure thus has $n + 1$ skeletal bonding orbitals, leading to $2n + 2$ skeletal electrons for $B_nH_n^{2-}$. For example, octahedral $B_6H_n^{2-}$ has seven $(= 6 + 1)$ bonding orbitals, leading to $14\ (= (2)(6) + 2)$ skeletal electrons.

B_nH_{n+4} derived from the hypothetical $B_nH_n^{4-}$: The preferred structure is a so-called *nido* structure (from the Latin, *nidus*, nest) derived from an $(n + 1)$-vertex deltahedron with $n + 2$ skeletal bonding orbitals, leading to $2n + 4$ skeletal electrons upon completely filling these bonding orbitals. Neural boron hydrides, which are nido compounds, include B_5H_9 derived from an octahedron by removal of one vertex, B_6H_{10} derived from a pentagonal bipyramid by removal of one vertex, and $B_{10}H_{14}$ derived from an 11-vertex, edge-coalesced icosahedron by removal of one vertex (Figure 8.9).

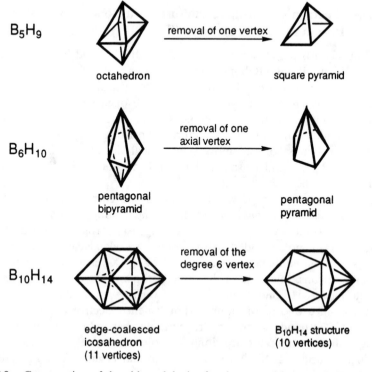

Figure 8.9. Construction of the nido polyhedra for the neutral boranes B_5H_9, B_6H_{10}, and $B_{10}H_{14}$ by removal of a vertex from the next larger deltahedron.

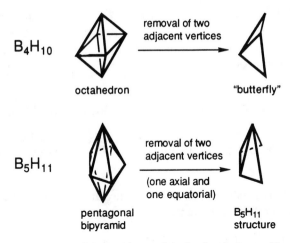

Figure 8.10. Construction of the arachno polyhedra for the neutral boranes B_4H_{10} and B_5H_{11} by removal of two vertices from a larger deltahedron.

B_nH_{n+6} derived from the hypothetical $B_nH_n^{6-}$: The preferred structure is a so-called arachno structure (from the Greek, αραχνη, *arachné* spiderweb) derived from an $(n + 2)$-vertex deltahedron with $n + 3$ skeletal bonding orbitals, leading to $2n + 6$ skeletal electrons upon completely filling these bonding orbitals. Neutral boron hydrides that are arachno compounds include B_4H_{10} derived from an octahedron by removal of two adjacent vertices and B_5H_{11} derived from a pentagonal bipyramid by removal of two adjacent vertices (Figure 8.10).

The special stability of three-dimensional deltahedral boranes and carboranes has led chemists to compare them with two-dimensional planar polygonal aromatic hydrocarbons such as benzene.[27,28] In both types of compound the vertex atoms have four valence orbitals, three of which are used for skeletal bonding, leaving only one orbital for external bonding (e.g., to hydrogen atoms in either benzene or $C_2B_{10}H_{12}$). Furthermore, in both types of compound the three skeletal orbitals are divided into a set of two equivalent orbitals (generically called *twin internal orbitals*) and a third unique orbital. In the case of planar polygonal molecules such as benzene, the twin internal orbitals of a vertex atom are used to form two-electron, two-center bonds (i.e., σ bonds) to its neighbors on each side. The unique internal orbitals in a planar polygonal hydrocarbon are the p orbitals perpendicular to the plane of the polygon; they are used for π-bonding interactions leading to aromatic stabilization. In deltahedral boranes and carboranes, the twin internal orbitals are used for bonding in the deltahedral surface and the unique internal orbitals are used for a multicenter two-electron "core" bond (i.e., the "2" of the "$2n + 2$ rule") in the center of the deltahedron. The positions of the nodes and the relative energies of the molecular orbitals obtained by these interactions of atomic orbitals in deltahedral boranes and carboranes are shown by *tensor surface harmonic theory*[29] to be similar to atomic orbitals (s, p, d, f, etc.), since the vertices of the borane deltahedra can be placed on the surface of a sphere with relatively little distortion.

8.3.5 Metal Complexes of Polyhedral Boranes and Carboranes

A deltahedral carborane can be converted to a nido compound in one of two general ways such as depicted in Figure 8.11 for the icosahedral carborane $1,2\text{-}C_2B_{10}H_{12}$:

1. *Polyhedral degradation*: removal of a vertex and all its attached edges by treatment with a base, for example,

$$1,2\text{-}C_2B_{10}H_{12} + MeO^- + 2MeOH \rightarrow C_2B_9H_{12}^-$$
$$+ (MeO)_3B + H_2\uparrow \qquad (8.17a)$$

$$C_2B_9H_{12}^- + H^- \rightarrow C_2B_9H_{11}^{2-} + H_2\uparrow \qquad (8.17b)$$

The anion $C_2B_9H_{11}^{2-}$ is given the trivial name *dicarbollide*, derived from the trivial name *ollide* (Spanish *olla*, pot) for the hypothetical 11-vertex nido anion $B_{11}H_{11}^{4-}$. Note that removal of the boron vertex from $1,2\text{-}C_2B_{10}H_{12}$ leads to an open pentagonal face in $C_2B_9H_{11}^{2-}$.

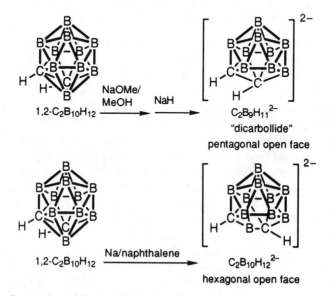

Figure 8.11. Conversion of the icosahedral carborane $1,2\text{-}C_2B_{10}H_{12}$ to the dicarbollide $C_2B_9H_{11}^{2-}$ (top) and to $C_2B_{10}H_{12}^{2-}$ (bottom).

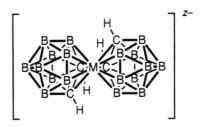

Figure 8.12. The "peanut-shaped" structure of the bis(dicarbollyl) derivatives of transition metals $[(C_2B_9H_{11})_2M]^{z-}$ (M = Cr, Fe, Co, Ni, Cu, etc., z = 0, 1, and/or 2). Hydrogen atoms bonded to boron are omitted for clarity.

2. *Polyhedral expansion*: addition of electrons to a deltahedral carborane with $2n + 2$ skeletal electrons to form the corresponding nido compound with $2n + 4$ skeletal electrons, for example,

$$1,2\text{-}C_2B_{10}H_{12} + 2Na \xrightarrow{C_{10}H_8} 2Na^+ + C_2B_{10}H_{12}^{2-} \qquad (8.18)$$

In the latter reaction, the $C_2B_{10}H_{12}^{2-}$ nido structure is derived from a 13-vertex deltahedron by removal of a vertex of degree 6; such structures derived from deltahedra larger than an icosahedron are called *supraicosahedral structures*. Note that reduction of $C_2B_{10}H_{12}$ opens up a hexagonal face in the nido dianion $C_2B_{10}H_{12}^{2-}$.

The open polygonal faces in the anions $C_2B_9H_{11}^{2-}$ and $C_2B_{10}H_{12}^{2-}$, as well as other related anions, can complex with transition metals much in the manner of cyclopentadienyl and benzene. The resulting metal complexes are even more stable than the corresponding cyclopentadienyl derivatives.[30,31] Thus reactions of $C_2B_9H_{11}^{2-}$ or its precursor $C_2B_9H_{12}^-$ (equation 8.17a) with transition metal compounds leads to bis(dicarbollyl)metal derivatives $(C_2B_9H_{11})_2M^{z-}$ (z = 0, 1, and/or 2, depending on the electronic configuration of the metal), with a characteristic "peanut-shaped" structure (Figure 8.12). The anionic cobalt derivative $(C_2B_9H_{11})_2Co^-$ (Figure 8.11: z = 1) is particularly stable because the cobalt atom has the favored 18-electron noble gas electronic configuration. Metal salts of the $(C_2B_9H_{11})_2Co^-$ anion, even the alkali metal salts, are unusually soluble in organic solvents for ionic compounds because of the very low charge density of this anion. For this reason, addition of this anion to an aqueous solution containing alkali metals allows the alkali metals to be extracted from aqueous solution with organic solvents such as nitrobenzene.[32]

8.4 Boron Halides and Their Derivatives

The binary boron halides BX_3 and B_2X_4 (X = F, Cl, Br, I) are known, as well as a few halides with X/B ratios lower than 2, such as neutral B_nH_n derivatives.

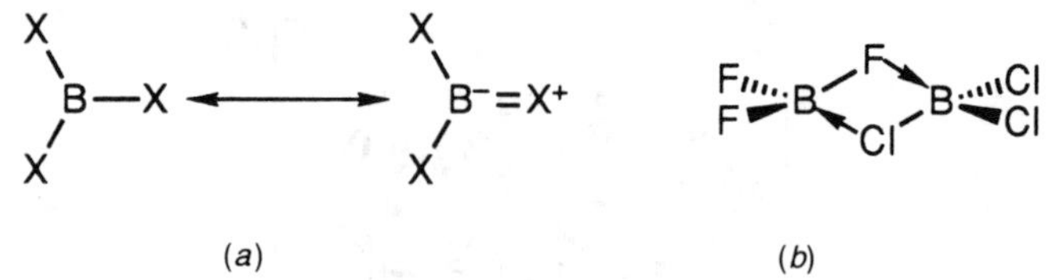

(a)   (b)

Figure 8.13. (a) $p\pi$–$p\pi$ bonding in boron trihalides, BX_3 and (b) the $F_2B(\mu\text{-F})(\mu\text{-}Cl)BCl_2$ intermediate in the BF_3/BCl_3 scrambling reaction.

The B_nH_n "boron subhalides" contain boron deltahedra similar to the deltahedral boranes and carboranes already discussed. The boron trihalides have a trigonal planar structure using sp^2 hybrids for the bonds to the three halogen atoms; there is also an empty p orbital perpendicular to the BX_3 plane (Figure 8.1). The electron deficiency arising from this empty p orbital can be relieved by partial B≡X $p\pi$–$p\pi$ bonding involving donation of a halogen lone pair into this empty boron p orbital (Figure 8.13a). Since, however, such B≡X $p\pi$–$p\pi$ bonding is very limited, the boron trihalides are among the strongest known Lewis acids. Nevertheless, the decrease in the Lewis acidity of boron trihalides in the sequence $BBr_3 > BCl_3 \geqslant BF_3$ can be related to an increase in the strength of partial B≡X $p\pi$–$p\pi$ bonding in the sequence B≡Br < B≡Cl < B≡F. This partial B≡X π-bonding is essentially lost in going from trigonal boron in a boron trihalide, BX_3, to a tetrahedral adduct $L{\rightarrow}BX_3$. The high Lewis acidity of the boron trihalides accounts for rapid halogen scrambling when different boron trihalides are mixed, for example,

$$BF_3 + BCl_3 \rightleftarrows BF_2Cl + BFCl_2 \tag{8.19}$$

Such scrambling reactions, which can be followed by ^{11}B and ^{19}F NMR spectroscopy, involve intermediates of the type $F_2B(\mu\text{-F})(\mu\text{-Cl})BF_2$ (Figure 8.13b), in which the boron octets are completed by a lone pair from a bridging halogen atom.

8.4.1 Boron Trifluoride

Boron trifluoride (mp $-127.1°C$, bp $-99.9°C$), is a pungent gas that can be prepared either by heating B_2O_3 with NH_4BF_4 or by treatment of B_2O_3 or borax with a mixture of CaF_2 and sulfuric acid; the latter method is less expensive for the preparation of BF_3 in large quantities.

Boron trifluoride is a strong Lewis acid and forms adducts with water, ethers, alcohols, amines, phosphines, and so on. The boron trifluoride diethyl ether adduct, $Et_2O{\rightarrow}BF_3$, is a liquid, boiling at $126°C$; it is readily available commercially and is a conveniently handled source of BF_3 for many purposes. The high Lewis acidity of BF_3 makes it a useful catalyst for acid-catalyzed organic reactions including esterification, alkylation of aromatic hydrocarbons with alcohols, and polymerizations of olefins and their expoxides. Friedel–Crafts acylations and alkylations of aromatic hydrocarbons with acyl and alkyl halides, respectively,[33] are promoted by BF_3 but often require stoichiometric amounts of BF_3 for reaction with the HX liberated in the Friedel–Crafts reaction to give BF_3X^-.

The B—F bonds in BF_3 are not readily hydrolyzed by water, in contrast to the other boron trihalides. Instead BF_3 reacts with water at low temperatures to form two hydrates, $BF_3 \cdot H_2O$ (mp 10.2°C) and $BF_3 \cdot 2H_2O$ (mp 6.4°C), which decompose extensively above $+20$°C, regenerating BF_3. The structures of each of these hydrates contain a tetrahedral $H_2O \rightarrow BF_3$ adduct; the dihydrate contains an additional hydrogen-bonded H_2O molecule.[34] Self-ionization of the BF_3 hydrates generates the tetrahedral BF_3OH^- ion:

$$2BF_3 \cdot H_2O \rightarrow [H_3O \rightarrow BF_3]^+ + BF_3OH^- \tag{8.20}$$

At room temperature, the reaction of BF_3 with a limited amount of water results in the formation of boric acid, $B(OH)_3$, and the tetrafluoroborate ion, BF_4^-:

$$4BF_3 + 6H_2O \rightarrow 3H_3O^+ + 3BF_4^- + B(OH)_3 \tag{8.21}$$

8.4.2 Tetrafluoroboric Acid and the Tetrafluoroborate Ion

Unsolvated HBF_4 does not exist; it is an "impossible" structure because of the apparent need for five-coordinate boron, which is not feasible because of the absence of d orbitals in the boron valence shell. However, a number of solvates of HBF_4 are known. The commercially available etherates $R_2O \cdot HBF_4$ (R = Me, Et) are actually the oxonium salts $R_2OH^+BF_4^-$. The BF_4^- ion is tetrahedral. Salts of the BF_4^- ion resemble the corresponding salts of the likewise tetrahedral ClO_4^-. For this reason, tetrafluoroborates frequently can provide safe substitutes for potentially explosive perchlorates of organic cations. The BF_4^- ion is a noncoordinating anion in most systems.

8.4.3 Other Boron Trihalides and Tetrahaloborates

Boron trichloride, BCl_3 (mp -102°C, bp 12.5°C), and *boron tribromide*, BBr_3 (mp -46°C, bp 91.3°C), are fuming, strongly acidic liquids, which are rapidly and violently hydrolyzed by water, in contrast to BF_3. They can be prepared by reactions of B_2O_3/C mixtures with the elemental halogen at elevated temperatures:

$$B_2O_3 + 3C + 3X_2 \xrightarrow{530°C} 3CO + 2BX_3 \tag{8.22}$$

Boron triiodide, BI_3 (mp 49.9°C, bp 210°C), is a rather unstable solid, which can be made by reaction of $NaBH_4$ with elemental iodine:

$$NaBH_4 + 2I_2 \xrightarrow{130°C} BI_3 + NaI + 2H_2 \qquad (8.23)$$

The tetrahaloborates, BX_4^- (X = Cl, Br, I), are much less stable than the corresponding tetrafluoroborates but can be obtained as salts of large monopositive cations. With a given monopositive cation M^+, the order of stability of the $M^+BX_4^-$ salts is X = F ≫ Cl > Br > I.

8.4.4 Halides with Boron–Boron Bonds[35]

The *diboron tetrahalides*, B_2X_4 (X = F, Cl, Br, and I), have B—B distances of 1.70–1.75 Å, suggesting a two-center, two-electron B—B σ bond. The structure of B_2F_4 (mp $-56°C$, bp $-34°C$) has been shown to be planar in the solid and gas phases, consistent with a structure with two linked trigonal boron atoms. Diboron tetrahalides can be obtained by a number of methods, including passing BCl_3 or BBr_3 through an electric discharge between mercury or copper electrodes or reaction of BCl_3 with the boron–boron bonded compound $(Me_2N)_2B—B(NMe_2)_2$ (Section 8.7.5).[36] Reactions of B_2Cl_4 with ethylene and acetylene result in additions of BCl_2 groups across the carbon–carbon multiple bonds. Solvolysis of B_2X_4 with water and alcohols, ROH, gives the boron–boron bonded compounds $B_2(OH)_4$ and $B_2(OR)_4$, respectively.

Pyrolysis of B_2X_4 results in disproportionation to give BX_3 and *neutral* B_nX_n derivatives (X = Cl, n = 4, 8–12; X = Br, n = 7–10), including yellow tetrahedral B_4Cl_4 and red tricapped trigonal prismatic B_9Cl_9. These B_nX_n derivatives are unusual because they are neutral compounds with $2n$ skeletal electrons, in contrast to the $B_nH_n^{2-}$ dianions with $2n + 2$ skeletal electrons.

8.5 Boron–Oxygen Compounds

8.5.1 Boron(III) Oxide, Boric Acid, and Borate Esters

Boron(III) oxide, B_2O_3 (mp 450°C), is obtained by fusing boric acid. This compound, which is very difficult to crystallize and usually forms a glass, is a key ingredient in borosilicate glass. The ability to incorporate boron into silicate glasses is a consequence of the similarity of boron and silicon resulting from the diagonal relationship of these elements (Section 8.1). Crystalline B_2O_3 can be obtained by the careful dehydration of boric acid.

Boron(III) oxide is an acidic oxide and reacts with water to form white flaky crystalline *boric acid*, $B(OH)_3$. The structure of boric acid consists of trigonal $B(OH)_3$ units linked by hydrogen bonding to form infinite layers of nearly hexagonal symmetry separated by 3.18 Å. Boric acid is a weak monobasic acid that functions as a Lewis acid rather than as a proton donor:

$$B(OH)_3 + 2H_2O \rightleftarrows H_3O^+ + B(OH)_4^- \qquad K_a = 6 \times 10^{-10} \qquad (8.24)$$

The $B(OH)_4^-$ anion occurs in several minerals. The acidity of boric acid is increased drastically by the addition of bidentate oxygen donor chelate agents such as ethylene glycol through the formation of tetrahedral chelates such as $(C_2H_4O_2)_2B^-$ (**8.2**). Boric acid functions as a strong acid in anhydrous H_2SO_4 owing to the formation of the tetrahedral $B(OSO_3H)_4^-$ ion in this medium.

Pyrolysis of boric acid ultimately results in dehydration to B_2O_3, as noted earlier. Intermediates in this pyrolysis process include *cyclotriboric acid*, $B_3O_3(OH)_3$, containing a six-membered B_3O_3 ring (**8.3**), and polymeric "metaboric acid," $(HBO_2)_n$. Solid cyclotriboric acid has a hydrogen-bonded layer structure.

8.2   8.3

Boric acid is readily esterified by treatment with alcohols in the presence of a strong acid such as sulfuric acid. The resulting *borate esters*, $(RO)_3B$, are readily distillable liquids—for example, trimethyl borate, $(MeO)_3B$ (bp 69°C). The borate esters are considerably weaker Lewis acids than the boron trihalides (Section 8.4) owing to considerable $B{=}O$ $p\pi{-}p\pi$ bonding.

8.5.2 Polynuclear Borate Structures[37,38]

Boric acid and its anion $B(OH)_4^-$ readily condense to polynuclear structures with B—O—B bonds as illustrated by the formation of cyclotriboric acids, **8.3**, in the pyrolysis of boric acid. Other polynuclear borate anions containing B—O—B units are depicted in Figure 8.14; most of these structures contain six-membered B_3O_3 rings. The terminal O^- groups are readily protonated in aqueous borate solutions to give terminal hydroxy groups leading to *pentaborate*, $B_5O_6(OH)_4^-$, with a charge/boron ratio of $-\frac{1}{5}$, *triborate* $B_3O_3(OH)_4^-$, with a charge/boron ratio of $-\frac{1}{3}$, and *tetraborate* $B_4O_5(OH)_4^{2-}$, with a charge/boron ratio of $-\frac{1}{2}$ (Figure 8.14). Such polynuclear borate structures contain both three-coordinate and four-coordinate boron atoms with each four-coordinate boron atom bearing a negative charge.

A large number of crystalline borates are known containing the structural units depicted in Figure 8.14. In most cases the stoichiometry provides little clue as to the anion structure. The important boron mineral *borax* is commonly written

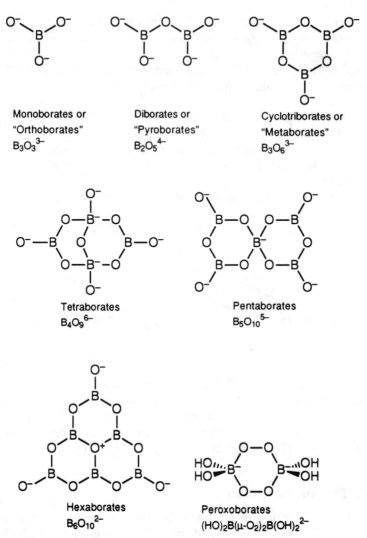

Figure 8.14. Structures of some borate anions. The terminal O^- groups in these structures are generally protonated to terminal OH groups in aqueous media. The peroxoborate ion $(HO)_2B(\mu\text{-}O_2)_2B(OH)_2^{2-}$ is also shown.

as "$Na_2B_4O_7 \cdot 4H_2O$" but is actually the salt $Na_2[B_4O_5(OH)_4] \cdot 2H_2O$ containing the tetraborate ion depicted in Figure 8.14 but with the four terminal O^- groups protonated to OH groups. The same tetraborate ion is present in the so-called octaborates, $M_2^IM^{II}B_8O_{14} \cdot 12H_2O = M_2^IM^{II}[B_4O_5(OH)_4]_2] \cdot 8H_2O$. The hydrated triborate dianion $B_3O_3(OH)_5^{2-}$ is commonly found as 1:1 salts with Ca^{2+} and Mg^{2+}.

The main systematics of the structures of crystalline metal borates can be summarized as follows.

1. Borons can link either three oxygens to form a triangle or four oxygens to form a tetrahedron.
2. Polynuclear anions are formed by vertex sharing only of these boron–oxygen triangles or tetrahedra to give compact insular groups.
3. In the hydrated borates, protonatable oxygen atoms will be protonated in the following sequence: (a) available protons are first assigned to free O^{2-} ions to convert them to OH^- ions; (b) additional protons are then assigned to oxygens first on tetrahedral boron atoms and then on trigonal boron atoms to give B—OH units; (c) any remaining protons are assigned to free OH^- groups to give H_2O molecules.
4. The hydrated insular groups may polymerize in various ways to eliminate water; there may be B—O bond formation within the polyanion framework.
5. Complicated borate polyanions may be modified by the attachment of individual side groups such as a borate tetrahedron or triangle, two linked borate triangles, and/or heteroatom tetrahedron.

8.5.3 Peroxoborates

Reactions of borates with hydrogen peroxide or boric acid with sodium peroxide give the peroxoborate variously formulated as "$NaBO_3 \cdot 4H_2O$" or "$NaBO_2 \cdot H_2O_2 \cdot 3H_2O$." X-Ray diffraction indicates the structure $[(OH)_2B(\mu\text{-}O_2)_2B(OH)_2]^{2-}$ (Figure 8.14) with two bridging peroxo groups. In solution, boric acid and hydrogen peroxide are in equilibrium with a mononuclear peroxoborate anion:

$$B(OH)_3(O_2H)^- + H_2O \rightleftarrows B(OH)_4^- + H_2O_2 \qquad K \approx 0.04 \qquad (8.25)$$

Peroxoborates are useful in bleaches because of their ability to release hydrogen peroxide by such processes, giving only innocuous boric acid or borates as by-products.

8.6 Organoboron Compounds

In addition to the carboranes discussed in Section 8.3.3, a variety of other compounds with boron–carbon bonds are known.[39] The *trialkylboranes* and *triarylboranes*, R_3B, contain three B—C bonds; they can be made by reactions of $Et_2O \cdot BF_3$ with the corresponding alkylmagnesium derivative RMgX or alkyllithium derivative RLi. They have monomeric structures with trigonal planar boron (i.e., as in Figure 8.1). The lower alkyls such as Me_3B and Et_3B burn spontaneously in air with a green flame. Triarylboron derivatives are significantly less oxygen sensitive; Ph_3B oxidizes in air but not highly exothermically and trimesitylboron, $(2,4,6\text{-}Me_3C_6H_2)_3B$, is stable in air because the boron is shielded by the bulky mesityl groups.

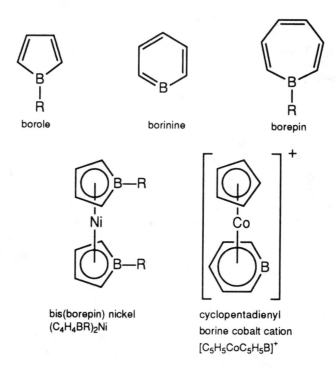

Figure 8.15. Planar unsaturated boron heterocycles and examples of their transition metal complexes.

Boron forms unsaturated heterocyclic planar ring systems such as boroles (boracyclopentadienes), borinins (borabenzenes), and borepins (Figure 8.15). These are rather reactive and unstable unless they are highly substituted and have no direct B—H bonds. Even borepins are not particularly stable, although they have the favored 6π electrons for a planar polygonal structure like the relatively stable cyclopentadienide, benzene, and tropylium ions. The syntheses of unsaturated boron heterocycles are rather complicated and beyond the scope of this book. Stable transition metal complexes are formed by boroles and borabenzenes.[40] For example, bis(borole)nickel complexes and cyclopentadienylborabenzenecobalt cations (Figure 8.15) are very stable because the central transition metals have the favored 18-electron noble gas configuration.

Organoboron compounds containing a single B—C carbon bond are also known. *Alkyl-* and *arylboronic acids*, $RB(OH)_2$, can be obtained by alkylation of $Et_2O \cdot BF_3$ or $(RO)_3B$ with organomagnesium halides, $RMgX$, followed by hydrolysis. They undergo facile dehydration to trimeric cyclic anhydrides, $R_3B_3O_3$, known as *alkyl-* or *arylboroxines*, which contain the stable B_3O_3 ring similar to that found in the cyclic borates (Figure 8.14).

One of the most important reactions in boron chemistry is the *hydroboration* of olefins, which was developed extensively by H. C. Brown and his co-workers as an important method in symethic organic chemistry.[41,42] Hydroboration involves the addition of the B—H bonds from a borane source such as $Me_2S{\rightarrow}BH_3$, $C_4H_8O{\rightarrow}BH_3$, or diborane across the C=C double bond of olefins to generate one or more boron–carbon bonds. This addition proceeds in the anti-Markovnikov sense, with the boron attaching to the least substituted carbon atom. The number of boron–carbon bonds that can be formed by olefin hydroboration depends on the steric hindrance of the olefin. Thus hydroboration of tetrasubstituted olefins stops after forming a single boron–carbon bond:

$$R^1R^2C{=}CR^3R^4 + BH_3 \rightarrow R^1R^2CH{-}C(BH_2)R^3R^4 \qquad (8.26)$$

Less hindered disubstituted and trisubstituted olefins can form two boron–carbon bonds in the hydroboration reaction:

$$2R^1R^2C{=}CHR^3 + BH_3 \rightarrow (R^1R^2CH{-}CHR^3)_2BH \qquad (8.27)$$

Hydroboration of unhindered monosubstituted alkenes can proceed all the way to the corresponding trialkylboron with the formation of three boron–carbon bonds:

$$3RCH{=}CH_2 + BH_3 \rightarrow (RCH_2CH_2)_3B \qquad (8.28)$$

The value of the hydroboration reaction lies in the ability to convert the organoboron compounds to organic compounds of interest such as hydrocarbons and alcohols through boron–carbon bond cleavage reactions such as the following:

$$R_3B + 3H_2O \xrightarrow{\;H_3O^+\;} 3R{-}H + B(OH)_3 \qquad (8.29a)$$

$$R_3B + 3H_2O_2 \xrightarrow{\;NaOH\;} 3R{-}OH + B(OH)_3 \qquad (8.29b)$$

$$R_3B + CO \xrightarrow[110^\circ C]{diglyme} \frac{1}{x}[R_3CBO]_x \qquad (8.29c)$$

$$\frac{1}{x}[R_3CBO]_x + H_2O_2 + H_2O \xrightarrow{\;NaOH\;} R_3COH + B(OH)_3 \qquad (8.29c')$$

Other compounds with B—H bonds may be used instead of BH_3 sources for the hydroboration reaction. A very regiospecific hydroborating agent is 9-borabicyclo[3.3.1]nonane or "9-BBN," which is obtained as a liquid (bp $195^\circ C/12$ mm) by the hydroboration of 1,5-cyclooctadiene. 9-Borabicyclo[3,3,1]nonane has a dimeric structure that is closely related to the diborane structure, with two three-center B—H—B bonds between a pair of boron atoms (Figure 8.16).

The ultimate products of the alkylation or arylation of boron compounds are the tetraalkyl- and tetraarylborate anions, R_4B^-, containing an anionic tetrahedral boron atom; they are formally alkyl anion (R^-) adducts of the

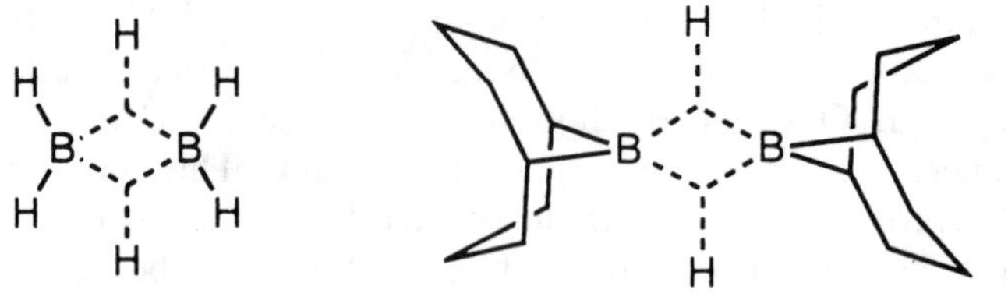

Figure 8.16. The related structures of diborane and 9-borabicyclo[3.3.1]nonane.

Lewis acids R_3B. The most important R_4B^- anion is tetraphenylborate. Sodium tetraphenylborate, $Na^+BPh_4^-$, is soluble in water and stable in neutral aqueous solution without decomposition. However, the tetraphenylborate ion decomposes in acid solution to give benzene and triphenylboron:

$$(C_6H_5)_4B^- + H^+ \rightarrow C_6H_6 + (C_6H_5)_3B \qquad (8.30)$$

Tetraphenylborates of larger monovalent cations, including K^+, Rb^+, Cs^+, and R_4N^+, are so insoluble in water that these large monovalent cations can be precipitated quantitatively with tetraphenylborate anion.

8.7 Boron–Nitrogen Derivatives

A B—N unit is isoelectronic with a C—C unit. For this reason R_2B—$NR_2 \leftrightarrow R_2B^-$=$N^+R_2'$ derivatives are isoelectronic with the corresponding olefins R_2C=CR_2' and RB=NR' $\leftrightarrow$ RB$^-$≡N^+R' derivatives are isoelectronic with the corresponding acetylene RC≡CR'. Such boron–nitrogen bonds have appreciable π character. Since, however, the polarity in the π-bonding network implied by the charge separation is approximately balanced by a reverse polarity in the σ-bonding network relating to the higher electronegativity of nitrogen relative to boron, the B—N bonds in such molecules are nearly nonpolar. The properties of such boron–nitrogen compounds resemble those of the corresponding carbon–carbon compounds in many ways.

8.7.1 Borazenes: Analogues of Benzenes

The boron–nitrogen analogue of benzene, C_6H_6, is *borazene*, $B_3N_3H_6$ (Figure 8.17), which is a colorless liquid (mp $-57°C$, bp $80.5°C$). Borazene has a regular planar hexagonal ring structure with physical properties resembling the isoelectronic benzene including an "aromatic" odor.

The original synthesis of borazene (Figure 8.17: R = X = H) used the reaction of diborane with ammonia at elevated temperatures:

$$3B_2H_6 + 6NH_3 \xrightarrow{180°C} 2B_3H_3N_6 + 15H_2 \qquad (8.31)$$

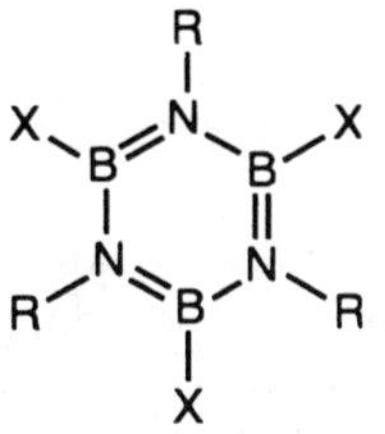

Figure 8.17. The structure of borazenes, "inorganic" analogues of benzene.

More convenient preparations of borazene and substituted borazenes use reactions of boron halides with N–H compounds, for example,

$$3NH_4Cl + 3BCl_3 \xrightarrow[150°C]{PhCl} (-NH-BCl-)_3 + 9HCl\uparrow \qquad (8.32a)$$

$$RNH_2 + BCl_3 \xrightarrow{PhCl} RNH_2 \cdot BCl_3 \qquad (8.32b)$$

$$3RNH_2 \cdot BCl_3 + 6Me_3N \rightarrow (-NRBCl-)_3 + 6Me_3NH^+Cl^- \qquad (8.32b')$$

Thus N-substituents in borazenes (R in Figure 8.17) can be introduced by choice of a suitable primary amine, RNH_2 (equation 8.32b). The B—Cl bonds in B-chloroborazene derivatives (X = Cl in Figure 8.17) are very reactive and can be converted to other groups by metathesis reactions, allowing the introduction of diverse B-substituents in borazenes. Unsubstituted borazene is thus best obtained by the hydride reduction of B,B',B''-trichloroborazene obtained from NH_4Cl and BCl_3 (equation 8.32a):

$$2(-NH-BCl-)_3 + 6NaBH_4 \rightarrow 2B_3N_3H_6 + 6NaCl + 3B_2H_6 \qquad (8.33)$$

Despite the structural and physical resemblance of borazene to benzene, the chemical properties of borazene suggest only weak aromaticity. Thus, borazene is considerably more reactive than benzene and in contrast to benzene readily undergoes addition reactions to give derivatives of "inorganic cyclohexane":

$$(-NH-BH-)_3 + 3HX \rightarrow (-NH_2-BHX-)_3 \qquad (8.34)$$

Unsubstituted borazene decomposes slowly upon storage at ambient conditions and is hydrolyzed at elevated temperatures to NH_3 and $B(OH)_3$. Polyalkylborazenes resemble polyalkylbenzenes in forming arene–transition metal complexes such as $Me_6B_3N_3Cr(CO)_3$, although such hexahaptoborazene metal complexes are much less stable than their hexahaptobenzene analogues.

8.7.2 Boron Nitride: Analogue of Elemental Carbon

Boron nitride, $(BN)_x$, is isoelectronic with elemental carbon and exists in the following allotropic forms, corresponding to two of the three allotropic forms of carbon (Section 2.2.2):

1. *Hexagonal boron nitride* has a layer structure similar to that of graphite, but the layers of hexagonal boron nitride are stacked in an eclipsed manner. Hexagonal boron nitride is colorless and nonconducting, in contrast to graphite. The B—N distances in hexagonal boron nitride of 1.45 Å are similar to the 1.415 Å C—C distances in graphite.
2. *Cubic boron nitride* has an infinite three-dimensional structure similar to diamond with sp^3-hybridized boron and nitrogen atoms and B—N distances of 1.56 Å, which are comparable to the C—C distances of 1.514 Å in diamond. Cubic boron nitride is as hard as diamond.

Hexagonal boron nitride can be prepared by a variety of methods,[43] including treatment of borax with ammonium chloride at elevated temperatures, reaction of urea with boric acid in an ammonia atmosphere at 500–950°C followed by annealing at 1800°C, and reduction of B_2O_3 with KCN at 1100°C in a graphite crucible. Cubic boron nitride is obtained from hexagonal boron nitride at elevated temperatures and very high pressures (1800°C/85,000 atm), analogous to the preparation of diamond from graphite.

8.7.3 Iminoboranes: Analogues of Acetylenes

Iminoboranes, $RB^-\equiv N^+R'$, are analogues of alkynes and are isolable only when the substituents R and R' are bulky groups (e.g., R = tetramethyl-piperidine, R' = *tert*-butyl).[44,45] They can be obtained by reactions involving R_3SiX elimination, for example,

$$RB(X)NR'(SiMe_3) \xrightarrow[\text{vacuum}]{\Delta} R{-}B^-\equiv N^+{-}R' + Me_3SiX \qquad (8.35)$$

8.7.4 Pyrazole Derivatives of Boron

Pyrazole forms a variety of boron derivatives[46] (Figure 8.18) some of which, namely the polypyrazolylborates, are of interest as ligands in coordination chemistry. Reaction of a borane adduct, $L{\rightarrow}BH_3$, or a trialkylboron with pyrazole or an alkyl-substituted pyrazole at elevated temperatures, results in elimenation of hydrogen or alkane to give a binuclear derivative with two four-coordinate boron atoms known as a *pyrazabole*, for example,

$$2R_3B + 2C_3H_3N_2H \xrightarrow{\Delta} [C_3H_3N_2BR_2]_2 + 2RH\uparrow \qquad (8.36)$$

Similarly, reaction of an alkali metal borohydride with excess pyrazole or an alkylpyrazole at elevated temperatures results in successive replacement of B—H bonds with B—pyrazolyl bonds to give successively dihydrobis-(pyrazolyl)borates, $H_2B(C_3H_3N_2)_2^-$; hydrotris(pyrazolyl)borates, $HB(C_3H_3N_2)_3^-$,

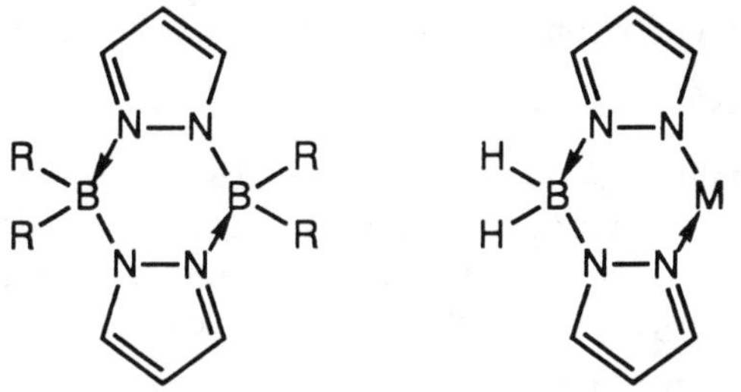

Pyrazabole
[C₃H₃N₂BR₂]₂
R = H, alkyl, etc.

A bis(pyrazolyl)borate
metal complex:
H₂B(C₃H₃N₂)₂M

Figure 8.18. Boron-pyrazole derivatives: (a) the pyrazabole structure and (b) bis-(pyrazolyl)borate as a bidentate ligand.

and tetrakis(pyrazolyl)borates, $B(C_3H_3N_2)_2^-$, for example,

$$KBH_4 + 2C_3H_3N_2H \xrightarrow{\sim 90°C} K^+[H_2B(C_3H_3N_2)_2]^- + 2H_2\uparrow \qquad (8.37a)$$

$$KBH_4 + 3C_3H_3N_2H \xrightarrow{\sim 170°C} K^+[HB(C_3H_3N_2)_3]^- + 3H_2\uparrow \qquad (8.37b)$$

$$KBH_4 + 4C_3H_3N_2H \xrightarrow{> 200°C} K^+[B(C_3H_3N_2)_4]^- + 4H_2\uparrow \qquad (8.37c)$$

The dihydrobis(pyrazolyl)borates are uninegative bidentate chelate ligands, and both the hydrotris(pyrazolyl)borates and tetrakis(pyrazolyl)borates are uninegative tridentate chelate ligands forming BN_2MN_2 six-membered chelate rings in all cases (Figure 8.18). Divalent metals form tetrahedral metal chelates (e.g., $[H_2B(C_3H_3N_2)_2]_2M^{II}$) with bis(pyrazolyl)borate ligands and octahedral metal chelates (e.g., $[HB(C_3H_3N_2)_3]_2M^{II}$ and $[B(C_3H_3N_2)_4]_2M^{II}$) with tris-(pyrazolyl)borate and tetrakis(pyrazolyl)borate ligands, respectively. Note that the tetrahedral configuration of the boron atom prevents the fourth pyrazolyl ring of the tetrakis(pyrazolyl)borate anion from coordinating to the same metal atom as the first three pyrazolyl rings, with the result that the tetrakis-(pyrazolyl)borate anion functions only as a tridentate rather than a tetradentate ligand to a single metal atom.

8.7.5 Other Dialkylaminoboron Derivatives

Several types of B—B bonded network are stabilized by external dialkylamino groups owing to relief of the electron deficiency of the boron atoms through partial B=N double bonding related to that found in borazenes and boron nitride (see Section 8.7.1 and 8.7.2). Dialkylaminoboron halides are readily obtained by treatment of boron trihalides with secondary amines, for example,

$$BCl_3 + 4Me_2NH \rightarrow (Me_2N)_2BCl + 2Me_2NH_2^+Cl^- \tag{8.38}$$

Low molecular weight dialkylaminoboron halides are distillable liquids that are readily hydrolyzed. A boron–boron bond can be obtained by treatment of $(Me_2N)_2BCl$ with an alkali metal in an inorganic analogue of the Wurtz coupling reaction in organic chemistry, for example,

$$2(Me_2N)_2BCl + 2K \rightarrow (Me_2N)_2B—B(NMe_2)_2 + 2KCl \tag{8.39}$$

The product is a distillable liquid (mp $-33°C$, bp $42.5°C/0.05$ mm), which can readily be converted to other B_2X_4 derivatives including B_2Cl_4 (Section 8.4.4). A minor product from the reaction of $(Me_2N)_2BCl$ with Na/K alloy is orange-red hexakis(dimethylamino)cyclohexaborane, $(Me_2N)_6B_6$, which contains a B_6 ring in a chair conformation and trigonal planar boron and nitrogen coordination.[47]

References

1. Kölle, P.; Nöth, H., The Chemistry of Borinium and Borenium Ions, *Chem. Rev.*, 1985, **85**, 399–418.

2. Matkovich, V. I., Ed., *Boron and Refractory Borides*, Springer-Verlag, Berlin, 1977.

3. Vlasse, M.; Naslain, R.; Kaspar, J. S.; Ploog, K., *J. Solid State Chem.*, 1979, **28**, 289–301.

4. Decker, B. F.; Kasper, J. S., *Acta Crystallogr.*, 1959, **12**, 503–506.

5. Hoard, J. L.; Sullenger, D. B.; Kennard, C. H. L.; Hughes, R. E., *J. Solid State Chem.*, 1970, **1**, 268–277.

6. Rogl, P.; Nowotny, H., Structural Chemistry of Ternary Metal Borides, *J. Less Common Met.*, 1978, **61**, 39–45.

7. Étourneau, J.; Hagenmuller, P., Structural Features of Rare Earth Borides, *Phil. Mag. B.* 1985, **52**, 589–610.

8. Lipscomb, W. N.; Britton, D., Valence Structures of the Higher Borides, *J. Chem. Phys.*, 1960, **33**, 275–280.

9. Lipscomb, W. N., Borides and Boranes, *J. Less Common Met.*, 1981, **82**, 1–20.

10. Heřmánek, S., ^{11}B NMR Spectra of Boranes, Main-Group Heteroboranes, and Substituted Derivatives, Factors Influencing Chemical Shifts of Skeletal Atoms, *Chem. Rev.*, 1992, **92**, 325.

11. Dickerson, R. E.; Lipscomb, W. N., Semitopological Approach to Boron Hydride Structures, *J. Chem. Phys.*, 1957, **27**, 212–217.

12. Lipscomb, W. N., *Boron Hydrides*, Benjamin, New York, 1963.

13. Lipscomb, W. N., *Advances in Theoretical Studies of Boron Hydrides and Carboranes*, in E. L. Muetterties, Ed., *Boron Hydride Chemistry*, Academic Press, New York, 1975, pp. 30–78.

14. Long, L. H., Recent Studies of Diborane, *Prog. Inorg. Chem.*, 1972, **15**, 1–99.

15. Stock, A., *Hydrides of Boron and Silicon*, Cornell University Press, Ithaca, NY, 1933.

16. Gaines, D. F., Recent Advances in the Chemistry of Pentaborane (9), in *Boron Chemistry*, Vol. 4, W. R. Parry and G. Kodama, Eds., Pergamon Press, Oxford, 1980.

17. Wade, R., *Sodium Borohydride and Its Derivatives*, in *Specialty Inorganic Chemicals*, R. Thompson, Ed., Royal Society of Chemistry, London, 1981, pp. 25–58.

18. Hajós, A., *Complex Hydrides and Related Reducing Agents in Organic Synthesis*, Elsevier, Amsterdam, 1979.

19. Muetterties, E. L.; Knoth, W. H., *Polyhedral Boranes*, Dekker, New York, 1968.

20. King, R. B.; Duijvestijn, A. J. W., *Inorg. Chim. Acta*, 1990, **178**, 55–57.

21. Grimes, R. N., *Carboranes*, Academic Press, New York, 1970.

22. Bregadze, V. I., Dicarba-*closo*-dodecaboranes $C_2B_{10}H_{12}$ and their Derivatives, *Chem. Rev.*, 1992, **92**, 209.

23. Štíbr, B., Carboranes Other than $C_2B_{10}H_{12}$, *Chem. Rev.*, 1992, **92**, 225–250.

24. Williams, R. E., Coordination Number–Pattern Recognition Theory of Carborane Structures; *Adv. Inorg. Chem. Radiochem.*, 1976, **18**, 67–142.

25. Wade, K., Structural and Bonding Patterns in Cluster Chemistry, *Adv. Inorg. Chem. Radiochem.*, 1976, **18**, 1–66.

26. Mingos, D. M. P., Polyhedral Skeletal Electron Pair Approach, *Acc. Chem. Res.*, 1992, **17**, 311–320.

27. King, R. B.; Rouvray, D. H., *J. Am. Chem. Soc.*, 1977, **99**, 7834–7840.

28. King, R. B., *Applications of Graph Theory and Topology in Inorganic Cluster and Coordination Chemistry*, CRC Press, Boca Raton, FL, 1993, Chapter 4.

29. Stone, A. J., The Bonding in Boron and Transition-Metal Cluster Compounds, *Polyhedron*, 1984, **12**, 1299–1306.

30. Callahan, K. P.; Hawthorne, M. F., Ten Years of Metallocarboranes, *Adv. Organomet. Chem.*, 1976, **14**, 145–186.

31. Grimes, R. N., The Role of Metals in Borane Clusters, *Acc. Chem. Res.*, 1983, **16**, 22–26.

32. Rais, J.; Selucký, P.; Kyrš, M., *J. Inorg. Nuclear Chem.*, 1976, **38**, 1376–1378.

33. Olah, G., Ed., *Friedel–Crafts and Related Reactions*, Wiley-Interscience, New York, 1963.

34. Mootz, D.; Steffen, M., *Acta Crystallogr.*, 1981, **B37**, 1110–1112.

35. Morrison, J. A., Chemistry of the Polyhedral Boron Halides and the Diboron Tetrahalides, *Chem. Rev.*, 1991, **91**, 35–48.

36. Nöth, H.; Meister, W., *Chem. Ber.*, 1961, **94**, 509–514.

37. Farmer, J. B., Metal Borates, *Adv. Inorg. Chem. Radiochem.*, 1982, **25**, 187–237.

38. Heller, G., A Survey of Structural Types of Borates and Polyborates, *Top. Curr. Chem.*, 1986, **131**, 39–98.

39. Onak, T., *Organoborane Chemistry*, Academic Press, New York, 1975.

40. Herberich, G. E.; Ohst, H., Borabenzene Metal Complexes, *Adv. Organomet. Chem.*, 1986, **25**, 199–236.

41. Brown, H. C., *Boranes in Organic Chemistry*, Cornell University Press, Ithaca, NY, 1972.

42. Brown, H. C., *Organic Synthesis via Boranes*, Wiley, New York, 1975.

43. Paine, R. T.; Narula, C. K., Synthetic Routes to Boron Nitride, *Chem. Rev.*, 1990, **90**, 73–91.

44. Paetzold, P., Iminoboranes, *Adv. Inorg. Chem.*, 1987, **31**, 123–170.

45. Nöth, H., The Chemistry of Amino Iminoboranes, *Angew. Chem. Int. Ed.* (Engl., 1988, **27**, 1603–1623.

46. Niedenzu, K.; Trofimenko, S., Pyrazole Derivatives of Boron, *Top. Curr. Chem.*, 1986, **131**, 1–37.

47. Nöth, H., Pommerening, H., *Angew. Chem. Int. Ed. Engl.*, 1980, **19**, 482–483.

CHAPTER

9

Aluminum, Gallium, Indium, and Thallium

9.1 General Aspects of the Chemistry of Aluminum, Gallium, Indium, and Thallium

The trivalent oxidation state is favored for aluminum, gallium, indium, and thallium, all of which have aqueous trivalent cationic chemistry, in contrast to boron. The hydrous oxides of trivalent aluminum and gallium are amphoteric and thus dissolve both in aqueous acid, to form the corresponding $M(H_2O)_6^{3+}$ cations, and in aqueous alkali, to form the corresponding $M(OH)_4^-$ anions. Trivalent compounds of the type MX_3 are often Lewis acids, which combine with Lewis bases to form tetrahedral species of the type $L \rightarrow MX_3$. The acceptor strengths of such trivalent compounds to form the tetrahedral adducts decrease in the sequence Al > Ga > In. Coordination numbers higher than 4 are possible for Al, Ga, In, and Tl because of the presence of d orbitals in the valence shells, in contrast to boron. Thus aluminium hydride forms not only the monoadducts $LAlH_3$ but also the diadducts L_2AlH_3, as exemplified by $(Me_3N)_2AlH_3$. Binary hydrides are stable only in the case of $[AlH_3]_x = [AlH_{6/2}]_x$, which is non-volatile owing to polymerization through Al—H—Al bridge bonding and the ability of aluminum to exhibit octahedral coordination because of accessible d orbitals for d^2sp^3 hybridization. The stability of metal–hydrogen bonds decreases rapidly in the sequence Al > Ga > In > Tl, with the result that no compounds of any type with Tl—H bonds are stable at room temperature.

These elements exhibit the so-called *inert pair effect*, which relates to the stability of an electronic configuration retaining an s^2 pair. This effect is particularly marked with thallium, leading to a $+1$ oxidation state even more

233

stable than the $+3$ oxidation state and similar to the greater stability of the $+2$ oxidation state relative to the $+4$ oxidation state for lead (Chapter 3). Thus Tl^+ is stable in aqueous solution and exibits chemistry similar to that of K^+ and Ag^+, whereas Tl^{3+} is a strong oxidizing agent:

$$Tl_3^+ + 2e^- = Tl^+ \qquad E° = +1.25 \text{ V} \qquad (9.1)$$

The oxidation state $+2$ is very rare for aluminum and its congeners. Some compounds with formulas corresponding to the $+2$ oxidation state are actually mixed oxidation state derivatives containing equal amounts of metals in the $+1$ and $+3$ oxidation states: for example, "$GaCl_2$" $= Ga_2Cl_4 = Ga^+GaCl_4^-$ with Ga(I) in the cation and Ga(III) in the anion. The mixed metal oxidation state in such compounds is indicated by their diamagnetism, since "true" M(II) (M = Al, Ga, In, Tl) must have an unpaired electron and thus be paramagnetic.

9.2 The Elements and Their Alloys

Aluminum is the most abundant metallic element in the earth's crust (8.8 mass %) and is an important constituent of silicate minerals (Section 3.5), where frequently an AlO_2^- unit can replace an SiO_2 unit in the structure. Aluminum minerals can be classified into three general types:

1. Aluminum-containing silicates such as micas and feldspars (Section 3.5.3). Aluminum is an important constituent in clay-type minerals such as kaolinite, montmorillonite, and vermiculite.
2. Aluminum oxide minerals such as the hydroxo-oxide *bauxite*.
3. Aluminum fluoride minerals, notably *cryolite*, Na_3AlF_6.

Gallium, indium, and thallium are relative rare elements in comparison to aluminum, occurring in the earth's crust in concentrations of 19, 0.21, and 0.7 ppm, respectively. Gallium and indium are found in aluminum and zinc ores. Indium is also found in low concentrations in sulfide minerals. Thallium is widely distributed; it can be recovered from flue dusts obtained from roasting pyrites. Thallium and its compounds are very toxic; skin contact, ingestion, and inhalation are all dangerous.

All four metals (Al, Ga, In, Tl) have isotopes suitable for NMR studies (Table 9.1).[1] However, the quadrupole moments of Al, Ga, and particularly In can lead to relatively broad spectra. Thallium-205, which has a spin of $\frac{1}{2}$ and thus no quadrupole moment, is the most sensitive heavy metal nucleus; its NMR sensitivity is more than twice that of [31]P (Section 5.1).

Aluminum metal is prepared on a gigantic scale by the *Hall process*: first the bauxite ore is purified by dissolving in sodium hydroxide followed by reprecipitation with carbon dioxide, then the purified bauxite is dissolved in molten cryolite at 800–1000°C and the melt is electrolyzed. Synthetic cryolite, Na_3AlF_6, prepared for aluminum oxide, sodium hydroxide, and hydrofluoric acid, can be used as a flux for this melt.

Table 9.1 Isotopes of Aluminum, Gallium, Indium, and Thallium for NMR Spectroscopy

Isotope	Spin	Abundance (%)	Quadrupole Moment ($\times 10^{-28}$ m^2)	NMR Sensitivity*
^{27}Al	$\frac{5}{2}$	100	0.149	0.206
^{69}Ga	$\frac{3}{2}$	60.2	0.19	0.069
^{71}Ga	$\frac{3}{2}$	39.8	0.12	0.142
^{115}In	$\frac{9}{2}$	95.7	0.83	0.332
^{203}Tl	$\frac{1}{2}$	29.5	0	0.187
^{205}Tl	$\frac{1}{2}$	70.5	0	0.192

* Relative to the proton for equal numbers of nuclei.

Aluminum is a hard, strong, white, highly electropositive metal (mp 660.2°C, bp 2467°C), which is resistant to corrosion because of a hard tough film of oxide on its surface. The surface of aluminum can be protected further by *anodization*, which involves electrolysis in 15–20% sulfuric acid. The protective oxide film on aluminum can be removed by amalgamation with traces of mercury; aluminum amalgam is attacked even by pure water, with evolution of hydrogen. Aluminum is attacked by dilute mineral acids but passivated by concentrated nitric acid. Aluminum is also attacked under ordinary conditions by hot alkali hydroxide and free halogens.

Gallium, indium, and thallium metals can be liberated from aqueous solutions of their salts by electrolysis because of a large overvoltage for hydrogen evolution, in contrast to aluminum. Thus gallium is obtained by electrolysis of a sodium gallate solution using a stainless steel cathode. Gallium (mp 29.8°C, bp 2070°C), indium (mp 157°C, bp 2000°C), and thallium (mp 303°C, bp 1457°C) are soft, white, relatively reactive metals with low melting points. The exceptionally low melting point and typical boiling point of gallium give it the longest liquid range of any known substance; it thus is used as a thermometer liquid. In general, gallium, indium, and thallium metals all dissolve readily in acids. However, the dissolution of thallium in hydrochloric or sulfuric acid is relatively slow because of the limited solubilities of the thallium(I) salts of these acids. Gallium, like aluminum, is soluble in sodium or potassium hydroxides. Gallium, indium, and thallium react readily with many nonmetals including halogens and sulfur.

A characteristic of boron chemistry is the stability of B_{12} icosahedra, which are found in the allotropes of elemental boron (Section 8.2). Metallic aluminum, gallium, indium, and thallium have typical metallic structures without discrete metal icosahedra or other deltahedra. However, aluminum is an important component in many icosahedral quasicrystal alloys,[2-4] which include the following types:

1. The i(Al—Mt) class (Mt = transition metal) such as $Al_{80}Mn_{20}$, $Al_{74}Mn_{20}Si_6$, and $Al_{79}Cr_{17}Ru_4$.

2. The Al—Zn—Mg class such as $Al_{25}Zn_{38}Mg_{37}$, Al_{44}, Zn_{15}, Cu_5, Mg_{36}, and $Al_{60}Cu_{10}Li_{30}$.

Icosahedral quasicrystals exhibit diffraction patterns (e.g., X-ray) with apparently sharp spots containing fivefold symmetry axes despite the incompatibility of fivefold axes, with the periodicity of normal crystals; for this reason these materials have been called quasicrystals. The structure and bonding of quasi-crystals is obviously very complicated, but it can be related to the likewise complicated structure and bonding of the boron allotropes (Section 8.2).[5]

Deltahedral structures (Section 8.3.2.3: Figure 8.7) are found in a number of alloys of gallium with alkali metals.[6] Thus Ga_{12} icosahedra are found in MGa_7 (M = Rb, Cs) = $M_2(Ga_{12})(Ga_2)$ and $Li_3Ga_{14} = Li_3(Ga_{12})(Ga_2)$, as well as the more complicated phases $K_3Ga_{13} = K_6(Ga_{11})(Ga_{12})(Ga)_3$ and $Na_{22}Ga_{39} = Na_{22}(Ga_{12})_2(Ga_{15})$. Similarly a Ga_8 bisdisphenoid, the eight-vertex deltahedron, which has vertices of degrees 4 and 5 only, is found in MGa_3 (M = K, Rb) = $M_3(Ga_8)(Ga)$. Electron-counting rules similar to those used for the deltahedral boranes (Section 8.3.2.3) can be applied to the deltahedra in these gallium alloys.[7] The structure of many of these gallium alloys have groups of one to 15 (in $Na_{22}Ga_{39}$) gallium atoms interspersed between the gallium deltahedra in the crystal lattice, which are chemically bonded to the vertex atoms of the deltahedra.

9.3 Aqueous Solution Chemistry of Trivalent Aluminum, Gallium, Indium, and Thallium

The octahedral aqua ions $[M(H_2O)_6]^{3+}$ (M = Al, Ga, In, Tl) exist in aqueous solutions and in crystalline hydrated salts. The octahedron in $[Tl(H_2O)_6]^{3+}$ is somewhat distorted, with two water molecules in *trans* positions bonded more strongly to the central thallium atom than the remaining four water molecules. These octahedral aqua ions are acidic by hydrolysis reactions (compare equations 1.22 in Section 1.4):

$$[M(H_2O)_6]^{3+} \rightleftarrows [M(H_2O)_5OH]^{2+} + H^+ \qquad (9.2)$$

The dissociation constants K_a in the absence of noncomplexing anions are $\sim 10^{-5}$, 10^{-3}, 10^{-4}, and 10^{-1} for Al, Ga, In, and Tl, respectively, indicating that Tl(III) salts are the most extensively hydrolyzed.

The hydrolysis of Al(III) solutions can be studied by ^{27}Al NMR spectros-copy, which distinguishes between aluminum in tetrahedral AlO_4 and octa-hedral AlO_6 environments.[8,9] Hydrolysis products of aluminum include monomeric $Al(OH)^{2+}$, $Al(OH)_3$, and $Al(OH)_4^-$ and oligomeric $Al_3(OH)_{11}^{2-}$, $Al_6(OH)_{15}^{3+}$, and $Al_8(OH)_{22}^{2+}$. The most complicated aluminum hydrolysis product that has been characterized is $[Al^{tet}(\mu_4\text{-}O)_4Al_{12}^{oct}(\mu_2\text{-}OH)_{24}(H_2O)_{12}]^{7+}$ (Figure 9.1), which has been isolated as crystalline $Na[Al^{tet}(\mu_4\text{-}O)_4Al_{12}^{oct}(\mu_2\text{-}OH)_{24}(H_2O)_{12}](SO_4)_4 \cdot 13H_2O$. This hydrolysis product contains a tetra-

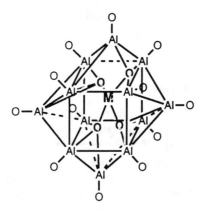

Figure 9.1. A schematic diagram of the structure of the hydrolysis product $[M^{tet}(\mu_4\text{-}O)_4Al^{oct}_{12}(\mu_2\text{-}OH)_{24}(H_2O)_{12}]^{7+}$ (M^{tet} = Al, Ga). The central tetrahedral atom and the μ_4-O groups connecting it to the peripheral Al atoms are shown in bold type. For clarity, neither the OH groups bridging the 24 edges of the cuboctahedron nor the hydrogen atoms of the 12 terminal water ligands are shown.

hedral aluminum (Al^{tet}) bridged by μ_4-oxygen atoms to 12 octahedral aluminum atoms (Al^{oct}) at the vertices of a cuboctahedron. Each of the 24 edges of the cuboctahedron contains a bridging hydroxyl group. The central tetra-hedral aluminum atom can be replaced by a gallium atom to give $[Ga^{tet}(\mu_4\text{-}O)_4Al^{oct}_{12}(\mu_2\text{-}OH)_{24}(H_2O)_{12}]^{7+}$.

An important class of crystalline salts of these elements are the *alums*, which are double sulfates of the general formula $M^IM^{III}(SO_4)_2 \cdot 12H_2O$. In the alum structure, M^I can be almost any monopositive ion except Li^+, which is too small to be accommodated in the structure, and M^{III} can be not only the tripositive ions Al^{3+}, Ga^{3+}, and In^{3+} but also most tripositive transition metal ions of similar size, including Ti^{3+}, V^{3+}, Cr^{3+}, Mn^{3+}, Fe^{3+}, Co^{3+}, and Ir^{3+}.

Aluminum and gallium hydroxides are amphoteric as indicated by the following equilibria:

$$Al(OH)_3 \text{ (s)} \rightleftarrows AlO_2^- + H^+ + H_2O \qquad K \approx 5 \times 10^{-33} \qquad (9.3a)$$

$$Al(OH)_3 \text{ (s)} \rightleftarrows Al^{3+} + 3OH^- \qquad K \approx 4 \times 10^{-13} \qquad (9.3b)$$

$$Ga(OH)_3 \text{ (s)} \rightleftarrows GaO_2^- + H^+ + H_2O \qquad K \approx 5 \times 10^{-33} \qquad (9.3b')$$

$$Ga(OH)_3 \text{ (s)} \rightleftarrows Ga^{3+} + 3OH^- \qquad K \approx 4 \times 10^{-13} \qquad (9.3b'')$$

Therefore, $Al(OH)_3$ and $Ga(OH)_3$ as well as the corresponding metals dis-solve in aqueous alkali (MOH: M = Li, Na, K, Rb, Cs) to form *aluminates* and *gallates*. However, the oxides of indium and thallium are strictly basic. Raman and ^{27}Al NMR studies indicate that the principal aluminate species is $Al(OH)_4^-$.

9.4 Oxygen Derivatives of Trivalent Aluminum, Gallium, Indium, and Thallium

9.4.1 Metal Oxides

9.4.1.1 Aluminum Oxides

The only aluminum oxide stoichiometry is *alumina*, Al_2O_3, but this compound exists in two anhydrous forms. In α-alumina the oxide ions form a hexagonal close-packed array with the Al^{3+} ions distributed in the octahedral interstices. α-Alumina occurs in nature as *corundum* and can be obtained by heating γ-alumina or any hydrous oxide above 1000°C. Corundum is stable at high temperatures and indefinitely metastable at low temperatures, as well as very hard and resistant to hydration and acid attack; it is used as an abrasive. Incorporation of trace quantites of colored metal ions into the α-alumina structure leads to several important gemstones such as red *ruby* (Cr^{3+}), blue *sapphire* (Fe^{2+}, Fe^{3+}, Ti^{4+}), and yellow *Oriental topaz* (Fe^{3+}). The second form of anhydrous aluminum oxide is γ-alumina, which has a "defect" spinel structure (i.e., a spinel structure with a deficit of cations). γ-Alumina is obtained by dehydration of hydrous oxides at relatively low temperatures ($\sim 450°C$). There are significant differences in the density and chemical reactivity between α-alumina and γ-alumina, with γ-alumina being significantly less dense (3.4 g/cm^3 for γ-alumina v. 4.0 g/cm^3 for α-alumina) and much more susceptible to hydration and acid attack.

A number of hydrated forms of alumina are known. Hydrated alumina of the stoichiometry $AlO \cdot OH$ occurs in two forms: γ-$AlO \cdot OH$ = *boehmite*, which can be precipitated by adding ammonia to boiling aqueous Al^{3+}, and α-$AlO \cdot OH$ = *diaspore*, which is found in nature. The "true" aluminum hydroxide, $Al(OH)_3$, which occurs in nature as *gibbsite*, is obtained by acidifying aluminate solutions with CO_2:

$$Al(OH)_4^- + CO_2 \rightarrow Al(OH)_3\downarrow + HCO_3^- \tag{9.4}$$

Aluminum hydroxide has a layer structure, $Al(OH)_{6/2}$, in which each hydroxide group bridges two aluminum atoms.

A third form of "alumina," namely "β-alumina," is not pure Al_2O_3 but contains other metal ions such as Na^+ and Mg^{2+}; a typical stoichiometry of β-alumina is $Na_{1.67}Mg_{0.67}Al_{10.33}O_{17}$. For example, sodium β-alumina can be made by heating Na_2CO_3, $NaNO_3$, or $NaOH$ with any modification of Al_2O_3 or its hydrates to 1500°C in a sealed platinum vessel. β-Aluminas are fast ion solid state conductors with ion exchange properties and high electrical conductivity.[10,11] They are useful in solid state, permeable membrane electrolytes for sodium/sulfur batteries (Section 6.4.2.1).

Another important class of mixed aluminum oxides containing other metals are the *spinels*, of general stoichiometry $M^{II}Al_2O_4$ and containing a lattice of close-packed oxygen atoms in which the divalent metal occurs in tetrahedral sites and aluminum in octahedral sites. Important spinels include *spinel*, $MgAl_2O_4$, and *chrysoberyl*, $BeAl_2O_4$. *Inverse spinels* are also known in which half of the trivalent metal occupies the tetrahedral sites and the other half some of the octahedral sites, with the remaining octahedral sites occupied by the divalent metal. For this reason, inverse spinels are favored when the divalent metal prefers an octahedral oxygen environment.

Aluminas are widely used as supports for catalysts and absorbents for chromatography. Such aluminas are made by heating hydrated aluminum oxides to various temperatures and may have surfaces that are partially or completely dehydrated. The activity of such alumina supports depends on their method of preparation and the extent of subsequent exposure to moist air.

9.4.1.2 Gallium Oxides

The oxides of gallium are rather similar to those of aluminum. Thus gallium oxide has a high temperature form α-Ga_2O_3 and a low temperature form γ-Ga_2O_3. Gallium oxides are prepared by heating gallium(III) nitrate, gallium(III) sulfate, or the hydrous oxides obtained by precipitating aqueous gallium(III) solutions with ammonia. In addition, β-Ga_2O_3 is known which contains both tetrahedrally and octahedrally coordinated gallium.[12] The hydrated gallium oxides $GaO \cdot OH$ and $Ga(OH)_3$ are similar to their aluminum analogues.

9.4.1.3 Indium Oxides

Yellow In_2O_3 is only known in a single form. The hydroxide $In(OH)_3$ can be obtained by addition of ammonia to aqueous $InCl_3$ at 100°C and aging the precipitate a few hours at this temperature.

9.4.1.4 Thallium Oxides

The only known thallium(III) oxide is brown-black Tl_2O_3 (mp 716°C), which is relatively dense (10.04 g/cm^3) and has a very low electrical resistivity (7×10^{-5} $\Omega \cdot$ cm at room temperature). Heating Tl_2O_3 to ~ 100°C results in oxygen evolution to give thallium(I) oxide, Tl_2O, indicative of the high stability of thallium(I).

9.4.2 Alkoxides and Complexes with Oxygen Ligands

The alkoxides of aluminum are much more important than those of gallium, indium, and thallium. Aluminum alkoxides may be obtained by reaction of aluminum metal with the alcohol in the presence of a mercury catalyst to remove the protective oxide layer or by reaction of aluminum chloride with

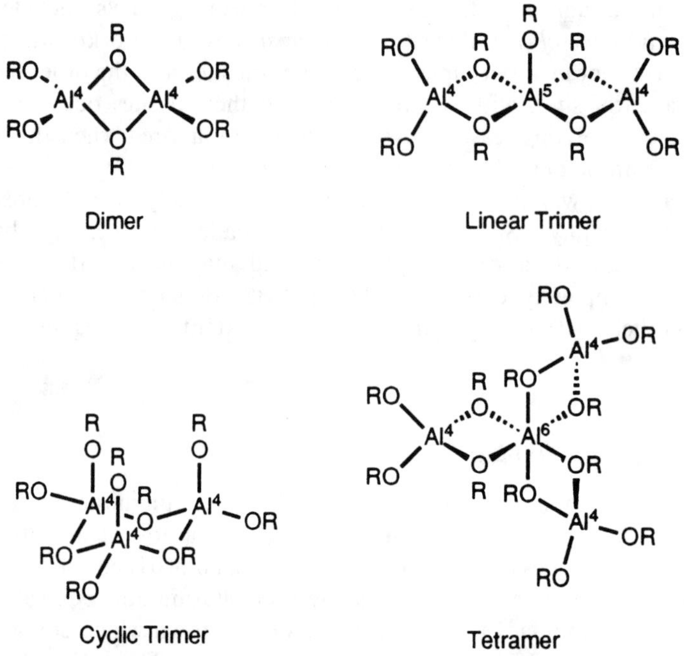

Dimer

Linear Trimer

Cyclic Trimer

Tetramer

Figure 9.2. Some structures of aluminum alkoxides. The coordination numbers of the aluminum atoms are indicated by the superscripts 4, 5, and 6.

the corresponding alkali metal alkoxide, for example,

$$2Al + 6ROH \xrightarrow[\Delta]{1\% \; HgCl_2} 2(RO)_3Al + 3H_2\uparrow \tag{9.5a}$$

$$AlCl_3 + 3RONa \rightarrow (RO)_3Al + 3NaCl \tag{9.5b}$$

Aluminum alkoxides hydrolyze vigorously in water. Aluminum alkoxides are useful precursors for deposition of alumina in various forms. Thus the *Yoldas process* is used for the preparation of alumina fibers from aluminum *n*-butoxide using sol–gel chemistry.[13] Aluminum isopropoxide is a useful reagent for the Meerwein–Pondorf–Verley reduction of aldehydes and ketones to the corresponding alcohols, with the isopropoxy groups being oxidized to acetone in the process.[14]

Aluminum alkoxides generally form oligomers of the general type $[Al(OR)_3]_n$ with the complexity, n, decreasing with increasing bulk of the alkoxide group. A variety of structures (Figure 9.2) are possible for aluminum alkoxides in which the aluminum atoms have coordination numbers of 4a, 5a, and 6. Such structures include the dimer, in which both aluminum atoms are four-coordinate, the linear trimer, in which the central aluminum atom is five-coordinate, the cyclic trimer, in which all three aluminum atoms are four-coordinate, and the tetramer, in which the center aluminum atom is six-coordinate and the outer aluminum atoms are four-coordinate. The structures of aluminum alkoxides have been elucidated by ^{27}Al NMR spectroscopy[15] and a tetrameric structure for aluminum isopropoxide $[Al(OPr^i)_3]_4$ has been determined by X-ray crystallography.[16]

Trivalent aluminum, gallium, indium, and thallium all form numerous complexes with uninegative bidentate, chelating oxygen ligands such as β-diketones, catechol (1,2-dihydroxybenzene), dicarboxylates (e.g., oxalate), salicylate, and 8-hydroxyquinolinate. The acetylacetonates, $(MeCOCHCOMe)_3M$ (M = Al, Ga, In, Tl) melt below 200°C and vaporize without decomposition.

9.5 Halogen Derivatives of Trivalent Aluminum, Gallium, Indium, and Thallium

All four halides of each of the four metals are known. However, black TlI_3 is not thallium(III) iodide, $Tl^{3+}(I^-)_3$, but the thallium(I) salt of the triiodide ion (Section 7.5.3.1), $Tl^+(I_3^-)$. Note that Tl^{3+} has a sufficiently high oxidation potential ($E° = +1.25$ V: equation 9.1) to oxidize iodide to elemental iodine ($E°$ for $I_2/2I^- = +0.54$ V). The other thallium(III) halides, $TlCl_3$ and $TlBr_3$, are unstable with respect to halogen loss to give TlX (X = Cl, Br) and free halogen; such decomposition of $TlCl_3$ occurs above $\sim 40°C$.

The metal trifluorides, MF_3, are high-melting solids containing six-coordinate metal atoms. Anhydrous AlF_3 is obtained by reaction of Al_2O_3 with hydrogen fluoride at 700°C. Dissolution of aluminum metal in aqueous hydrofluoric acid gives the hydrated fluorides $AlF_3 \cdot 3H_2O$ and $AlF_3 \cdot 9H_2O$; the latter hydrate is very soluble in water. Fluorine-19 NMR spectra of the resulting solution indicates the presence of $AlF_3(H_2O)_3$ as well as AlF_4^-, $AlF_2(H_2O)_4^+$, and $AlF(H_2O)_5^{2+}$. Addition of excess fluoride to the solutions gives octahedral AlF_6^{3-}. The most important hexafluoroaluminate(III) salt is *cryolite*, Na_3AlF_6.

The other metal halides, MX_3 (M = Al, Ga, In, Tl; X = Cl, Br, I), are made by direct combination of the elements. Solid $AlCl_3$ has a layer lattice with six-coordinate aluminum, which expands 85% in volume upon melting (mp 192°C) as the compound takes on a dimeric structure with four-coordinate metal atoms (Figure 9.3). Similar dimeric structures are formed by $[AlBr_3]_2$, $[AlI_3]_2$, and $[GaX_3]_2$ in all phases. The dimers are split by reactions with donor molecules to give tetrahedral complexes such as $R_3N \rightarrow AlCl_3$. In the presence of excess chloride, aluminum and gallium form only tetrahedral MCl_4^- with no evidence for octahedral MCl_6^{3-}, in contrast to the smaller fluoride ion. The AlX_4^- ions are hydrolyzed by water but $GaCl_4^-$ can be extracted from 8 M aqueous hydrochloric acid into ethers. Aluminum also forms the oxohalide ions $[Al_3(\mu_3\text{-}O)Cl_8]^-$ and $[Cl_2Al(\mu_2\text{-}OAlCl_3)_2AlCl_2]^{2-}$ (Figure 9.3). Reactions of MCl_3 with HCl in ethers give oxonium salts such as $[(Et_2O)_nH]^+MCl_4^-$, which often form viscous oils.

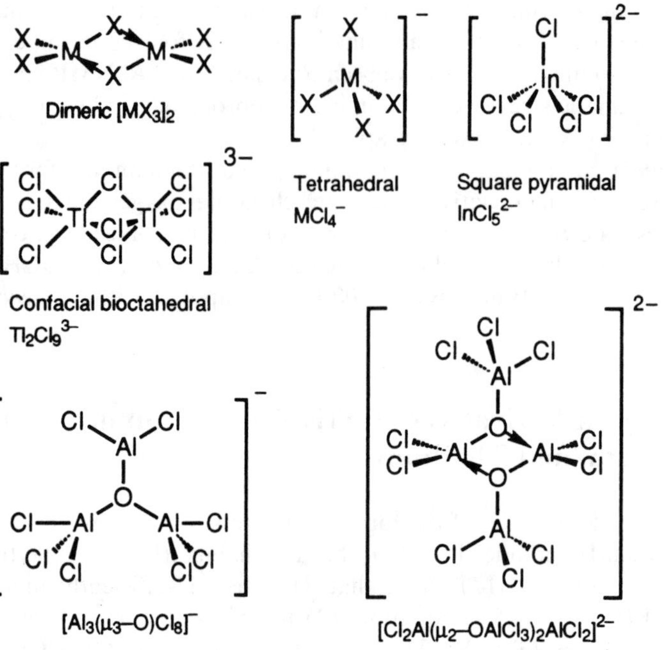

Figure 9.3. Structures of some halide complexes of aluminum, gallium, indium, and thallium.

The heavier metals of this group, namely indium and thallium, can have coordination numbers higher than 4 in their chloride complexes (Figure 9.3). Thus indium forms square pyramidal complexes[17] such as $[Et_4N]_2[InCl_5]$ and thallium forms square pyramidal $TlCl_5^{2-}$, octahedral $TlCl_6^{3-}$, and the binuclear confacial bioctahedral $Tl_2Cl_9^{3-}$.

Aluminum chloride is of interest as an ingredient in fused salt media. Thus the $AlCl_3/NaCl$ eutectic (mp 173°C) is much used as a molten salt medium for electrolytic and other reactions such as those with elemental sulfur and metal halides. The equilibria in the $NaCl/AlCl_3$ eutectic in the 175–300°C temperature range have been shown by potentiometric and vapor pressure methods to include the following[18]:

$$2AlCl_4^- \rightleftarrows Al_2Cl_7^- + Cl^- \tag{9.6a}$$

$$3Al_2Cl_7^- \rightleftarrows 2Al_3Cl_{10}^- + Cl^- \tag{9.6b}$$

$$2Al_3Cl_{10}^- \rightleftarrows 3Al_2Cl_6 + 2Cl^- \tag{9.6c}$$

Interaction of $AlCl_3$ with *N*-butylpyridinium chloride or 1-methyl-3-ethyl-imidazolium chloride (RCl) at room temperature gives conducting ionic liquids that are good solvents for inorganic compounds. These room temperature ionic liquids have a number of properties of interest, including an extended electro-

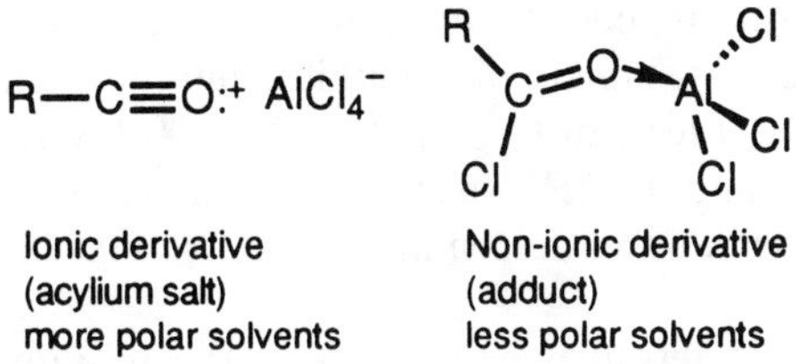

$R—C\equiv O:^+ AlCl_4^-$	R, $C=O$, Al, Cl structure
Ionic derivative	Non-ionic derivative
(acylium salt)	(adduct)
more polar solvents	less polar solvents

Figure 9.4. Ionic and nonionic derivatives formed by interaction of acyl chlorides with aluminum chloride: intermediates in the Friedel–Crafts reaction.

chemical window and readily controlled addition of excess chloride (base) or aluminum chloride (acid); moreover, the physical properties of the "neutral" melt do not change around the 1:1 mole ratio of $AlCl_3$ to RCl.[19]

The formation of $AlCl_4^-$ and $AlBr_4^-$ is essential to the role of $[AlCl_3]_2$ and $[AlBr_3]_2$ as Friedel–Crafts catalysts for the alkylation and acylation of benzenoid derivatives,[20,21] since in this way the carbonium ions required as electrophiles for aromatic substitution are concurrently formed. Reactions of aluminum chloride with acyl chlorides gives the acylium salts $[R—C\equiv O]^+AlCl_4^-$ in polar aprotic solvents and the nonionic adduct $R(X)C=O \rightarrow AlCl_3$ in less polar solvents such as dichloromethane (Figure 9.4). The acylium cation, $R—C\equiv O^+$, can then act as an electrophile for the Friedel–Crafts reaction.

9.6 Hydride Derivatives of Trivalent Aluminum, Gallium, Indium, and Thallium

The lithium tetrahydrometallates $LiMH_4$ (M = Al, Ga, In) can be obtained from lithium hydride and the corresponding halide:

$$4LiH + MCl_3 \xrightarrow{Et_2O} LiMH_4 + 3LiCl \tag{9.7}$$

The tetrahydroaluminate ion can be made more economically as the sodium salt $NaAlH_4$ by direct combination of the elements at elevated temperatures and pressure:

$$Na + Al + 2H_2 \xrightarrow[\text{tetrahydrofuran}]{150°C/130\ atm/24\ hr} NaAlH_4 \tag{9.8}$$

The $NaAlH_4$ product can be precipitated from its tetrahydrofuran solution by addition of toluene and converted efficiently by metathesis to the more useful lithium salt:

$$NaAlH_4 + LiCl \xrightarrow{Et_2O} LiAlH_4 + NaCl \tag{9.9}$$

Tetrahydroaluminates are much more vigorously hydrolyzed by water than the corresponding tetrahydroborates (Section 8.3.2.1). This relates to the ability for aluminum to increase its coordination number to 5 and 6 by d-orbital participation. The thermal stability of the lithium tetrahydrometallates decreases in the sequence $LiBH_4$ (dec 380°C), $LiAlH_4$ (dec 100°C), $LiGaH_4$ (dec 50°C), $LiInH_4$ (dec ~0°C), and $LiTlH_4$ (dec ~0°C), indicating a decrease in the stability of M—H bonds upon descending the B, Al, Ga, In, Tl column of the periodic table.

The most important tetrahydroaluminate is the lithium salt $LiAlH_4$,[22] a nonvolatile crystalline solid explosively hydrolyzed by water and soluble in diethyl ether (~29 g/100 g at room temperature) and many other ethers or even in benzene in the presence of crown ethers. In diethyl ether $LiAlH_4$ is extensively associated, but ion pairs are formed in dilute tetrahydrofuran solutions. X-Ray diffraction shows crystalline $LiAlH_4$ to contain the tetrahedral AlH_4^- ion with an average Al—H distance of 1.55 Å. The most important use of $LiAlH_4$ is as a reducing agent in organic chemistry[23]; it can reduce aldehydes, ketones, carboxylic acids, esters, and acid halides to the corresponding alcohols and nitriles (RC≡N) to the corresponding primary amines (RNH_2). Some substituted hydroaluminates are also useful reducing agents including the toluene-soluble $Na[AlH_2(OCH_2CH_2OMe)_2]$ ("Red-Al") and $Li[AlH(OBu^t)_3]$.

The octahedral hexahydroaluminate trianion, AlH_6^{3-}, is known in the form of alkali metal salts such as Li_3AlH_6 and Na_3AlH_6. These salts can be made by direct combination of the elements under hydrogen pressure or by reaction of $NaAlH_4$ with excess NaH.

No volatile binary aluminum hydrides are known. Aluminum(III) hydride[24] is a nonvolatile white solid, which can be made by reaction of $LiAlH_4$ either with $AlCl_3$ in diethyl ether or with 100% sulfuric acid in tetrahydrofuran:

$$3LiAlH_4 + AlCl_3 \xrightarrow{Et_2O} 4AlH_3 + 3LiCl \tag{9.10b}$$

$$2LiAlH_4 + H_2SO_4 \xrightarrow{tetrahydrofuran} 2AlH_3 + 2H_2\uparrow + Li_2SO_4 \tag{9.10b}$$

The structure of AlH_3 has been shown to be similar to that of AlF_3, consisting of octahedral $AlH_{6/2}$ units linked into a three-dimensional polymeric network through Al—H—Al three-center-bonds with an Al–H–Al angle 141°.[25] Aluminum hydride is a stronger reducing agent than $LiAlH_4$ in organic chemistry.

Monomeric $LAlH_3$ and L_2AlH_3 adducts can be prepared, even though AlH_3 itself is polymeric.[26] The four-coordinate trimethylamine adduct, $Me_3N{\rightarrow}AlH_3$ is a white volatile solid (mp 75°C), which is readily hydrolyzed by water and can be obtained by any of the following reactions:

$$Me_3NH^+Cl^- + LiAlH_4 \xrightarrow[-80°C]{Et_2O} Me_3N{\rightarrow}AlH_3 + LiCl + H_2\uparrow \tag{9.11a}$$

$$Me_3N{\rightarrow}AlCl_3 + 3LiH \xrightarrow{Et_2O} Me_3N{\rightarrow}AlH_3 + 3LiCl \tag{9.11b}$$

$$3LiAlH_4 + AlCl_3 + 4Me_3N \rightarrow 4Me_3N{\rightarrow}4AlH_3 + 3LiCl \tag{9.11c}$$

This adduct reacts reversibly with excess trimethylamine to form a five-coordinate adduct $(Me_3N)_2AlH_3$, which is shown to have a five-coordinate trigonal bipyramidal structure.[27] High purity aluminum films can be deposited in a hot wall, low pressure chemical vapor deposition reactor at temperatures as low as 100°C by using $(Me_3N)_2AlH_3$.[28]

Reaction of $AlCl_3$ with $LiBH_4$ gives the volatile $Al(BH_4)_3$ (bp 44°C), which is spontaneously flammable in air and reactive toward water. The structure of $Al(BH_4)_3$ is 9.1, containing a central six-coordinate aluminum atom surrounded by three bidentate BH_4 ligands (Figure 8.5).

9.1

9.7 Other Binary Compounds of Aluminum, Gallium, Indium, and Thallium

Aluminum carbide, Al_4C_3, is made by heating aluminum and carbon to 1100–2000°C. It is one of the saline metal carbides (Section 2.3.7.1) that reacts with water to give methane. The solid state structure contains discrete carbon atoms that are not within bonding distance of other carbon atoms (minimum $C \cdots C = 3.16$ Å), supporting formulations as a "methanide" with C^{4-} anions.

The *nitrides* AlN, GaN, and InN are fairly hard white solids having wurtzite structures with tetracoordinate M and N. Aluminum reacts directly with elemental nitrogen at elevated temperatures to form AlN, but GaN and InN must be made by indirect methods. Thus GaN can be obtained by reaction of gallium metal or Ga_2O_3 with ammonia at 600–1000°C and InN can be obtained by pyrolysis of $(NH_4)_3InF_6$. Aluminum nitride is of interest as a high temperature ceramic electrical insulator because of its large band gap (6.2 eV), high melting point (2400°C), higher thermal conductivity, and low chemical reactivity. Thin films of AlN can be deposited by chemical vapor deposition using ammonia as the nitrogen source and aluminum halides or alkyls as the aluminum source or by using a single-source precursor such as $Al_2(NMe_2)_6$, a volatile white solid (mp 88–89°C), which can be prepared from $LiNMe_2$ and $AlCl_3$.[29]

The phosphides and arsenides of gallium and indium are the so-called *III–V compounds* which are discussed elsewhere in this book (Section 5.3.1).

The chalcogenides of aluminum, Al_2E_3 (E=S, Se, Te), can be made by direct combination of the elements at 1000°C and hydrolyze readily to Al_2O_3 and H_2E.

9.8 Organometallic Compounds of Aluminum, Gallium, Indium, and Thallium

9.8.1 Organoaluminum Derivatives[30]

Trialkylaluminum compounds, R_3Al, are prepared on a large scale because of their use in Ziegler–Natta catalysts for olefin polymerization. This large-scale preparation uses the *hydroalumination* of terminal olefin at elevated temperature and pressures:

$$2Al + 3H_2 + 6RCH{=}CH_2 \rightarrow 2(RCH_2CH_2)_3Al \qquad (9.12)$$

This reaction requires some preformed AlR_3 as a catalyst, which can be made by methods such as the following:

$$2Al + 3R_2Hg \rightarrow 2R_3Al + 3Hg \qquad (9.13a)$$

$$3RMgCl + AlCl_3 \rightarrow R_3Al + 3MgCl_2 \qquad (9.13b)$$

These latter synthetic methods are also useful for the preparation of R_3Al derivatives that cannot be obtained by olefin hydroalumination such as Me_3Al and Ph_3Al.

Lower aluminum alkyls such as Me_3Al and Et_3Al are very reactive liquids, spontaneously flammable in air. All R_3Al derivates are very reactive toward air, water, alcohols, halocarbons, and so on. Trialkyluminum derivatives are Lewis acids, giving tetrahedral $L{\rightarrow}AlR_3$ derivatives with Lewis bases such as amines, phosphines, ethers, and sulfides. Trialkylaluminum derivatives function as alkylating agents but normally transfer only one alkyl group to the substrate, since R_2AlX are much less active alkylating agents than R_3Al.

Direct reaction of aluminum with many alkyl chlorides gives the "organo-aluminum sesquichlorides," $R_3Al_2Cl_3$:

$$2Al + 3RCl \rightarrow R_3Al_2Cl_3 \qquad (9.14)$$

This reaction fails for propyl and higher chlorides because such alkyl decompose in the presence of the alkylaluminum chlorides to give HCl, olefins, and so on.

The lower trialkylaluminums and triarylaluminums are dimeric in condensed phases, with alkyl carbon atoms bridging the aluminum atoms through three-centre Al—C—Al bonds (Figure 9.5). The proton NMR spectrum of trimethylaluminum $[AlMe_3]_2$ exhibits separate resonances at $-70°C$ for the terminal and bridging methyl groups, but these resonances coalesce upon warming, and a single sharp peak is seen at room temperature.[31] This indicates a *fluxional* process in which the bridging and terminal methyl groups exchange places over a small energy barrier. The mechanism for this methyl exchange in $[AlMe_3]_2$ involves reversible dissociation of the dimer to the monomer. However, the extent of dissociation of $[AlMe_3]_2$ is very small (only $\sim0.005\%$ at $20°C$).

Formation of dimers with bridging alkyl or aryl groups and three-center Al—C—Al bonds is inhibited in R_3Al derivatives if the R groups are sufficiently

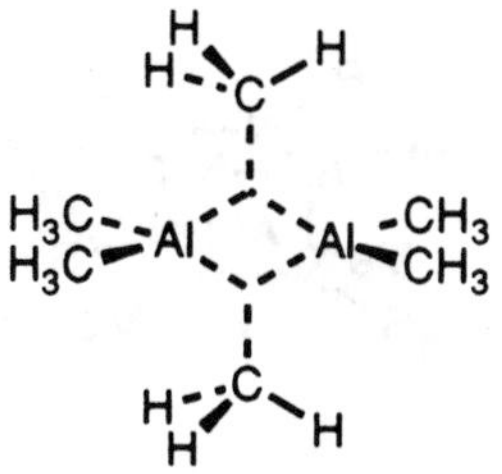

Figure 9.5. The dimeric structure of trimethylaluminum, showing the bridging methyl groups and the three-center Al—C—Al bonding.

bulky. For example, trimesitylaluminum, $(2,4,6\text{-}Me_3C_6H_2)_3Al$, is monomeric with planar three-cordinate aluminum both in the solid state and in solution.

The most important application of organoaluminum compounds such as Et_3Al, $Et_3Al_2Cl_3$, and R_2AlH is their use with transition metal halides, alkoxides, or organometallic compounds as catalysts (e.g., Ziegler–Natta catalysts) for the stereoregular polymerization of ethylene, propylene, and other unsaturated compounds.[32] Thus the polymerization of ethylene at 50–150°C and 10 atm with an $[Et_3Al]_2/TiCl_4$ mixture in heptane gives 85–95% crystalline polymer with a density of 0.95–0.98 g/cm^3.

Organoaluminum compounds with aluminum–nitrogen bonds are of interest as single-source precursors to aluminum nitride (Section 9.7). Such compounds can be obtained by reactions of aluminum alkyls with primary or secondary amines, for example,

$$2Me_3Al + 2R_2NH \rightarrow [R_2NAlMe_2]_2 + 2CH_4\uparrow \qquad (9.15a)$$

$$4Me_3Al + 4RNH_2 \rightarrow [RNAlMe]_4 + 8CH_4\uparrow \qquad (9.15b)$$

Adducts of aluminum alkyls with primary and secondary amines are not stable because of such reactions. The dimers $[R_2NAlMe_2]_2$ have a four-membered Al_2N_2 ring, whereas the tetramers $[RNAlMe]_4$ have an Al_4N_4 cubic cage (Figure 9.6). Poly(n-alkyliminoalanes) can also be obtained by reactions such as the following:

$$nMe_3N \cdot AlH_3 + nRNH_2 \rightarrow [RNAlH]_n + 2nH_2\uparrow + nMe_3N\uparrow \qquad (9.16a)$$

$$nLiAlH_4 + nRNH_3^+Cl^- \rightarrow [RNAlH]_n + 3nH_2\uparrow + nLiCl \qquad (9.16b)$$

$$2nAl + 2nRNH_2 \rightarrow 2[RNAlH]_n + nH_2\uparrow \qquad (9.16c)$$

The products are oligomers with cage structures, where n is commonly 4, 6, or 8 but can range from 2 to 35. Reaction of Me_3Al with the very bulky primary amine $2,6\text{-}Pr_2^iC_6H_3NH_2$ at 110°C followed by pyrolysis at 170°C

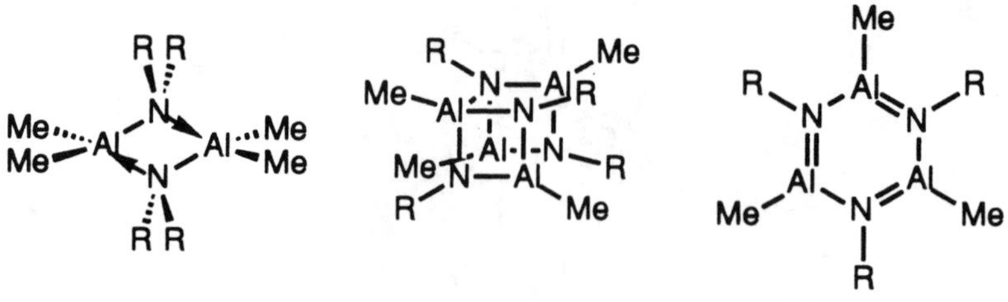

[Me₂AlNR₂]₂ dimer   [MeAlNR]₄ "cubane" cage   [MeAlNR]₃ planar ring
R = 2,6-Pr$^i_2$C₆H₃

Figure 9.6. Examples of ring and cage structures obtained from reactions of trimethyl-aluminum with secondary and primary amines.

gives $(2,6\text{-Pr}^i_2\text{C}_6\text{H}_3)_3\text{N}_3\text{Al}_3\text{Me}_3$ with a planar Al_3N_3 ring (Figure 9.6)[33] analogous to the planar B_3N_3 ring found in borazenes (Section 8.7.1). The very reactive compound $(2,6\text{-Pr}^i_2\text{C}_6\text{H}_3)_3\text{N}_3\text{Al}_3\text{Me}_3$ is instantly decomposed by traces of air and moisture.

9.8.2 Organometallic Derivatives of Gallium, Indium, and Thallium

The trialkyl derivatives R_3M (M = Ga, In, Tl) do not dimerize, in contrast to trialkylaluminums. However, Me_3In and Me_3Tl are tetramers in the solid state. Dialkyl derivatives R_2MX (M = Ga, In, Tl) are known, some of which are even stable in aqueous solution, in contrast to their aluminum analogues. Dimethyl-gallium hydroxide tetramer can be made by treatment of $Me_3Ga \cdot OEt_2$ with water:

$$4Me_3Ga \cdot OEt_2 + 4H_2O \rightarrow [Me_2GaOH]_4 + 4CH_4\uparrow + 4Et_2O \quad (9.17)$$

This hydroxide, like simple $Ga(OH)_3$ (Section 9.3), is amphoteric, forming the hydrated Me_2Ga^+ cation with acids and the $Me_2Ga(OH)_2^-$ anion upon treatment with alkali:

$$[Me_2GaOH]_4 + 4H_3O^+ \rightarrow 4[Me_2Ga(H_2O)_2]^+ \quad (9.18a)$$

$$[Me_2GaOH]_4 + 4OH^- \rightarrow 4[Me_2Ga(OH)_2]^- \quad (9.18b)$$

Thallium(III) gives very stable ionic dialkylthallium derivatives $R_2Tl^+X^-$ (X = halide, $\frac{1}{2}SO_4^{2-}$, CN^-, NO_3^-, etc.) which, like the isoelectronic dialkyl-mercury derivatives (R_2Hg), are unaffected by air or water. The Me_2Tl^+ ion is linear in aqueous solution and in its salts.

9.9 Aluminum, Gallium, Indium, and Thallium with Oxidation States Below 3

9.9.1 Aluminum

Simple aluminum(I) compounds are not stable under ordinary conditions. However, gaseous AlCl, Al_2O, and AlO have been observed at elevated temperatures. The disproportionation of aluminum(I) halides to $Al + AlCl_3$ is an exothermic process with enthalpies ranging from -105 kJ/mol for AlF to -46 kJ/mol for AlCl.

Several organoaluminum derivatives are known in which the aluminum has a formal oxidation state below $+3$. Reduction of $AlBu_3^i$ with potassium metal in hexane gives the brown dimeric dianion $[Al_2Bu_6^i]^{2-}$, which is isoelectronic with hexakis(isobutyl)disilane (Section 3.7.1) and has a direct Al—Al bond.[34] Similarly, reaction of Bu_2^iAlCl with potassium metal gives Bu_2^iAl—$AlBu_2^i$ which also has an Al—Al bond.[35]

A few examples of polyhedral aluminum clusters are known, but these are rather rare. Aluminum monochloride, although unstable, has been alkylated by bis(pentamethylcyclopentadienyl)magnesium to give $[Me_5C_5Al]_4$, which was shown by X-ray diffraction to contain an Al_4 tetrahedron with Al—Al edges of 2.77 Å.[36] The only aluminum analogue to the deltahedral borane anions $B_nH_n^{2-}$ (Section 8.3.2.3) is dark red icosahedral $K_2[Al_{12}Bu_{12}^i]$, which has been obtained as a by-product in 1.5% yield from the reaction of Bu_2^iAlCl with potassium metal in hexane solution.[37] In contrast to most other organoaluminum derivatives, this icosahedral aluminum derivative survives in air for about 2 hours in the solid state, suggestive of stabilization through the three-dimensional aromaticity in such deltahedral cage compounds (Section 8.3.4).

9.9.2 Gallium

Some simple binary gallium(I) compounds have been made by reactions such as the following:

$$Ga_2O_3 \text{ (s)} + 4Ga \text{ (l)} \xrightarrow{700°C} 3Ga_2O \text{ (g)} \tag{9.19a}$$

$$4Ga \text{ (l)} + SiO_2 \text{ (s)} \rightleftharpoons Si \text{ (in Ga)} + 2Ga_2O \text{ (g)} \tag{9.19b}$$

$$GaCl_3 \text{ (g)} \xrightleftharpoons{1100°C} GaCl \text{ (g)} + Cl_2 \text{ (g)} \tag{9.19c}$$

Pure Ga_2O and nonstoichiometric Ga_2S have been isolated, but GaCl has not been isolated in the pure state.

Several derivatives of the Ga^+ ion are known. The "gallium dihalides" $GaX_2 = Ga^+GaX_4^-$ can be made by treatment of gallium metal with $GaCl_3$, $HgCl_2$, or Hg_2Cl_2. The Ga^+ ion is known in a few other salts such as $Ga^+AlCl_4^-$.

A number of gallium(II) derivatives are known which contain direct Ga—Ga bonds. White stable diamagnetic $[Me_4N]_2[Ga_2X_6]$ is obtained by anodic dissolution of gallium metal in 6 M hydrohalic acid followed by addition of $Me_4N^+X^-$. X-ray diffraction of the chloride derivative (X = Cl) indicates the presence of a direct Ga—Ga bond (2.39 Å). Reaction of $Ga^+GaCl_4^-$ with dioxane gives the complex $(C_4H_8O_2)Cl_2Ga—GaCl_2(C_4H_8O_2)$ with a Ga–Ga bonding distance of 2.41 Å. Direct combination of gallium metal with the elemental chalcogens at elevated temperatures in the correct ratios gives GaE (E = S, Se, Te). Yellow GaS (mp 970°C) has been shown to have a hexagonal layer structure with direct Ga–Ga bonds (2.48 Å).

9.9.3 Indium

Aqueous solutions of indium(I) are obtained in low concentration from an indium metal anode in 0.01 M $HClO_4$. Such solutions are rapidly oxidized by both H^+ and air and are unstable with respect to disproportionation into indium metal and In(III). Solutions of indium(I) are more stable in acetonitrile than in water. Thus solid $In^+ClO_4^-$, $In^+BF_4^-$, and $In^+PF_6^-$ have all been isolated from reactions of indium amalgam with the corresponding silver salts in acetonitrile solution. The indium(I) halides InX (X = Cl, Br, I) are all isostructural, with the low temperature form of TlI. They are unstable to water but can be made by solid state reactions such as treatment of indium metal with the corresponding mercury(II) halide at 320–350°C.

"*Indium dihalides,*" $In^+InX_4^-$, can be made by direct combination of indium metal with the halogen in boiling xylene. They react with tetraalkylammonium halides to form $In_2X_6^{2-}$, which has a structure similar to that of $Ga_2X_6^{2-}$ with a direct metal–metal bond. Unlike their gallium analogues, the $In_2X_6^{2-}$ dianions disporportionate in nonaqueous solvents to give $In^IX_2^-$ and $In^{III}X_4^-$.

Cyclopentadienylindium(I) derivatives can be obtained from reactions of InX with the corresponding cyclopentadienide anion. The unsubstituted cyclopentadienyl derivative, C_5H_5In, contains zigzag polymeric chains. However, the volatile yellow pentamethylcyclopentadienyl derivative $[Me_5C_5In]_6$ contains an octahedral In_6 cluster, but the In–In distances are far too long (3.94–3.96 Å) for direct In—In bonding.[38]

9.9.4 Thallium

Thallium(I) is so stable that thallium(III) is a strong oxidant in aqueous solution. Electron exchange between Tl(I)/Tl(III) appears to involve two-electron transfer processes. The Tl^+ ion is not very sensitive to pH, but Tl^{3+} is extensively hydrolyzed to $TlOH^{2+}$ and Tl_2O_3 even at pH 1–2.5. For this reason the Tl^{3+}/Tl^+ redox potential of +1.25 V (equation 9.1) is very dependent on pH as well as the presence of complexing anions (e.g., this potential becomes +0.77 V in 1 M hydrochloric acid).

The colorless Tl^+ ion has a radius of 1.54 Å, which compares with 1.44 Å for K^+, 1.58 Å for Rb^+, and 1.27 Å. The aqueous chemistry of Tl^+ thus resembles that of Ag^+ and the alkali metal cations. Unlike other heavy metal hydroxides, the hydroxide TlOH is soluble in water and is a strong base, even absorbing CO_2 from the atmosphere like the alkali metal hydroxides. The solubilities of Tl^+ salts are lower than the corresponding alkali metal salts. Thallium(I) sulfate, nitrate, fluoride, and acetate are soluble in water, but the heavier thallium halides (TlCl, TlBr, TlI), sulfide, and chromate are only sparingly soluble in water.

The Tl^+ ion can be used to probe the behaviour of K^+ in biological systems because these ions are similar in size. In this connection ^{203}Tl and ^{205}Tl NMR spectroscopy (Section 9.2 and Table 9.1) can be used to investigate the environment of Tl^+ in aqueous media including models of biological systems.

Thallium(I) alkoxides are tetramers, $[TlOR]_4$, containing a Tl_4O_4 cube.[39] They are liquids except for the methoxide and can be obtained by treatment of thallium metal with the corresponding alcohol:

$$4Tl + 4ROH \rightarrow [TlOR]_4 + 2H_2\uparrow \tag{9.20}$$

Pale yellow crystalline monomeric thallium(I) cyclopentadienide can be precipitated by treatment of a basic solution of Tl^+ with cyclopentadiene:

$$Tl^+ + OH^- + C_5H_6 \rightarrow TlC_5H_5\downarrow + H_2O \tag{9.21}$$

It is useful as a relatively air-stable source of the cyclopentadienide ion in preparative reactions, especially for the synthesis of cyclopentadienyl derivatives of diverse metals.

No stable compounds of thallium(II) are known. Thallium(II) has been postulated as an intermediate in reactions of Tl^+ with one-electron oxidants or Tl^{3+} with one-electron reductants. Although Tl^{2+} has been detected in flash photolysis studies, its lifetime is only ~ 0.5 ms in aqueous media under ambient conditions.

References

1. Hinton, J. F.; Briggs, R. W., in *NMR and the Periodic Table*, R. K. Harris and B. E. Mann, Eds., Academic Press, London, 1978.

2. Henley, C. L., Quasicrystal Order, Its Origins and Consequences: A Survey of Current Models, *Comments Condens. Matter Phys.*, 1987, **13**, 59–117.

3. Janot, C., Dubois, J. M., Quasicrystals, *J. Phys. F; Met. Phys.*, 1988, **18**, 2303–2343.

4. Andersson, S.; Lidin, S.; Jacob, M.; Terasaki, O., On the Quasicrystalline State, *Angew. Chem. Int. Ed. Engl.*, 1991, **30**, 754–758.

5. King, R. B., The Chemical and Geometrical Genesis of Quasicrstals: Relationship of Icosahedral Aluminum Alloy Quasicrystal Structures to Icosahedral Structures in Elemental Boron, *Inorg. Chim. Acta*, 1991, **181**, 217–225.

6. Belin, C.; Ling, R. G., The Intermetallic Phases of Gallium and Alkali Metals. Interpretation of the Structures According to Wade's Electron-Counting Methods, *J. Solid State Chem.*, 1983, **40**, 40–48.

7. King, R. B., *Inorg. Chem.*, 1989, **28**, 2796–2799.

8. Haraguchi, H.; Fujiwara, S., Aluminum Complexes in Solution as Studied by Aluminum-27 Nuclear Magnetic Resonance, *J. Phys. Chem.*, 1969, **73**, 3467–3473.

9. Delpuech, J. J., Aluminum-27 in *NMR of Newly Accessible Nuclei*, Vol. 2, P. Laszlo, Ed., Academic Press, 1983, pp. 153–195.

10. Kummer, J. T., β-Alumina Electrolytes, *Prog. Solid State Chem.*, 1972, **7**, 141–175.

11. Kennedy, J. H., The β-Aluminas, *Top. Appl. Phys.*, 1977, **21**, 105–141.

12. Geller, S., *J. Chem. Phys.*, 1960, **33**, 674–676.

13. Brinker, C. J.; Scherer, G. W., *Sol–Gel Science*, Academic Press, New York, 1990, pp. 67–78.

14. Wilds, A. L., Reduction with Aluminum Alkoxides, *Org. React.*, 1944, **2**, 178–223.

15. Kříž, Čásenský, B.; Lyčka, A.; Fusek, J.; Heřmánek, S., ²⁷Al NMR Behavior of Aluminum Alkoxides, *J. Magn. Resonance*, 1984, **60**, 375–381.

16. Turova, N. Ya.; Kozunov, V. A.; Yanovskii, A. I.; Bokii, N. G.; Struchkov, Yu. T.; Tarnopol'skii, B. L., Physico-Chemical and Structural Investigation of Aluminium Isopropoxide, *J. Inorg. Nuclear Chem.*, 1979, **41**, 5–11.

17. Carty, A. J.; Tuck, D. J., Coordination Chemistry of Indium, *Prog. Inorg. Chem.*, 1975, **19**, 243–337.

18. Hjuler, H. A.; Mahan, J. H.; von Barner, J. H.; Bjerrum, N. J., *Inorg. Chem.*, 1982, **21**, 402–406.

19. Lipsztajn, M.; Osteryoung, R. A., *Inorg. Chem.*, 1985, **24**, 716–719.

20. Olah, G. A., *Friedel–Crafts Chemistry*, Wiley, New York, 1963.

21. Roberts, R. M.; Khalaf, A. A., *Friedel–Crafts Alkylation Chemistry*, Dekker, New York, 1984.

22. Finhold, A. E.; Bond, A. C.; Schlesinger, H. J., *J. Am. Chem. Soc.*, 1947, **69**, 1199–1203.

23. Gaylord, N. G., *Reduction with Complex Metal Hydrides*, Wiley-Interscience, New York, 1956.

24. Brower, F. M.; Matzek, N. E.; Reigler, P. F.; Rinn, H. W.; Roberts, C. B.; Schmidt, D. L.; Shover, J. A.; Terada, K., *J. Am. Chem. Soc.*, 1976, **98**, 2450–2453.

25. Turley, J. W.; Rinn, H. W., *Inorg. Chem.*, 1969, **8**, 18–22.

26. Ruff, J. K.; Hawthorne, M. F., The Amine Complexes of Aluminum Hydride, *J. Am. Chem. Soc.*, 1960, **82**, 2141–2144.

27. Heitsch, C. W.; Nordman, C. E.; Parry, R. W., *Inorg. Chem.*, 1963, **2**, 508–512.

28. Gladfelter, W. L.; Boyd, D. C.; Jensen, K. F., *Chem. Mater.*, 1989, **1**, 339–343.

29. Gordon, R. G.; Hoffman, D. M.; Riaz, U., Atmospheric Pressure Chemical Vapor Deposition of Aluminum Nitride Thin Films at 200–250°C, *J. Mater. Res.*, 1991, **6**, 5–7.

30. Mole, T.; Jeffrey, E. A., *Organoaluminum Compounds*, Elsevier, Amsterdam, 1972.

31. Ramey, K. C.; O'Brien, J. F.; Hasegawa, I.; Borchert, A. E., Nuclear Magnetic Resonance Study of Aluminum Alkyls, *J. Phys. Chem.*, 1965, **69**, 3418–3423.

32. Quirk, R. P., Ed., *Transition Metal Catalyzed Polymerizations: Ziegler–Natta and Metathesis Polymerizations*, Cambridge University Press, Cambridge, 1988.

33. Waggoner, K. M.; Hope, H.; Power, P. P., Synthesis and Structure of [MeAlN(2,6-i-Pr₂C₆H₃)]₃, *Angew. Chem. Int. Ed. Engl.*, 1988, **27**, 1699–1700.

34. Hoberg, H.; Krause, S., Dipotassium Hexaisobutylaluminate, a Complex Containing an Al–Al Bond, *Angew. Chem. Int. Ed. Engl.*, 1979, **17**, 949–950.

35. Miller, M. A.; Schram, E. P., *Organometallics*, 1985, **4**, 1362–1364.

36. Dohmeier, C.; Robl, C.; Tacke, M.; Schnöckel, H., The Tetrameric Aluminum(I) Compound [{Al(η^5-C$_4$Me$_5$)}$_4$], *Angew. Chem. Int. Ed. Engl.*, 1991, **30**, 564–565.

37. Hiller, W.; Klinkhammer, K.-W.; Uhl, W.; Wagner, J., K$_2$[Al$_{12}$t-Bu$_{12}$], a Compound with Al$_{12}$ Icosahedra, *Angew. Chem. Int. Ed. Engl.*, 1991, **30**, 179–180.

38. Beachley, O. T., Jr.; Churchill, M. R.; Fettinger, J. C.; Pazik, J. C.; Victoriano, L., *J. Am. Chem. Soc.*, 1986, **108**, 4666–4668.

39. Burke, P. J.; Matthews, R. W.; Gilles, D. G., Thallium-205 Nuclear Magnetic Resonance of Thallium(I), Alkoxides, *J. Chem. Soc. Dalton Trans.*, 1980, 1439–1442.

The Alkali and Alkaline Earth Metals

10.1 General Aspects of the Chemistry of the Alkali and Alkaline Earth Metals

The alkali metals (Li, Na, K, Rb, Cs, Fr) and alkaline earth metals (Be, Mg, Ca, Sr, Ba, Ra) are the two most electropositive families of elements. Their chemistry is largely ionic except for the lightest and least electropositive alkaline earth metals, namely beryllium and magnesium. Furthermore, their chemistry is restricted to the $+1$ oxidation state for the alkali metals and the $+2$ oxidation state for the alkaline earth metals, with the corresponding ions, M^+ and M^{2+}, respectively, having the electronic configurations of the immediately preceding noble gas. Because of these many general similarities, the chemistry of these two families of metals is conveniently treated together.

10.1.1 Alkali Metals

Alkali metal chemistry is principally that of the M^+ cations, which give ionic compounds in the solid state as well as solvated cations. The following properties *decrease* with increasing atomic weight of the alkali metals (i.e., Li $\rightarrow$ Na $\rightarrow$ K $\rightarrow$ Rb $\rightarrow$ Cs):

1. Melting points and heats of sublimation of the metals,
2. Effective *hydrated* radii and hydration energies of the M^+ cations,
3. Lattice energies of salts except those with the very smallest anions,
4. Heats of formation of fluorides, hydrides, oxides, and carbides,

5. Ease of thermal decomposition of nitrates and carbonates,
6. Strengths of covalent bonds in M_2 molecules.

The Li^+ ion is exceptionally small and has a charge/radius ratio comparable to Mg^{2+}, in accord with the "diagonal relationship" (Section 8.1).

A number of other monopositive ions with similar chemical behavior to the alkali metal cations are known. Salts of the quaternary ammonium cations NH_4^+, RNH_3^+, ..., R_4N^+ are similar in many respects to those of the heavier alkali metals. The properties of many NH_4^+ salts are similar to those of the corresponding K^+ salts. Tetraalkylammonium hydroxides, $R_4N^+OH^-$, are strong bases like alkali metal hydroxides and absorb CO_2 from the atmosphere to form the corresponding carbonates. The Tl^+ ion (Section 9.9.4) has an ionic radius (1.64 Å) similar to that of Rb^+ (1.66 Å) but is more polarizable. The $(C_5H_5)_2Co^+$ ion, which is very stable because of the 18-electron noble gas valence electron for the central cobalt atom, gives precipitation reactions similar to those of Cs^+. The corresponding hydroxide $(C_5H_5)_2Co^+OH^-$ is a strong base, absorbing CO_2 from the atmosphere.

10.1.2 Alkaline Earth Metals

The alkaline earth metals, like the alkali metals, are highly electropositive metals as shown by the high chemical reactivities, ionization enthalpies, and standard electrode potentials of the free metals as well as the ionic nature of the compounds of the heavier alkaline earth metals. The atomic radii of the alkaline earth metals are smaller than those of the alkali metals owing to increasing nuclear charge. The free alkaline earth metals have twice the number of bonding electrons as the alkali metals, leading to higher melting points, boiling points, and densities (Table 10.1). The alkaline earth metal cations M^{2+} are smaller and considerably less polarizable than the alkali metal cations M^+. The divalent lanthanide ions, namely Eu^{2+}, Sm^{2+}, and Yb^{2+} (Section 12.4.2), resemble Sr^{2+} and Ba^{2+} except for the ease of oxidation to the trivalent state. Beryllium has unique chemical behavior with predominantly covalent chemistry but forms the tetrahedrally coordinated hydrated ion $Be(H_2O)_4^{2+}$. Beryllium compounds are highly toxic, especially as dusts or smokes. This toxicity appears to arise from the ability for Be^{2+}, with its stronger coordinating ability, to displace Mg^{2+} from Mg-activated enzymes.

The four heaviest alkaline earth metals (Ca, Sr, Ba, and Ra) form a closely allied series. The following properties *increase* with increasing atomic weight of these alkaline earth metals (i.e., Ca → Sr → Ba → Ra):

1. Hydration tendencies of crystalline salts,
2. Solubilities of fluorides,
3. Thermal stabilities of carbonates, nitrates, and peroxides,
4. Rates of reactions of metals with hydrogen.

The solubilities of sulfates, nitrates, and halides decrease in the series

Table 10.1   Properties of the Free Alkali and Alkaline Earth Metals

Metal	Melting Point (°C)	Boiling Point (°C)	Density (g/cm³)*	Redox Potential (V)†
Alkali Metals				
Lithium	180.5	1326	0.534	−3.02
Sodium	97.8	883	0.97	−2.71
Potassium	63.7	756	0.86	−2.92
Rubidium	39.0	688	1.532	−2.99
Cesium	28.6	690	1.8785	−3.02
Alkaline earth metals				
Beryllium	1278	2970	1.85	−1.70
Magnesium	651	1107	1.74	−2.37
Calcium	843	1487	1.54	−2.87
Strontium	769	1384	2.6	−2.89
Barium	725	1140	3.51	−2.90
Radium	700	<1737	~5	−2.92

* At ambient temperature (15–25°C).
† In aqueous solution.

$Ca \rightarrow Sr \rightarrow Ba \rightarrow Ra$. The sulfates $SrSO_4$, $BaSO_4$, and $RaSO_4$ are precipitated essentially quantitatively from aqueous solutions.

10.1.3 NMR Properties of the Alkali and Alkaline Earth Metals

Table 10.2 summarizes the NMR properties of all the stable magnetic isotopes of the alkali and alkaline earth metals. The most favorable such isotope is ^{133}Cs, which has the advantages of both 100% natural abundance and a very low quadrupole moment. However, many of the other alkali and alkaline earth metal isotopes, such as ^7Li, ^{23}Na, and ^{25}Mg, have also been used for NMR studies despite their significant quadrupole moments. The symmetrical environments in solutions of the alkali and alkaline earth metals in many of their compounds allow reasonably sharp NMR spectra to be obtained from many of the quadrupolar isotopes.

10.2 The Free Alkali and Alkaline Earth Metals

All the alkali and alkaline earth metals except for the lightest alkaline earth metals (Be and Mg) are highly reactive toward air and water, evolving hydrogen in the latter case. Beryllium and magnesium metals are stable to air and water under ambient conditions with the aid of adherent protective coatings of the corresponding oxides.

Table 10.2   Isotopes of the Alkali and Alkaline Earth Metals for NMR

Isotope	Spin	Abundance (%)	Quadrupole Moment (10^{-28} m^2)	NMR Sensitivity*
Alkali metals				
^{6}Li	1	7.4	−0.0008	0.008 51
^{7}Li	$\frac{3}{2}$	92.6	−0.045	0.294
^{23}Na	$\frac{3}{2}$	100	0.12	0.092 7
^{39}K	$\frac{3}{2}$	93.1	0.055	0.000 051
^{41}K	$\frac{3}{2}$	6.9	0.067	0.000 084
^{85}Rb	$\frac{5}{2}$	72.8	0.25	0.010 5
^{87}Rb	$\frac{3}{2}$	27.2	0.12	0.177
^{133}Cs	$\frac{7}{2}$	100	−0.003	0.047
Alkaline earth metals				
^{9}Be	$\frac{3}{2}$	100	0.052	0.014
^{25}Mg	$\frac{5}{2}$	10.0	0.022	0.027
^{43}Ca	$\frac{7}{2}$	0.13	−0.05	0.064
^{87}Sr	$\frac{9}{2}$	7.0	0.36	0.027
^{135}Ba	$\frac{3}{2}$	6.6	0.18	0.005
^{137}Ba	$\frac{3}{2}$	11.3	0.28	0.007

* Relative to ^{1}H for equal numbers of nuclei; does not consider the natural abundance.

10.2.1 Alkali Metals

Sodium and potassium are relatively abundant elements (2.6 and 2.4%, respectively, in the earth's crust) and occur in large deposits of *rock salt*, NaCl, and *carnallite*, $KCl \cdot MgCl_2 \cdot 6H_2O$. Lithium, rubidium, and cesium are much rarer in nature and occur mainly in a few silicate minerals. All isotopes of francium are radioactive with short half-lives; the longest lived francium isotopes are ^{212}Fr and ^{223}Fr with half-lives of 19 and 22 minutes, respectively. For this reason, macroscopic amounts of francium compounds cannot be isolated. Francium chemistry is known only through limited tracer studies and appears to resemble that of cesium, in accord with expectation.

The lighter alkali metals (Li, Na) are obtained by electrolysis of suitable fused salt mixtures at elevated temperatures in an inert atmosphere. Electrolysis of a 55:45 melt of LiCl/KCl at 450°C can be used for the preparation of lithium metal, which is liberated in preference to the more electropositive potassium. Electrolysis of a 40:60 NaCl/CaCl$_2$ mixture at 580°C can be used for the preparation of sodium metal. The more volatile heavier alkali metals (K, Rb, Cs) can be obtained by reaction of their molten chlorides with sodium vapor at 850°C and are purified by distillation. At their boiling points, alkali metal vapors contain ∼1% of the corresponding diatomic molecules, M_2, in which the single-valence electrons of the two alkali metal atoms form a normal two-electron, two-center bond.

The alkali metals have relatively low melting points (Table 10.1). Some alloys containing two or more different alkali metals are liquids at room temperature.[1] Thus Na/K alloys are liquids at room temperature over a wide range of compositions with the eutectic melting point of $-12.3°C$ for the alloy containing 77% potassium. Sodium and potassium metals react vigorously with mercury to form the corresponding amalgams. Sodium amalgams with less than 7% sodium are liquids at room temperature.

All alkali metals react readily with oxygen; the O/M ratio in the favored product increases with increasing atomic weight of the alkali metal, and the oxides ultimately produced are Li_2O, Na_2O_2, KO_2, RbO_2, and CsO_2. Lithium metal is unusual in reacting directly with molecular nitrogen to give Li_3N; this reaction is slow under ambient conditions but rapid at 400°C. The heavier alkali metals are unreactive toward pure molecular N_2. Lithium and sodium react with carbon at elevated temperatures to form the corresponding acetylides, M_2C_2 (M = Li, Na). The heavier alkali metals give nonstoichiometric intercalation compounds (Section 2.2.3) upon reaction with graphite.

10.2.2 Alkaline Earth Metals

Beryllium, like its neighbors lithium and boron in the periodic table, is relatively rare in the earth's crust (~ 2 ppm). The most important beryllium mineral is *beryl*, $Be_3Al_2(SiO_3)_6$, which is an example of a cyclic silicate (Section 3.5.3). *Emerald* is also nominally $Be_3Al_2(SiO_3)_6$ but with some of the Al^{III} replaced by Cr^{III} to impart the characteristic green color. Except for radium, the remaining alkaline earth metals are much more abundant and are widely distributed in minerals and in seawater; calcium is the third most abundant metal in nature. Examples of alkaline earth minerals include *dolomite*, $CaCO_3 \cdot MgCO_3$, *carnallite*, $KCl \cdot MgCl_2 \cdot 6H_2O$, and *barytes*, $BaSO_4$. All isotopes of radium are radioactive. The radium isotope with the longest half-life, ^{226}Ra ($t_{1/2} = \sim 1600$ years, α-emitter), is found in the natural decay series of the common uranium isotope ^{238}U. Radium can be isolated from pitchblende, although it takes 10 tons of pitchblende to produce a milligram of radium.

Beryllium metal is obtained by electrolysis of a $BeCl_2$ melt with the addition of some NaCl to increase the conductivity of the melt. Beryllium metal is light (1.86 g/cm^3) and quite hard and brittle. Beryllium metal has the lowest electron density, hence the lowest stopping power of electromagnetic radiation of any metal that can be used for construction. This property of beryllium metal makes it useful for windows in X-ray apparatus and for nuclear technology. Alloying copper with $\sim 2\%$ beryllium increases its strength approximately sixfold.

Magnesium metal can be obtained by the following methods:

1. Electrolysis of fused halide mixtures (e.g., $MaCl_2/CaCl_2/NaCl$), from which the least electropositive metal, namely magnesium, is liberated,
2. Reduction of MgO or calcined dolomite (CaO·MgO) with ferrosilicon, followed by distillation of the magnesium metal from the reaction mixture.

Bulk magnesium metal, like aluminum metal, is protected by an oxide film and thus can be used as a construction material; it is the lightest construction metal used in industry.

Calcium metal and its heavier congeners are made on a relatively small scale by high temperature reduction of their oxides with aluminum metal or their halides with sodium metal or by electrolysis of their fused chlorides. They resemble the alkali metals by being reactive toward water under ambient conditions and by forming blue solutions in liquid ammonia.

10.2.3 Solutions of the Alkali and Alkaline Earth Metals in Liquid Ammonia[2,3]

All the alkali metals, the heavier alkaline earth metals (Ca, Sr, Ba), and the lanthanide metals exhibiting stable divalent oxidation states (Eu, Yb) dissolve in liquid ammonia to give blue solutions that conduct electricity electrolytically. Transport number measurements indicate that the main current carrier is the solvated electron, which has a high mobility. The rate of decomposition of the solvated electron in pure liquid ammonia is $\sim 1\%$ per day, resulting in hydrogen evolution. Decomposition of the solvated electron in liquid ammonia is accelerated by the addition of even small amounts of water. The half-life of the solvated electron in pure neutral water is only $\sim 10^{-4}$ second. Similar blue alkali metal solutions cann be obtained in a number of other polar solvents including amines (e.g., $MeNH_2$ and $H_2NCH_2CH_2NH_2$), relatively basic ethers (Me_2O, tetrahydrofuran, $MeOCH_2CH_2OCH_2CH_2OMe$), and hexamethylphosphoramide, $(Me_2N)_3P{=}O$ (Section 5.10.3).

More concentrated solutions of alkali metals in liquid ammonia are copper-colored with a metallic luster and high electrical conductivity. These solutions have been shown to contain the alkali metal anions, or *alkalides*, M^-, by the disproportionation reaction

$$2M \rightarrow M^+ + M^- \tag{10.1}$$

The lithium/ammonia system has an extraordinary deep eutectic at 20 mol% lithium and $-184°C$. X-Ray and neutron diffraction studies on this eutectic suggests the presence of the neutral "expanded-metal compound" $Li(NH_3)_4$, containing four-coordinate lithium(0).[4]

Addition of ligands that strongly coordinate alkali metal cations, M^+, such as crown ethers or cryptates, allow the isolation of crystalline derivatives from solutions of alkali metals in liquid ammonia. Structures of some of these crystalline derivatives can be determined by X-ray diffraction. These crystalline products are of two types:

1. *Alkalides.* The alkalides form yellow-bronze diamagnetic solids; examples of alkalides are $[Na(2,2,2\text{-crypt})]^+Na^-$ and $[Cs(18\text{-crown-}6)]^+Cs^-$. NMR studies indicate the presence of both M^+ and M^-.
2. *Electrides.* The electrides form dark paramagnetic, very reactive solids that have trapped electrons in their structures. Electrides are less stable, hence

more difficult to isolate, than the alkalides. An example of an electride is black $[Cs(19\text{-crown-}6)_2]^+e^-$.

The blue solutions of alkali metals in liquid ammonia and amines are useful reducing agents in organic chemistry.[5] They are thus potent enough reducing agents to destroy the aromaticity of benzene derivatives forming the corresponding cyclohexadiene derivatives; such reactions undoubtedly proceed through benzenoid radical anion intermediates formed by the solvated electrons. Solutions of sodium in liquid ammonia are most commonly used for this purpose. However, such solutions are unstable with respect to hydrogen evolution and give $NaNH_2$ in the presence of iron or other transition metals.

10.3 Compounds of the Alkali Metals

10.3.1 Oxides and Hydroxides

Alkali metal oxides include the compounds M_2O, which are derivatives of the oxide ion O^{2-} (Section 6.1.1.1), M_2O_2, which are derivatives of the peroxide ion O_2^{2-} (Section 6.3.3), and MO_2, which are derivatives of the superoxide ion O_2^- (Section 6.3.5). The M_2O alkali metal oxides have the antifluorite structure in which the metal ions and oxide ions occupy the positions occupied by the fluoride and calcium ions, respectively, in fluorite, CaF_2. The stability of alkali metal peroxides and superoxides increases with increasing size of the alkali metal. Sodium peroxide, Na_2O_2, is made on a large scale for use as a bleaching agent for fabrics, paper, pulp, wood, and so on. The alkali metal oxides react instantly with water to give the corresponding alkali metal hydroxide; the peroxides and superoxides also liberate oxygen with water.

Rubidium and cesium form highly colored suboxides containing fused octahedral metal clusters with oxygen atoms in the center of metal octahedra (Figure 10.1).[6] These include bioctahedral Rb_9O_2 and trioctahedral $Cs_{11}O_3$.

The alkali metal hydroxides, MOH, are very soluble in water and alcohols; their solutions readily absorb CO_2 from the atmosphere to form the corresponding carbonates. They sublime unchanged at 350–400°C. The gas phase

Bioctahedron
Rb_9O_2

Trioctahedron
$Cs_{11}O_3$

Figure 10.1. Alkali metal octahedral clusters in suboxides of the heavier alkali metals.

basicity of the alkali metal hydroxides increases in the sequence LiOH <
NaOH < KOH < RbOH < CsOH, but solvation can affect this sequence
in solution.

10.2.2 Alkali Metal Ions and Their Complexes

Most alkali metal salts are soluble in water. There are very few water-insoluble
sodium salts; $NaSb(OH)_6$ (Section 6.5.9.2) and $NaM(UO_2)_3(CO_2Me)_9 \cdot 6H_2O$
can be precipitated from aqueous solutions under very carefully controlled
conditions. The perchlorates, $MClO_4$, the hexanitrometallates, $M_3[M'(NO_2)_6]$
(M = Co, Rh, Ir), and the tetraphenylborates $M[B(C_6H_5)_4]$ of the heavier
alkali metals (M = K, Rb, Cs) can be precipitated from aqueous solution.

The alkali metal cations M^+ are only very weakly complexed by simple
anions.[7] However, β-diketones form alkali metal complexes. Such complexes
can be the basis for the separation of lithium from the other alkali metal
complexes in aqueous solution by extraction into xylene as the tri-n-octyl-
phosphine oxide complex $Li(PhCOCHCOPh)[OP(n\text{-}C_8H_{17})_3]_2$.

Alkali metals are complexed with a variety of chelating ligands containing
multiple ether or carbonyl donors, including the "glyme-type" acyclic ethers
$MeO(CH_2CH_2O)_nMe$, macrocyclic crown ethers and cryptates (Figure 10.2),

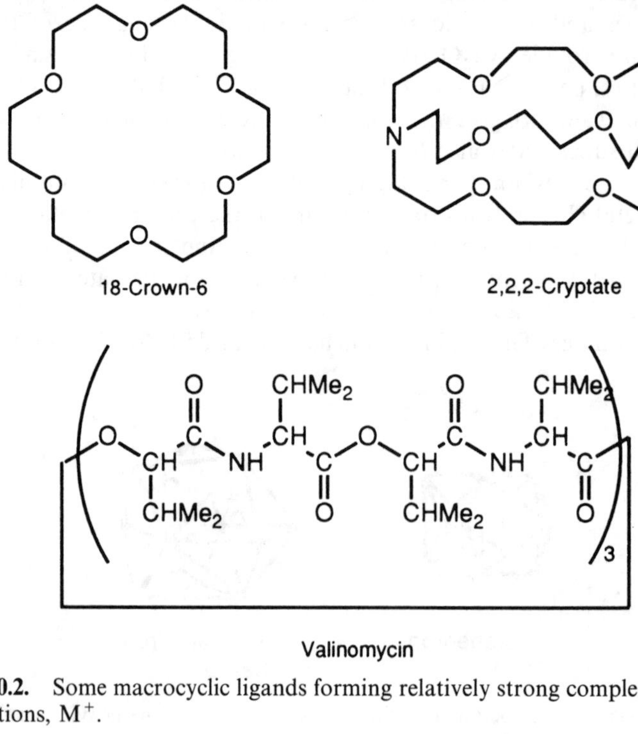

18-Crown-6   2,2,2-Cryptate

Valinomycin

Figure 10.2.   Some macrocyclic ligands forming relatively strong complexes with alkali
metal cations, M^+.

or certain polypeptides such as valinomycin, containing a 36-membered macrocyclic ring (Figure 10.2). The affinity of the macrocylic chelating ligands for alkali metals depends on the fit of the alkali metal cations into the ligand cavities.[8] Naturally occurring macrocyclic polypeptides and polyethers such as valinomycin and monesin A function as *ionophores* (i.e., alkali metal ion-selective transport agents) in biological systems.[9]

10.3.3 Organolithium Compounds[10,11]

Lithium alkyls and aryls can be obtained by reactions of the corresponding alkyl or aryl chloride with lithium metal in a suitable inert either or hydrocarbon solvent:

$$RCl + 2Li \rightarrow RLi + LiCl \qquad (10.2)$$

Reactions of the corresponding alkyl or aryl bromides or iodides with lithium metal are generally less desirable for the preparation of the corresponding lithium alkyls or aryls because of lower yields arising from more extensive coupling reactions of the organolithium compound with excess alkyl halide in a Wurtz reaction:

$$RLi + RX \rightarrow R—R + LiX \qquad (10.3)$$

Such coupling reactions can lead to significant lowering of the yield of the organolithium compound, particularly in the case of relatively reactive alkyl halides such as allyl and benzyl halides.

Organolithium compounds are very reactive toward water, air, and carbon dioxide by the following reactions:

$$RLi + H_2O \rightarrow RH + LiOH \qquad (10.4a)$$

$$RLi + D_2O \rightarrow RD + LiOD \qquad (10.4b)$$

$$2RLi + O_2 \rightarrow 2ROLi \qquad (10.4c)$$

$$RLi + CO_2 \rightarrow RCO_2^- Li^+ \qquad (10.4d)$$

The reaction with D_2O is useful for identifying the location of the lithium atom in the organolithium compound from the position of the deuterium atom in the deuterolysis product. The reaction with CO_2 can be useful for the preparation of carboxylic acids. Organolithium derivatives can cleave some ether solvents, particularly cyclic ethers, to form lithium alkoxides so that some care must be used in selecting suitable conditions for the preparation of organolithium, compounds in ether solvents. Organolithium derivatives are soluble in hydrocarbon solvents and sublimable in vacuum. They form complexes with lithium halides such as Me_3Li_4X.

Some organolithium compounds such as *n*-butyllithium are commercially available in large quantities as solutions in hexane or other inert solvents.

Such organolithium compounds are useful for the preparation of other organo-lithium compounds by reactions such as the following.

Metal–hydrogen exchange or lithiation, for example:

$$C_5H_5FeC_5H_5 + Bu^nLi \rightarrow C_5H_5FeC_5H_4Li + Bu^nH \qquad (10.5a)$$

$$C_5H_5FeC_5H_4Li + Bu^nLi \rightarrow LiC_5H_4FeC_5H_4Li + Bu^nH \qquad (10.5b)$$

Such lithiation reactions are sometimes promoted by addition of a chelating poly(tertiary amine) such as $N,N,N'N'$-tetramethylethylenediamine, Me_2NCH_2-CH_2NMe_2, to complex with the lithium. In some cases several hydrogen atoms in an organic compound can be replaced by lithium by this method (compare Section 2.3.1.3):

$$CH_3C{\equiv}CH + 4Bu^nLi \rightarrow Li_4C_3 + 4Bu^nH \qquad (10.6)$$

Metal–halogen exchange, for example:

$$C_6F_5Br + Bu^nLi \rightarrow C_6F_5Li + Bu^nBr \qquad (10.7)$$

Metal–halogen exchange reactions generally proceed under milder conditions than either direct reactions of lithium with alkyl halides or metal–hydrogen exchange reactions and thus are useful for preparing relatively unstable organolithium compounds, which must be handled under mild conditions. Thus C_6F_5Li (equation 10.7), like many fluorinated organolithium compounds, is rather unstable with respect to elimination of lithium fluoride.

Organolithium compounds exhibit a variety of oligomeric structures[12] with the following general features.

1. The lithium atom is bonded to the alkyl carbon atom by multicenter bonds but can interact with other atoms in the organic moiety including not only nitrogen and oxygen donor atoms but even hydrogen atoms in some cases.
2. The lithium atom can have apparent coordination numbers from 2 to 6, although 4 is the most common. Apparent lithium coordination numbers above 4 clearly must involve multicenter bonding, since lithium has no accessible d orbitals for five-coordinate sp^3d or six-coordinate sp^3d^2 hybridization.
3. Lithium is commonly found in tetrahedral Li_4 units, although other lithium polyhedra can occur.
4. Direct lithium–lithium bonding never occurs in organolithium compounds.

In accord with this general scheme, LiMe and LiEt are tetrameric $(LiR)_4$ derivatives based on an Li_4 tetrahedron with μ_3-alkyl groups in each of the four tetrahedron faces. However, cyclohexyllithium is a hexameric $(LiR)_6$ derivative with μ_3-cyclohexyl groups in six of the octahedron faces and benzene of solvation in the remaining two octahedron faces.

10.3.4 Organosodium and Organopotassium Compounds[13]

10.3.4.1 Alkyls and Aryls

Alkyl and aryl derivatives of sodium and potassium, MR (M = Na, K), are essentially ionic derivatives that are sparingly soluble at most in hydrocarbon solvents. They are very air-sensitive and reactive and therefore are difficult to handle. They cannot be used in the presence of ethers because of rapid cleavage of the ether, for example,

$$R_2O + MR' \rightarrow RO^-M^+ + ROR' \tag{10.8}$$

In addition, reactions of alkyl halides with sodium or potassium cannot be used to prepare the corresponding metal alkyls similar to the preparation of organolithium compounds (e.g., equation 10.2) because of Wurtz coupling (equation 10.3 with Na or K instead of Li). Thus the most general method for the preparation of sodium and potassium alkyls uses reactions of alkyl-mercury compounds with the alkali metal. Thus methylpotassium can be precipitated from the reaction of dimethylmercury with a liquid sodium/potassium alloy in an inert hydrocarbon solvent. Methylsodium has a tetrameric structure similar to that of methyllithium. However, methylpotassium has a structure similar to NiAs, with disordered isolated methyl anions and potassium cations.

Alkyl and aryl derivatives of sodium and potassium are of little practical value because of their insolubility. However, they can be solubilized by addition of suitable metal alkoxides such as $Mg(OCH_2CH_2OEt)_2$, which gives reactive alkylsodium or alkylpotassium derivatives soluble in benzene.

10.3.4.2 Derivatives of Acidic Hydrocarbons

Acidic hydrocarbons such as cyclopentadienes and terminal alkynes, $RC{\equiv}CH$, react with alkali metals or their hydrides in liquid ammonia, ethers, or hydrobcarbons to give the corresponding alkali metal derivatives, for example,

$$NaH + C_5H_6 \xrightarrow{\text{tetrahydrofuran}} Na^+C_5H_5^- + H_2\uparrow \tag{10.9a}$$

$$2Na + 2RC{\equiv}CH \xrightarrow{\text{tetrahydrofuran}} 2RC{\equiv}C^-Na^+ + H_2\uparrow \tag{10.9b}$$

The most important derivative of this type is sodium cyclopentadienide, $Na^+C_5H_5^-$, which is used extensively for the preparation of diverse cyclopentadienyl transition metal derivatives such as $(C_5H_5)_2M$ (M = Ti, V, Cr, Mn, Fe, Co, Ni, Ru, Os, etc.) and $Na^+M(CO)_3C_5H_5^-$ (M = Cr, Mo, W).[14] Although the cyclopentadienide anion, $C_5H_5^-$, is stabilized by an aromatic sextet similar to benzene, solid sodium cyclopentadienide is pyrophoric in air, and its tetrahydrofuran solutions must be handled in an inert atmosphere.

10.3.4.3 Hydrocarbon Dianions

The most important alkali metal derivatives of a hydrocarbon dianion are the cyclooctatetraene dianion derivatives, $(M^+)_2C_8H_8^{2-}$, which are made by reactions of the free alkali metal with neutral cyclooctatetraene in tetrahydrofuran or similar ether solvent. In this reaction, nonplamar neutral cyclooctatetraene becomes the planar octagonal $C_8H_8^{2-}$, with an aromatic 10π-electron system:

$$+ 2K \xrightarrow{\text{tetrahydrofuran}} 2K^+ \quad 2- \tag{10.10}$$

Despite this considerable aromatic stabilization, solid alkali metal salts of the $C_8H_8^{2-}$ anion react explosively with air. The $C_8H_8^{2-}$ derivatives such as $K_2C_8H_8$ are useful for the synthesis of cyclooctatetraene metal derivatives such as $(C_8H_8)_2U$ (Section 12.6.2).

10.3.4.4 Radical Anions

A variety of compounds with two or more benzene rings (e.g., naphthalene, anthracene, $(C_6H_5)_2C=O$, $(C_6H_5)_3PO$, $(C_6H_5)_3As$, $C_6H_5N=NC_6H_5$) form the corresponding radical anions upon reaction with alkali metals, generally in coordinating ether solvents such as tetrahydrofuran. Such radical anions are generally deep green, blue, or purple and require a structure in which the negative charge is delocalized over an aromatic system. The deep green naphthalene radical anion as the sodium salt ("sodium naphthalenide"), $Na^+C_{10}H_8^{\overline{\cdot}}$ is frequently used as a powerful reducing agent (see Section 8.3.5 for such an application in carborane chemistry). In such $Na^+C_{10}H_8^-$ reductions, the naphthalene can be regarded as an "electron carrier." The deep blue benzophenone radical anion $[(C_6H_5)_2CO]^{\overline{\cdot}}$ or "benzophenone ketyl" is a useful and rapid reagent for the removal of oxygen in the purification of solvents for handling highly oxygen-sensitive compounds.

10.4 Beryllium Chemistry

The X—Be—X unit exhibits linear sp hybridization, but beryllium has a strong tendency to achieve the maximum coordination number of 4 for sp^3 hybridization. Examples of association of BeX_2 units to give derivatives with three-coordinate and four-coordinate beryllium are given in Figure 10.3. In addition, beryllium halides, BeX_2, are dibasic Lewis acids, forming four-coordinate tetrahedral L_2BeX_2 derivatives such as $(Et_2O)_2BeCl_2$ and BeF_4^{2-}.

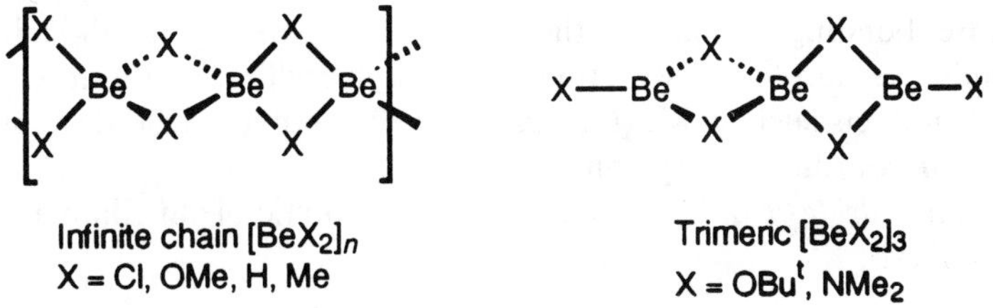

Infinite chain [BeX₂]ₙ
X = Cl, OMe, H, Me

Trimeric [BeX₂]₃
X = OBuᵗ, NMe₂

Figure 10.3. Ways of increasing the coordination number of beryllium from 2 in monomeric BeX_2 derivatives to 3 and 4.

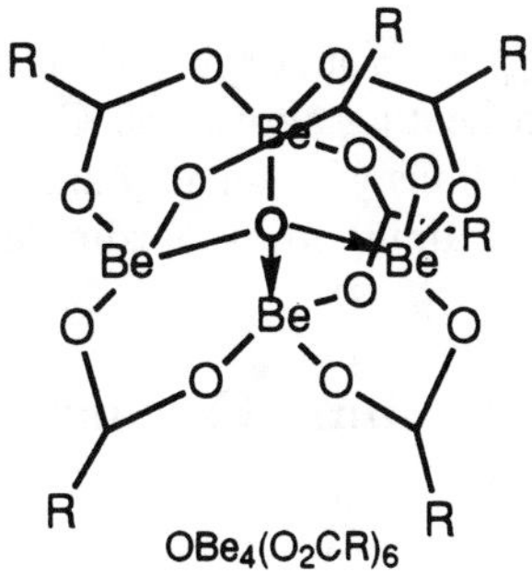

$OBe_4(O_2CR)_6$

Figure 10.4. The structure of the basic beryllium carboxylates $OBe_4(O_2CR)_6$; the central tetrahedral oxygen atom appears in bold type..

Linear two-coordinate beryllium is found in monomeric BeX_2 derivatives with bulky X groups such as $(2,4\text{-}Bu_2^t\ C_6H_3O)_2$ Be and $[(Me_3N)_2Si]_2$ Be.

White crystalline *beryllium oxide*, BeO, is amphoteric: it dissolves both in acids to from the $Be(H_2O)_4^{2+}$ cation and in aqueous alkali to form the $Be(OH)_4^{2-}$ anion. The white crystalline *beryllium halides*, BeX_2 (X = F, Cl, Br, I), are deliquescent and cannot be obtained from their hydrates by simple heating because of hydrolysis. Beryllium forms basic carboxylates, $OBe_4(O_2CR)_6$, containing a central tetrahedral OBe_4 unit with bidentate carboxylate groups bridging each of the six edges of the tetrahedron (Figure 10.4). These beryllium basic carboxylates are inert to water but are hydrolyzed by dilute acids.

Dialkylberylliums and *diarylberylliums* can be made by standard methods, for example,

$$Me_2Hg + Be \xrightarrow{110°C} Me_2Be + Hg \tag{10.11a}$$

$$2LiPh + BeCl_2 \xrightarrow[\text{solvent}]{\text{hydrocarbon}} Ph_2Be + 2LiCl \tag{10.11b}$$

Such organoberyllium compounds are highly reactive toward air and water and are associated into dimers or polymers through three-center, two-electron

Be—C—Be bonding similar to the Al—C—Al bonding in alkylaluminum compounds (Section 9.8.1). Reactions of dialkylberylliums with aprotic Lewis bases give adducts such as Me_3NBeMe_2 and (2,2'-bipyridyl)$BeMe_2$ containing three-or four-coordinate beryllium.

Beryllium hydrides can be obtained by the pyrolytic olefin elimination from beryllium alkyls:

$$[(CH_3)_2CH]_2Be \xrightarrow{\Delta} [(CH_3)_2CHBeH]_n + CH_3CH=CH_2\uparrow \quad (10.12a)$$

$$n[CH_3)_3C]_2Be \xrightarrow{>100^\circ C} [BeH_2]_n + 2n(CH_3)_2C=CH_2\uparrow \quad (10.12b)$$

Beryllium hydride, $[BeH_2]_n$, is an amorphous white polymeric solid (Figure 10.3) that evolves H_2 upon heating above 250°C. The polymeric structure of $[BeH_2]_n$ is broken upon reactions with tertiary amines to give dimeric adducts of the type $[R_3NBeH_2]_2$ containing four-coordinate beryllium.

10.5 Chemistry of Magnesium, Calcium, Strontium, and Barium

10.5.1 Binary Compounds

10.5.1.1 Oxides

The oxides MO (M = Mg, Ca, Sr, Ba) are white high-melting crystalline solids with cubic NaCl-type lattice structures. Calcium oxide is made on a vast scale by roasting *limestone*, $CaCO_3$:

$$CaCO_3 \xrightarrow{\Delta} CaO + CO_2 \qquad \Delta H_{298^\circ} = 178.1 \text{ kJ/mol} \quad (10.13)$$

The oxides CaO, SrO, and BaO react exothermically with water to form the corresponding hydroxides $M(OH)_2$. The water solubility of $M(OH)_2$ increases with increasing atomic number, and all $M(OH)_2$ compounds are relatively strong bases, reacting with CO_2 to produce the corresponding carbonates MCO_3.

10.5.1.2 Halides

The anhydrous halides can be made by dehydration of the corresponding hydrated salts without hydrolysis. Magnesium and calcium halides, MX_2 (M = Mg, Ca; X ≠ F) readily absorb water but SrX_2, BaX_2, and RaX_2 are normally anhydrous. The solubility of the fluorides increases in the sequence $MgF_2 < CaF_2 < SrF_2 < BaF_2$ because of the small size of F^- relative to M^{2+}; all these fluorides can be precipitated from their aqueous solutions. Calcium fluoride is used for prisms in spectrometers and optical cell windows because of its optical dispersion properties, transparency, and low water solubility.

10.5.1.3 Carbides (compare Section 2.3.1)

The carbides MC_2 (M = Ca, Sr, Ba) can be obtained by heating the metal or its oxide with carbon in an electric furnace. These carbides are ionic acetylides and hydrolyze readily to acetylene:

$$M^{2+}C_2^{2-} + 2H_2O \rightarrow M(OH)_2 + HC{\equiv}CH\uparrow \qquad (10.14)$$

Magnesium metal reacts with carbon at $\sim 500°C$ to give a similar acetylide, MgC_2, but at higher temperatures (500–700°C) it gives Mg_2C_3. Hydrolysis of Mg_2C_3 gives propyne, $CH_3C{\equiv}CH$, suggesting the presence of the C_3^{4-} anion.

10.5.2 Alkaline Earth Metal Complexes

The alkaline earth metals form more stable complexes than the alkali metals with a much greater variety of ligands.[15] The variation of stability constants within the alkaline earth metals depends on the type of ligand, as follows[16]:

1. For small or highly charged anions and certain uni- and bidentate ligands, the stability constants decrease with increasing crystal radii (i.e., Ma > Ca > Sr > Ba).
2. For oxoanions like NO_3^-, SO_4^{2-}, and IO_4^-, the stability constants increase with the hydrated radii (i.e., Mg < Ca < Sr < Ba).
3. For hydroxycarboxylic, polycarboxylic, and polyaminocarboxylic acids, the stability constants reach a maximum with calcium (i.e., Mg < Ca > Sr > Ba).

An important polydentate chelating ligand in alkaline earth coordination chemistry is ethylenediaminetetraacetic acid (EDTA), $(HO_2CCH_2)_2NCH_2$-$CH_2N(CH_2CO_2H)_2$, which readily forms complexes with the alkaline earths. Complexing of calcium with EDTA as well as with polyphosphates (Section 5.8.4.2) is important for the treatment of "hard" water by removal of calcium as well as for the volumetric determination of calcium using a metal indicator. A crystalline calcium complex of EDTA of the formula $Ca[CaEDTA]\cdot 7H_2O$ has been isolated and shown by X-ray diffraction to have eight-coordinate, distorted square antiprismatic calcium in the $CaEDTA^{2-}$ anion, with a hexadentate EDTA ligand and two coordinated water molecules.

The most important alkaline earth metal complex is *chlorophyll*, which is a porphyrin derivative of magnesium (Figure 10.5). In addition, polymerization through hydrogen bonding occurs in the chlorophyll structure. Chlorophyll provides a source of electrons for the photosynthetic reduction of CO_2 by water in plants.

10.5.3 Organomagnesium Compounds: Grignard Reagents

Organomagnesium compounds are much more important than the organometallic compounds of calcium and its heavier congeners. The most important organomagnesium compounds are *Grignard reagents*, which are obtained by

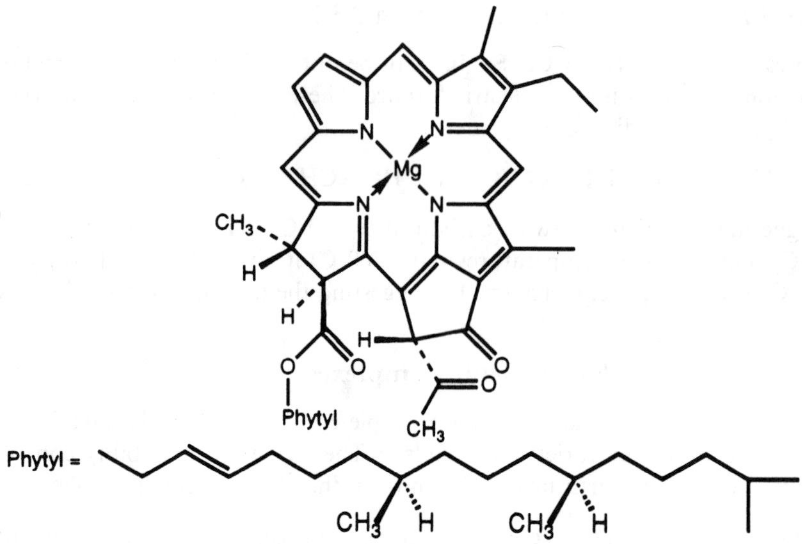

Figure 10.5. The structure of chlorophyll a.

reactions of alkyl or aryl halides with magnesium metal in an ether, most frequently diethyl ether or tetrahydrofuran.[17] The rates of the reactions of organic halides with magnesium decrease in the sequence I > Br > Cl. The reaction between organic halides and magnesium in an ether solvent often occurs only after an initial induction period and may become violent after it has started. This reaction can be initiated by addition of elemental iodine or 1,2-dibromoethane. Reactions of the dihalides XCH_2X, XCH_2CH_2X, and $X(CH_2)_3X$ with magnesium metal in diisopropyl ether gives the "di-Grignard reagent" $XMgCH_2MgX$, ethylene, and cyclopropane, respectively.[18]

The stoichiometry of Grignard reagents can be represented by the formula "RMgX", and the formation of Grignard reagents can be nominally represented by the following equation:

$$Mg + RX \rightarrow RMgX \qquad (10.15)$$

However, the actual Grignard reagent solutions contain a number of complicated equilibria (Figure 10.6). All three species MgX_2, $RMgX$, and R_2Mg

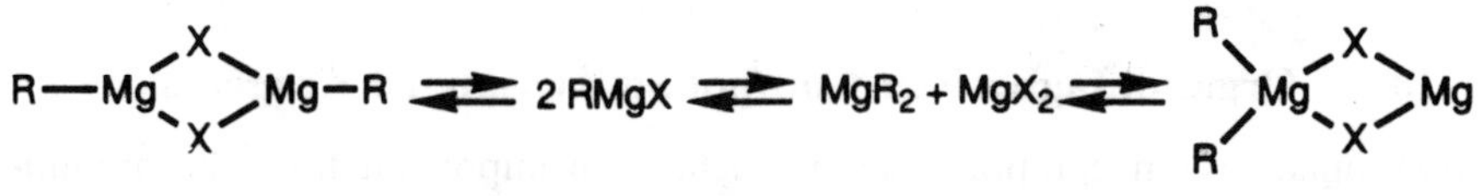

Figure 10.6. Some of the equilibria present in solutions of Grignard reagents. Co-ordinated ether molecules are omitted for clarity.

can be identified in the equilibrium mixture by ^{25}Mg NMR spectroscopy, which can be used to study the following equilbria (see so-called *Schlenk equilibria*) for different R groups[19]:

$$2RMgBr \rightleftarrows R_2Mg + MgBr_2 \tag{10.16}$$

The equilibrium constants for this reaction are ~ 30 for R = Me and ~ 5 for R = Et. Proton NMR spectroscopy cannot distinguish between R_2Mg and RMgX at ambient temperature because of rapid exchange of the R groups through some of the equilibria in Figure 10.6. However, ^{19}F NMR techniques can be used to distinguish between C_6F_5MgBr and $(C_6F_5)_2Mg$ at 25°C and proton NMR can distinguish between CH_3MgBr and $(CH_3)_2Mg$ at low temperatures. Both the etherates $RMgBr(OEt_2)_2$ (R = Et, Ph), with tetrahedral magnesium, and $MeMgBr (OC_4H_8)_3$, with five-coordinate magnesium have been crystallized from solutions of the corresponding Grignard reagents and their structures determined by X-ray diffraction.

A number of dialkylmagnesium derivatives, R_2Mg, are known. Thus, treatment of a Grignard reagent solution with *p*-dioxane precipitates the ether-insoluble dioxane magnesium halide complex, $C_4H_8O_2 \cdot MgX_2$:

$$2RMgX + 2C_4H_8O_2 \xrightarrow{Et_2O} 2C_4H_8O_2 \cdot MgX_2 \downarrow + R_2Mg \tag{10.17}$$

After removal of the precipitate, the dialkylmagnesium derivative can be isolated from the filtrate. Pure dialkylmagnesium compounds can also be obtained by reaction of magnesium metal with the corresponding dialkylmercury:

$$R_2Hg + Mg \rightarrow R_2Mg + Hg \tag{10.18}$$

Dialkylmagnesium compounds of the type $(RCH_2CH_2)_2Mg$ can be obtained by reaction of magnesium metal with the olefin under hydrogen pressure using anthracene to catalyze the formation of MgH_2 and 1% $ZrCl_4$ to catalyze the addition of MgH_2 to the olefin[20]:

$$Mg + H_2 + 2RCH{=}CH_2 \rightarrow (RCH_2CH_2)_2Mg \tag{10.19}$$

Dicyclopentadienylmagnesium (magnesium cyclopentadienide), $(C_5H_5)_2Mg$, is a pyrophoric white fluffy solid that can be made directly from magnesium metal and cyclopentadiene at elevated temperatures:

$$Mg + 2C_5H_6 \xrightarrow{500°C} (C_5H_5)_2Mg + H_2 \tag{10.20}$$

Dicyclopentadienylmagnesium is a useful reagent for the preparation of cyclopentadienyl transition metal derivatives.

References

1. Addison, C. C., *The Chemistry of the Liquid Alkali Metals*, Wiley, New York, 1984.

2. Dye, J. L., Compounds of Alkali Metal Anions, *Angew. Chem. Int. Ed. Engl.*, 1979, **18**, 587–598.

3. Dye, J. L., Electrides, Negatively Charged Metal Atoms, and Related Phenomena, *Prog. Inorg. Chem.*, 1984, **32**, 327–441.

4. Stacy, A. M.; Sienko, M. J., Reevaluation of the Crystal Structure Data on the Expanded-Metal Compounds $Li(NH_3)_4$ and $Li(ND_3)_4$, *Inorg. Chem.*, 1982, **21**, 2294–2297.

5. Birch, A. J., Reduction of Organic Compounds by Metal–Ammonia Solutions, *Q. Rev.*, 1950, **4**, 69–93.

6. Simon, A., Structure and Bonding in Alkali Metal Suboxides, *Struct. Bonding*, 1979, **36**, 81–127.

7. Poonia, N. S.; Bajaj, A. V., Coordination Chemistry of Alkali and Alkaline-Earth Cations, *Chem. Rev.*, 1979, **79**, 389–445.

8. Hiraoka, M., *Crown Compounds: Their Characteristics and Applications*, Elsevier, Amsterdam, 1982.

9. Pangborn, W.; Duax, W.; Langs, D., *J. Am. Chem. Soc.*, 1987, **109**, 2163–2165.

10. Wakefield, B. J., *The Chemistry of Organolithium Compounds*, Pergamon Press, Oxford, 1974.

11. Wakefield, B. J., *Organolithium Methods*, Academic Press, London, 1988.

12. Setzer, W. N.; Schleyer, P. von R., X-Ray Structural Analyses of Organolithium Compounds, *Adv. Organomet. Chem.*, 1985, **24**, 353–451.

13. Schade, C.; Schleyer, P. von R., Sodium, Potassium, Rubidium, and Cesium: X-Ray Structural Analysis of their Organic Compounds, *Adv. Organomet. Chem.*, 1987, **27**, 169–278.

14. Birmingham, J., Synthesis of Cyclopentadienyl Metal Compounds, *Adv. Organomet. Chem.*, 1964, **2**, 365–413.

15. Poonia, N. S.; Bajag, A. V., Coordination Chemistry of Alkali and Alkaline Earth Cations, *Chem. Rev.*, 1979, **79**, 389–445.

16. Aruga, R., *Inorg. Chem.*, 1980, **19**, 2895–2896.

17. Ashby, E. C., Grignard Reagents. Compositions and Mechanisms of Reaction, *Q. Rev.*, 1967, **21**, 259–285.

18. Bickelhaupt, F., Di-Grignard Reagents and Metallocycles, *Angew. Chem. Int. Ed. Engl.*, 1987, **26**, 990–1005.

19. Benn, R., Lehmkuhl, H., Mehler, K.; Rufinska, A., ^{25}Mg-NMR: A Method for the Characterization of Organomagnesium Compounds, Their Complexes, and Schlenk Equilibria, *Angew. Chem. Int. Ed. Engl.*, 1984, **23**, 534–535.

20. Bogdanović, B., Catalytic Synthesis of Organolithium and Organomagnesium Compounds and of Lithium and Magnesium Hydrides—Applications in Organic Synthesis and Hydrogen Storage, *Angew. Chem. Int. Ed. Engl.*, 1985, **24**, 262–273.

11

Zinc, Cadmium, and Mercury

11.1 General Aspects of the Chemistry of Zinc, Cadmium, and Mercury

The chemistry of zinc, cadmium, and mercury is conveniently treated with that of the main group elements because their filled d shells are retained in all stable derivatives. The maximum oxidation state of these elements in stable compounds is thus $+2$, arising from loss of the two s electrons. Examples of oxidation states below $+2$ for zinc, cadmium, and mercury include the following:

1. The $+1$ formal oxidation state in the M—M bonded ions M_2^{2+}, of which only Hg_2^{2+} is stable in water, and related polar solvents at ambient conditions.
2. Mercury ions with Hg—Hg bonds in fractional formal oxidation states such as $+\frac{2}{3}$ in $Hg_3[AsF_6]_2$ and $+\frac{1}{2}$ in $Hg_4[AsF_6]_2$.

In the $+2$ oxidation states zinc and cadmium have very similar chemistry, but that of mercury is markedly different, as indicated in the following examples:

1. Zinc ($E°$ for $Zn^{2+}/Zn^0 = -0.762$) and cadmium ($E°$ for $Cd^{2+}/Cd^0 = -0.402$) are electropositive metals, whereas mercury has a high positive standard potential ($E°$ for $Hg^{2+}/Hg^0 = +0.854$). For this reason zinc and cadmium metals dissolve in nonoxidizing acids with hydrogen evolution, whereas mercury metal dissolves only in oxidizing acids.

2. The ions Zn^{2+} and Cd^{2+} resemble Mg^{2+} (Chapter 10), whereas Hg^{2+} is much more polarizable than Mg^{2+}. Furthermore, $ZnCl_2$ and $CdCl_2$ are essentially ionic, whereas $HgCl_2$ gives a molecular crystal and Hg^{2+} complexes have stability constants orders of magnitude greater than those of Zn^{2+} and Cd^{2+}. In addition, mercury has a notoriously high affinity for sulfur ligands in metal complexes. The name *mercaptan for* organic RSH compounds comes from the Latin phrase *mercurium aptans*, related to the affinity of mercaptans for mercury.

3. $Zn(OH)_2$ is amphoteric, $Cd(OH)_2$ is more basic than $Zn(OH)_2$, but $Hg(OH)_2$ is an extremely weak base.

4. Linear two-coordinate derivatives, although possible for all three metals, are much more prevalent for mercury than for zinc and cadmium. All three metals commonly have coordination numbers 4, 5, and 6, with the higher coordination numbers being less favored for mercury because of the unusual stability of two-coordinate derivatives.

Cadmium and mercury are rather favorable elements for NMR studies because they have the spin $\frac{1}{2}$ isotopes ^{111}Cd, ^{113}Cd, and ^{199}Hg of natural abundances 12.8, 12.3, and 16.9%, respectively, with sensitivities of 0.0095, 0.011, and 0.0057, respectively, relative to the proton for equal numbers of nuclei. The only possible NMR isotope for zinc is ^{67}Zn, with a quadrupole moment ($I = \frac{5}{2}$), a natural abundance of 4.1%, and a sensitivity of 0.0029 relative to the proton for equal numbers of nuclei.

11.2 Elemental Zinc, Cadmium, and Mercury

Zinc, cadmium, and mercury have a low abundance in nature. The main source of zinc is *sphalerite*, a mixed zinc/iron sulfide. Cadmium occurs by isomorphism in almost all zinc ores. The most important mercury ore is *cinnabar*, HgS. Volcanic eruptions, such as those from Kilauea, Hawaii, generate mercury in the environment.[1]

The metals zinc (mp 419°C, bp 907°C), cadmium, (mp 321°C, bp 767°C), and particularly mercury (mp −38.9°C, bp 357°C) are characterized by relatively low melting and boiling points for free metals. Mercury is the only metal that is liquid even below room temperature and exhibits appreciable vapor pressure under ambient conditions (1.3×10^{-3} mm); mercury vapor, like that of the noble gases (Chapter 7), is monatomic. The volatility of these metals facilitates liberation from their ores by suitable roasting processes. Elemental zinc thus can be obtained by roasting the oxide with coke in a special blast furnace where the zinc vapor can be condensed. Element mercury can be obtained by simple roasting of cinnabar in air, since the initially produced HgO decomposes thermally to its elements at elevated temperatures.

Metallic zinc and cadmium react with nonoxidizing acids to evolve hydrogen, whereas metallic mercury is inert to nonoxidizing acids. Zinc but not cadmium

Figure 11.1. Hg_4^{6-} and Hg_8^{8-} cluster anions found in alkali metal amalgams.

dissolves in strong bases to form $Zn(OH)_4^{2-}$. Zinc and cadmium react readily when heated in oxygen to give the corresponding oxides MO (M = Zn, Cd). All three elements react directly with halogens and with nonmetals such as sulfur and selenium.

Zinc and cadmium form many alloys, including several of technical importance such as *brasses*, which are zinc–copper alloys (e.g., β-brass, CuZn, and γ-brass, Cu_5Zn_8). Mercury combines with most other metals to form amalgams. Amalgamation of mercury with alkali metals is rather violent and leads to species such as $NaHg_2$, NaHg, Na_3Hg_2, KHg_2, and KHg.[2,3] Some alkali metal amalgams contain definite mercury cluster anions such as square Hg_4^{6-} in Na_3Hg_2 $(=Na_6Hg_4)$[4] and cubic[5] Hg_8^{8-} in $Rb_{15}Hg_{16}$ (Figure 11.1). Electrolytic reduction of quaternary ammonium cations R_4N^+ (Section 6.4.1) at mercury electrodes give "quaternary ammonium amalgams," of approximate stoichiometry $R_4N^+Hg_4^-$, which appear to contain similar mercury cluster anions.[6]

11.3 Divalent Zinc and Cadmium Compounds

11.3.1 Oxides and Hydroxides

The oxides ZnO and CdO are formed by burning the metals in air or by pyrolysis of the carbonates or nitrates. Oxide smokes can be obtained by combustion of the corresponding metal alkyls; those of cadmium are highly toxic. The oxides ZnO and CdO sublime without decomposition at high temperatures. Zinc oxide is normally white but turns yellow upon heating, whereas cadmium oxide varies from greenish yellow to brown depending on its thermal history. Zinc oxide is an *n*-type semiconductor[7] because of non-stoichiometry (metal abundance) and impurities; it has many applications, including use in a catalyst as a ZnO/CuO mixture supported on Al_2O_3 or Cr_2O_3 for methanol synthesis for CO + H_2.

The hydroxides $Zn(OH)_2$ and $Cd(OH)_2$ are precipitated from aqueous solutions of the corresponding M^{2+} ions (M = Zn, Cd) by addition of bases. Excess aqueous NaOH readily dissolves $Zn(OH)_2$ to form zincate ions, whereas concentrated NaOH is necessary to dissolve $Cd(OH)_2$. Raman spectra indicate that the principal zincate ion in $Zn(OH)_2/NaOH/H_2O$ solutions is $Zn(OH)_3(H_2O)^-$, although crystalline $Zn(OH)_4^{2-}$ salts can be obtained. Excess concentrated aqueous ammonia dissolves the hydroxides $Zn(OH)_2$ and $Cd(OH)_2$ to form the corresponding ammines $M(NH_3)_4^{2+}$ (M = Zn, Cd).

11.3.2 Other Chalcogenides

All the possible six chalcogenides ME (M = Zn, Cd; E = S, Se, Te) are known. They can be made by direct combination of the elements at elevated temperatures. Very pure epitaxial layers of zinc selenide[8] can be made by the following reaction:

$$Me_2Zn + H_2Se \xrightarrow{200-350°C} ZnSe + 2CH_4\uparrow \qquad (11.1)$$

The structures of these chalcogenides consist of $ME_{4/4}$ units in which both the metal and chalcogen atoms have tetrahedral coordination. These chalcogenides are of considerable interest as semiconductors; they are isoelectronic with elemental silicon or germanium.

11.3.3 Halides

All eight possible halides MX_2 (M = Zn, Cd; X = F, Cl, Br, I) are known. The fluorides ZnF_2 and CdF_2 are considerably more ionic than the other halides, as indicated by their higher melting and boiling points, lower aqueous solubilities, and lower tendency to form complexes with excess halide, due to their higher lattice energies. The structures of the chlorides, bromides, and iodides consist of close-packed arrays of halide ions. The metal atoms occupy tetrahedral interstices in the zinc halides but octahedral interstices in the cadmium halides.

Zinc chloride is extremely soluble in water. The hydrates $ZnCl_2 \cdot nH_2O$ ($1 \leq n \leq 4$) are all liquid at room temperature. These liquids do not contain free chloride ions but do contain $Zn(\mu-Cl)_2Zn$ bridging units.[9] Concentrated aqueous solutions of $ZnCl_2$ dissolve starch, cellulose, and silk and thus are used in textile processing. More dilute aqueous solutions of $ZnCl_2$ contain the species $Zn(H_2O)_6^{2+}$, $ZnCl(H_2O)_5^+$, $ZnCl_4^{2-}$, $ZnCl_4(H_2O)_2^{2-}$, and $ZnCl_2(H_2O)_4$. Aqueous solutions of the cadmium halides are incompletely dissociated into Cd^{2+} and X^- ions; they are therefore weak electrolytes. The halides ZnX_2 and CdX_2 (X = Cl, Br, I) are readily soluble in alcohols, acetone, and other donor solvents.

Zinc and cadmium form only relatively weak complexes with fluoride ion. With the other halide ions the tetrahedral ions MX_4^{2-} are known; the cadmium derivatives are more stable than their zinc analogues. Pentahalometallate ions such as $CdCl_5^{3-}$ are also known.

11.3.4 Oxoanion Salts and Aqueous Solution Chemistry

The ions Zn^{2+} (radius 0.69 Å) and Cd^{2+} (radius 0.92 Å) are rather similar to the magnesium ion Mg^{2+} (radius 0.78 Å), and many of their salts are isomorphous. Zinc and cadmium salts of oxoacids such as nitrate, sulfate, sulfite, perchlorate, and acetate are soluble in water like the corresponding magnesium salts. The aqua ions of zinc and cadmium are normally the octahedral $M(H_2O)_6^{2+}$; these aqua ions are quite strong acids, and their salts are hydrolyzed by water (Section 1.4). Thus in perchlorate solutions the only species for zinc, cadmium, and mercury below 0.1 M are the hydrolyzed ions MOH^+. More concentrated Cd(II) solutions contain binuclear ions of the type $Cd_2(\mu_2\text{-}OH)^{3+}$. Vacuum distillation of zinc acetate gives $OZn_4(O_2CMe)_6$, which is isomorphous with its beryllium analogue (Section 10.4). In contrast to the water-stable $OBe_4(O_2CMe)_6$, however, $OZn_4(O_2CMe)_6$ is rapidly hydrolyzed by water, since zinc, unlike beryllium, can increase its coordination number above 4 because of available d orbitals in its valence shell.

11.3.5 Complexes of Zinc and Cadmium

Cadmium is a significantly softer acid than zinc, which means that Cd^{2+} is bound more strongly than Zn^{2+} to Cl, S, and P ligands, whereas Zn^{2+} favors bonding to F and O ligands. A number of important sulfur ligand complexes are known (Figure 11.2). Thus reactions of zinc or cadmium acetates with polysulfide ion gives $M(S_6)_2^{2-}$ ions with two MS_6 seven-membered rings sharing the metal atom.[10] A variety of thiolate complexes are known, including diverse polynuclear derivatives such as phenylthiolate complexes of the stoichiometries $[Zn(SPh)_2]_n$, $Cd_{10}(SPh)_{20}$, $Zn(SPh)_4^{2-}$, $Cd_4(SPh)_{10}^{2-}$, $Cd_{10}S_4(SPh)_{16}^{4-}$, and $Zn_4Cl_2(SPh)_8^{2-}$. The complexes $M_4(SPh)_{10}^{2-}$ and $Zn_4Cl_2(SPh)_8^{2-}$ have adamantane structures with bridges similar to the structure of P_4O_{10} (Section 5.6.1).

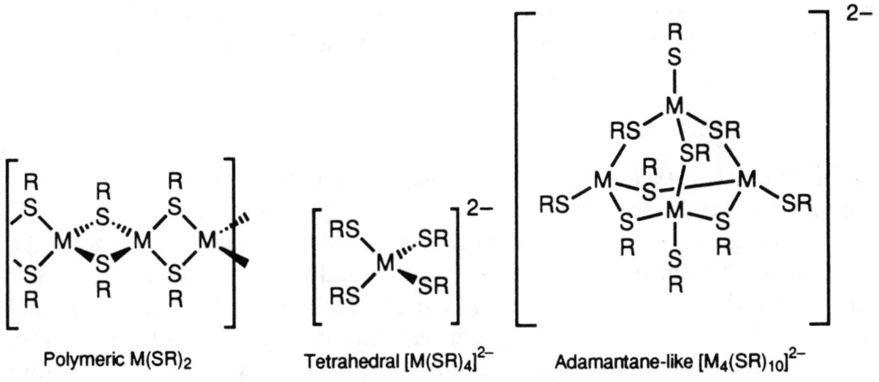

Figure 11.2. Some structure of thiolate complexes of zinc and cadmium (M = Zn, Cd; R = alkyl or aryl group).

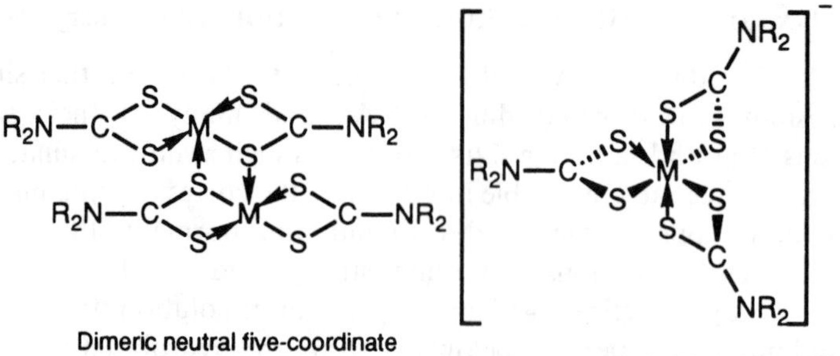

Dimeric neutral five-coordinate
[(R₂NCS₂)₂M]₂

Anionic six-coordinate [(R₂NCS₂)₃M]⁻

Figure 11.3. Structures of some dithiocarbamate complexes of zinc and cadmium (M = Zn, Cd).

Zinc complexes of small bite bidentate sulfur ligands such as dithiocarbamates and phosphorodithioates are made in large quantities for a number of applications such as antioxidants, antiwear lubricants, and accelerators for the vulcanization of rubber by sulfur. The structure of $\{Zn[S_2P(OR)_2]_2\}_2$ has five-coordinate zinc with sulfur bridges, whereas the anion $[(Me_2NCS_2)_3Zn]^-$ has six-coordinate zinc (Figure 11.3).

11.4 Divalent Mercury Compounds[11]

The Hg^{2+} cation (radius 0.93 Å), although roughly as large as Ca^{2+} (radius 1.06 Å), has a considerably greater polarizing power as a result of the inefficient shielding of the nucleus by completed $4f$ and $5d$ shells. For this reason, the only Hg(II) derivatives that have ionic structures and properties are those with highly electronegative anions such as fluoride, nitrate, and perchlorate. On the other hand, HgO, HgS, and HgX_2 (X = Cl, Br, I, CN, SCN) are all appreciably covalent. Water-soluble compounds of this type, such as $HgCl_2$ and $Hg(CN)_2$, exist largely as discrete un-ionized HgX_2 molecules in solution.

11.4.1 Mercury(II) Chalcogenides

Red mercury(II) oxide, HgO, is obtained by mild heating of mercury(I) or mercury(II) nitrates, by direct reaction of mercury with oxygen at 300–350°C, or by heating an alkaline solution of K_2HgI_4. *Yellow mercury (II) oxide*, HgO, is precipitated from solutions of Hg^{2+} salts by addition of aqueous base. Red and yellow HgO differ only in particle sizes. Heating HgO results in thermal decomposition into its elements with the equilibrium pressure of O_2 being 1 atm at 447°C. The structure of HgO contains zigzag —Hg—O—Hg—O— chains, with Hg—O distances of 2.82 Å, essentially linear O–Hg–O angles of 179°, and bent Hg—O—Hg bonds of 109°. The bonding between the chains in HgO is very weak. Linear O—Hg—O groups are also frequently found in mixed mercury oxides such as $Hg_2Nb_2O_7$.

Mercury(II) hydroxide, $Hg(OH)_2$, cannot be isolated because it dehydrates immediately to HgO. The Hg^{2+} ion in aqueous solution in the presence of a noncomplexing anion such as perchlorate undergoes the following hydrolysis reactions:

$$Hg^{2+} + HgO \rightleftarrows Hg(OH)^+ + H^+ \qquad K = 2.6 \times 10^{-4} \qquad (11.2a)$$

$$Hg(OH)^+ + H_2O \rightleftarrows Hg(OH)_2 + H^+ \qquad K = 2.6 \times 10^{-3} \qquad (11.2b)$$

No polynuclear hydrolysis products have been detected.

Black mercury(II) sulfide, HgS, is precipitated from aqueous solution (e.g., $Hg^{2+} + H_2S$) as a highly insoluble compound. The solubility product $[Hg^{2+}][S^{2-}]$ is only 10^{-54}, but HgS is actually slightly less insoluble because of hydrolysis of the Hg^{2+} and S^{2-} ions. Black HgS is unstable with respect to *red mercury(II) sulfide*, which is identical to the mineral cinnabar. Thus treatment of black HgS with alkali metal polysulfides or Hg_2Cl_2 converts it to red HgS.[12] Red HgS has a distorted NaCl structure with Hg—S chains.

11.4.2 Mercury(II) Halides and Pseudohalides

Mercuric(II) fluoride, HgF_2, is essentially ionic and crystallizes in the fluorite structure; it is almost completely decomposed even by cold water and is useful as a mild fluorinating agent.[13] *Mercury(II) chloride*, $HgCl_2$ "corrosive sublimate" (mp 280°C, bp 303°C, solubility 6.6 g/100 g water at 20°), and *mercury(II) bromide*, $HgBr_2$ (mp 238°C, bp 318°C, solubility 0.62 g/100 g water at 25°C), have molecular lattices with two short Hg—X bonds (Hg—Cl = 2.25 Å in $HgCl_2$ and Hg—Br = 2.48 Å in $HgBr_2$) and linear X—Hg—X coordination of the short Hg—X bonds. *Mercury(II) iodide*, HgI_2 (mp 257°C, bp 351°C, solubility only 0.006 g/100 g water at 25°C) exists in a red form with an infinite layer structure consisting of $HgI_{4/2}$ tetrahedra and a yellow form with a molecular HgI_2 structure. The red form of HgI_2 undergoes a sharp transition at 126°C to give the yellow form. Similar thermochromic properties are exhibited by the water-insoluble heavy metal tetraiodomercurates Cu_2HgI_4 (red → black) and Ag_2HgI_4 (yellow → red). The mercury(II) halides HgX_2 (X = Cl, Br) undergo relatively little ionization in aqueous solutions.

White *mercury(II) cyanide*, $Hg(CN)_2$, is obtained by heating stable cyano complexes such as $Fe(CN)_6^{4-}$ or Prussian blue with mercuric oxide at 90°C for several hours in aqueous solution. It dissolves in water to give a nonconducting aqueous solution, indicating the absence of dissociation into Hg^{2+} and CN^- ions and has a structure with a linear N≡C—Hg—C≡N unit. White *mercury(II) thiocyanate*, $Hg(SCN)_2$, can be precipitated from concentrated aqueous solutions of mercury (II) and thiocyanate ion; it burns in air to give a voluminous gray-black, spongy, snakelike ash known as *Pharaoh's serpents*. *Mercury(II) azide*, $Hg(N_3)_2$, is a very sensitive and dangerous explosive that will explode even under water. *Mercury(II) fulminate*, $Hg(CNO)_2$, can be obtained as a sparingly soluble (0.07 g/100 g water at 12°C) white solid from mercury metal, ethanol, and nitric acid; it is used as a detonator.

11.4.3 Mercury(II) Oxosalts

Mercury(II) nitrate, sulfate, and perchlorate are ionic and highly dissociated in aqueous solution. Such aqueous solutions must be acidified to prevent precipitation of HgO. The mercury(II) aqua ion is $Hg(H_2O)_6^{2+}$, but two Hg—O bonds in trans positions are shorter than the other four, in accord with the preference of mercury for linear O—Hg—O structural units. Mercury(II) carboxylates, $Hg(OCOR)_2$, can be obtained from HgO and the corresponding carboxylic acid. Mercury(II) phosphate and oxalate are insoluble in water.

11.4.4 Mercury(II) Complexes

Mercury(II) has a strong tendency to form complexes. Linear two-coordinate complexes are favored more for mercury than any other metal; tetrahedral four-coordinate mercury(II) complexes are also fairly common. Octahedral mercury(II) coordination is much rarer; a few three-coordinate and five-coordinate complexes are also known.[14]

The following equilibria are found in solutions of mercury halide complexes (X = Cl, Br, I):

$$HgX^+ \underset{-X^-}{\overset{+X^-}{\rightleftharpoons}} HgX_2 \underset{-X^-}{\overset{+X^-}{\rightleftharpoons}} HgX_3^- \underset{-X^-}{\overset{+X^-}{\rightleftharpoons}} HgX_4^{2-} \tag{11.3}$$

Dissolution of $Hg(CN)_2$ in excess cyanide solution gives the anions $Hg(CN)_3^-$ and $Hg(CN)_4^{2-}$. Similarly, the thiocyanate complex $Hg(SCN)_4^{2-}$ can be obtained from $Hg(SCN)_2$ and excess thiocyanate; its cobalt salt $CoHg(SCN)_4$, a blue water-insoluble solid, is useful as a magnetic susceptibility standard. Mercury(II) has a relatively low affinity for oxygen in complexes, but oxoanion complexes such as $Hg(SO_3)_2^{2-}$, $Hg(C_2O_4)_2^{2-}$, and $Hg(NO_2)_2^{2-}$ can be obtained.

Mercury(II) is a very soft acid with with a high affinity for sulfur and phosphorus ligands. Anionic thiol derivatives such as $Hg(SPh)_3^-$ and $Hg_2(SMe)_6^{2-}$ can be obtained, as well as the polysulfide complexes $Hg(S_4)_2^{2-}$ and $Hg(S_6)_2^{2-}$, which have two HgS_n chelate rings sharing the mercury atom. Tertiary phosphine mercury(II) complexes include monomeric $(R_3P)_2HgX_2$ and dimeric or polymeric $[R_3PHgX_2]_n$ with halide bridges; complexes of both types contain four-coordinate mercury(II).

Mercury salts give a variety of products upon treatment with aqueous ammonia under various conditions[15]:

$$HgCl_2 + 2NH_3 \rightarrow Hg(NH_3)_2Cl_2\downarrow \tag{11.4a}$$

$$nHgCl_2 + 2nNH_3 \rightarrow [HgNH_2Cl]_n\downarrow + nNH_4^+ + nCl^- \tag{11.4b}$$

$$2nHgCl_2 + 4nNH_3 \rightarrow [Hg_2N^+Cl^-]_n + 3nNH_4^+ + 3nCl^- \tag{11.4c}$$

The amido derivative $[HgNH_2Cl]_n$ (equation 11.4b) has a zigzag chain of NH_2 groups and linearly coordinated Hg units with Cl^- ions between the chains, that is, $[Hg(NH_2)_{2/2}^+Cl^-]_n$. The chloride $[Hg_2N^+Cl^-]_n$ (equation 11.4c) is the chloride of *Millon's base*, $[Hg_2N^+OH^-]_n$, which can be obtained as a bright yellow precipitate by treatment of yellow mercuric oxide with carbonate-free concentrated aqueous ammonia in the dark at room temperature. The structure of Millon's base consists of a three-dimensional silicalike $[Hg_2N^+]_n$ framework with tetrahedral four-coordinate nitrogen and linear two-coordinate mercury ($Hg-N = 2.04-2.09$ Å). The OH^- ions and water molecules occupy rather spacious cavities and channels in this network. Many salts of Millon's base of the general type $[Hg_2N^+X^-]_n \cdot mH_2O(X^- = NO_3^-, ClO_4^-, Cl^-, Br^-, I^-; m = 0-2)$ are known in which there are anions and water molecules in the cavities of the $[Hg_2N^+]_n$ framework. The $[Hg_2N^+]_n$ framework is unaltered upon anion exchange and may be regarded as a nonoxide analogue of zeolites (Section 3.5.3.5).

11.4.5 Complexes with Direct Mercury–Metal Bonds

Transition metal complexes frequently react with mercury or mercury compounds to give products with linear M—Hg—M units containing direct metal–mercury bonds and two-coordinate mercury. A simple example is $Hg[Co(CO)_4]_2$, which can be obtained by insertion of mercury metal into the Co—Co bond of $Co_2(CO)_8$. It is possible to obtain more complicated metal–mercury cluster derivatives containing large transition metal polyhedra with linear M—Hg—M edges lacking direct M—M bonds. Examples of such compounds include $Hg_6Rh_4(PMe_3)_{12}$, containing a large Rh_4 tetrahedron with six Rh—Hg—Rh edges, and $Hg_9Co_6(CO)_{18}$, containing a large Co_6 trigonal prism with nine Co—Hg—Co edges.[16] The mercury complex $Hg[Co(CO)_4]_2$ reacts with excess $Co(CO)_4^-$ to form the anion $Hg[Co(CO)_4]_3^-$, containing planar three-coordinate mercury.

11.5 Organometallic Compounds of Zinc, Cadmium, and Mercury[17]

11.5.1 Organozinc and Organocadmium Compounds

Organozinc compounds are of historical interest because they were the first metal alkyls to be prepared (Frankland, 1949). The original preparation of organozinc compounds used the reaction of zinc–copper couple with an alkyl iodide to give the corresponding alkylzinc iodide:

$$RI + Zn \xrightarrow{\ Cu\ } RZnI \tag{11.5}$$

Pyrolysis of alkylzinc iodides results in disproportionation to give the corresponding dialkylzinc derivative:

$$2RZnI \xrightarrow{\Delta} R_2Zn + ZnI_2 \tag{11.6}$$

Dialkylzinc and diarylzinc derivatives can also be made by treatment of the corresponding organomercury derivative with zinc metal or reaction of the corresponding organolithium, organomagnesium, or organoaluminum derivative with anhydrous zinc halides, for example,

$$R_2Hg + Zn \rightarrow R_2Zn + Hg \tag{11.7a}$$

$$ZnCl_2 + 2RLi \rightarrow R_2Zn + 2LiCl \tag{11.7b}$$

Dialkylcadmium and diarylcadmium derivatives are best obtained by alkylation of anhydrous cadmium halides with RLi or RMgX. Perfluoroalkylcadmium derivatives can be made by reactions such as the following[18]:

$$R_fX + Cd \xrightarrow{Me_2NCHO} R_fCdK \quad (R_fX = CF_3I, C_6F_5Br, etc.) \tag{11.8a}$$

$$(CF_3)_2Hg + Me_2Cd \xrightarrow{tetrahydrofuran} (CF_3)Cd \cdot OC_4H_8 + Me_2Hg \tag{18.8b}$$

$$2Cd + 2CF_2Br_2 + Me_2NCHO \longrightarrow$$
$$CF_3CdBr + CdBr_2 + CO\uparrow + Me_2N{=}CFH^+Br^- \tag{11.8c}$$

The last reaction is of interest because the relatively inexpensive CF_2Br_2 is used as the CF_3 source.

The dialkyl and diaryl derivatives of zinc and cadmium are nonpolar liquids or low-melting solids. The low molecular weight derivatives can be purified by distillation in an inert atmosphere (e.g., Me_2Zn, bp 46°; Et_2Zn, bp 118°C; Me_2Cd, bp 105.7°C). Ethylzinc iodide is a polymer $[EtZnI_{3/3}]_n$ with each iodine atom bridging three zinc atoms. Few RCdX compounds have been isolated and characterized.

The lower dialkylzincs such as Me_2Zn and Et_2Zn are spontaneously flammable in air, and all zinc alkyls react vigorously with oxygen and water. Dialkylcadmium derivatives are less sensitive toward oxygen but are also less thermally stable. Water and other active hydrogen compounds react with R_2Zn and R_2Cd to liberate the corresponding hydrocarbon (N = Zn, Cd):

$$R_2M + 2H_2O \rightarrow 2RH + M(OH)_2 \tag{11.9}$$

Dialkylzinc derivatives were used in the later nineteenth and early twentieth centuries for the conversion of aldehydes and ketones to the corresponding alcohols but were completely superseded by the more convenient and safer Grignard reagents, RMgX (Section 10.5.3), after their discovery by Grignard in the early twentieth century. However, organozinc derivatives are still used

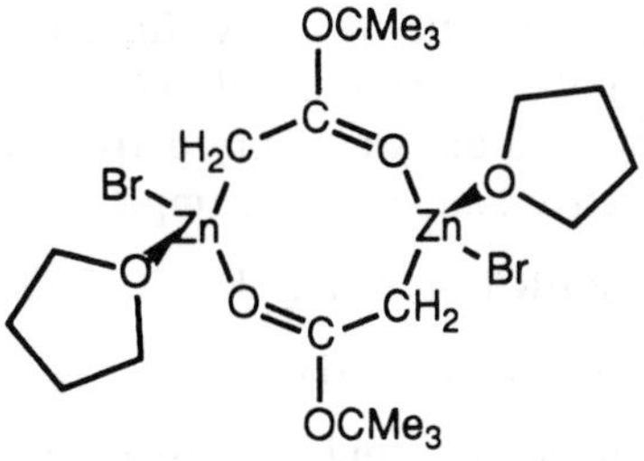

Figure 11.4. Structure of the Reformatsky reagent $[BrZn(OC_4H_8)CH_2CO_2Bu^t]_2$.

as intermediates in the *Reformatsky reaction*[19] for the conversion of esters to hydroxy esters, for example,

$$BrCH_2CO_2R' + Zn + R_2C{=}O \rightarrow R_2C(OZnBr)CH_2CO_2R' \qquad (11.10a)$$

$$2R_2C(OZnBr)CH_2CO_2R + 2H_2O \rightarrow$$
$$2R_2C(OH)CH_2CO_2R + Zn(OH)_2 + ZnBr_2 \qquad (11.10b)$$

This reaction involves an organozinc intermediate of the type $BrZnCH_2CO_2R'$. An intermediate of this type, $[BrZn(OC_4H_8)CH_2CO_2Bu^t]_2$, has been shown to have a dimeric structure with an eight-membered C_2OZnC_2OZn ring and four-coordinate zinc atoms (Figure 11.4). Dialkylcadmiums undergo self-exchange reactions and exchange reactions with alkyls of zinc, gallium, or indium through alkyl-bridged intermediates.

11.5.2 Organomercury Compounds[20]

The mercury–carbon bond is thermodynamically weak but has a low affinity for oxygen. For this reason. organomercury compounds are generally stable to air and water, in contrast to organozinc and organocadmium compounds. Dialkyl- and diarylmercury derivatives, R_2Hg, are nonpolar, volatile, toxic, colorless liquids or low-melting solids (e.g., bp of $Me_2Hg = 92°C$). A variety of compounds of the type RHgX (X = Cl, Br, I, CN, SCN, OH, etc.) are also known; these are nonpolar substances more soluble in organic liquids than water. However, $RHgNO_3$ and $(RHg)_2SO_4$ derivatives are saltlike substances that ionize in water to give $[RHgOH_2]^+$ ions. All R_2Hg and RHgX compounds are monomeric derivatives with two-coordinate mercury and linear R—Hg—X or R—Hg—R bonds.

Dialkylmercury and diarylmercury derivatives transfer alkyl groups to more electropositive metals or some other elements with the liberation of mercury:

$$nR_2Hg + 2M \rightleftarrows 2R_nM + nHg \qquad (11.11)$$

Elements for which reactions of this type go to completion include the alkali and alkaline earth metals, zinc, aluminum, gallium, tin, lead, antimony, bismuth, selenium, and tellurium. Reversible equilibria are obtained with metals of more similar electronegativity to mercury (e.g., Cd, In, Tl). Such reactions are useful for the synthesis of a number of pure metal and metalloid alkyls and aryls in small quantities. Dialkyl- and diarylmercury derivatives are also useful for the partial alkylation of reactive halides, for example,

$$BCl_3 + Ph_2Hg \rightarrow PhBCl_2 + PhHgCl \tag{11.12a}$$

$$AsCl_3 + Et_2Hg \rightarrow EtAsCl_2 + EtHgCl \tag{11.12b}$$

Alkylmercuric halides undergo rapid exchange reactions with mercury(II) halides of the following type:

$$RHgBr + {}^*HgBr_2 \rightleftarrows R^*HgBr + HgBr_2 \tag{11.13}$$

Such reactions can be studied using an isotopic mercury tracer.

Mercury perhaloalkyls, $RHgCX_3$, are convenient dihalocarbene generators[21] by the following reaction:

$$RHgCX_3 \rightarrow \{CX_2\} + RHgX \tag{11.14}$$

They can be obtained by dehydrohalogenation of haloforms, HCX_3, with a strong base in the presence of $RHgX$ (X = Cl, Br), for example,

$$PhHgCl + HCX_3 + KBu^t \xrightarrow{C_6H_6} PhHgCX_3 + KCl + Bu^tOH \tag{11.15}$$

Bis(trifluoromethyl)mercury, $(CF_3)_2Hg$, can be obtained by the thermal or photochemical reaction of CF_3I with cadmium amalgam or the thermal decaroxylation of $Hg(OCOCF_3)_2$ in the presence of sodium or potassium carbonate[22]; it is a colorless volatile solid that dissolves in water to form essentially nonconducting solutions.

Mercury–carbon bonds can be formed very easily. Thus Hg^{2+} can be methylated to $MeHg^+$ by diverse methyl sources including aluminum carbide, Al_4C_3 (Sections 2.3.1.1 and 9.7), and methylcobalt derivatives including vitamin B-12, which has a $Co—CH_3$ bond. In addition, aromatic hydrocarbons undergo electrophilic substitution upon treatment with mercury(II) carboxylates, for example,

$$C_6H_6 + Hg(O_2CMe)_2 \rightarrow C_6H_5HgOC(O)Me + MeCO_2H \tag{11.16}$$

Mercury compounds add reversibly to olefins and catalyze the hydration of acetylenes to give ketones through the corresponding enol intermediate, for example,

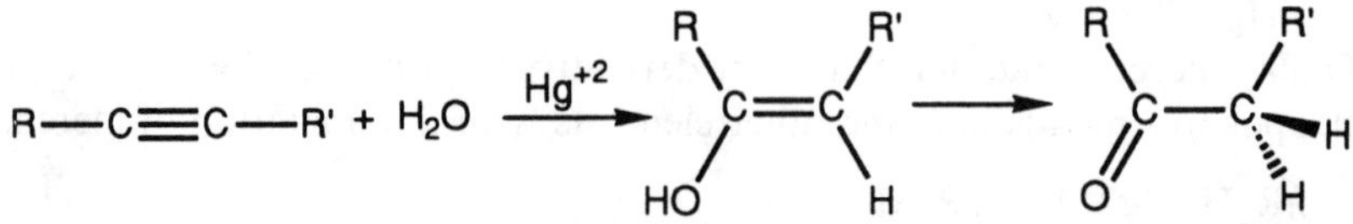

The methylmercury ion, CH_3Hg^+, which can be formed very readily by the biological alkylation of mercury, is the source of toxic effects of mercury and mercury compounds in the environment; it causes complex and irreversible

disturbances to the nervous system. The CH_3Hg^+ ion is hydrated in aqueous solution[23] and undergoes the following pH-dependent reactions leading to the formation of oxonium ions:

$$CH_3Hg(OH_2)^+ + OH^- \rightleftarrows CH_3HgOH + H_2O \tag{11.17a}$$

$$CH_3Hg(OH_2)^+ + CH_3HgOH \rightleftarrows (CH_3Hg)_2OH^+ + H_2O \tag{11.17b}$$

$$CH_3HgOH + (CH_3Hg)_2OH^+ \rightleftarrows (CH_3Hg)_3O^+ + H_2O \tag{11.17c}$$

Such oxonium ions bind sulfur and selenium very strongly; the formation constant of CH_3HgSR compounds are 10^8 times greater than those of CH_3Hg^+ complexes of the amino group. The toxicity of CH_3Hg^+ may arise from this high affinity for sulfur, which leads to its complexation with cystine and methionine units in peptides.

11.6 Zinc, Cadmium, and Mercury with Oxidation States Below 2

11.6.1 Zinc and Cadmium

Zinc and cadmium in formal oxidation states below 2 are not stable under normal conditions. The Zn_2^{2+} ion has been shown by Raman and other spectroscopic methods to be present in the yellow diamagnetic glass obtained by adding metallic zinc to molten $ZnCl_2$ at 500–700°C. Reaction of Cd^{2+} $[AlCl_4^-]_2$ with metallic cadmium at 340°C gives yellow solutions from which crystalline $Cd_2^{2+}[AlCl_4^-]_2$ can be isolated. The structure of $Cd_2^{2+}[AlCl_4^-]_2$ has been shown by X-ray diffraction to contain a Cd—Cd bond (2.576 Å). In addition, three chlorine atoms of each $AlCl_4^-$ anion form weak bridges to cadmium so that the cadmium coordination is distorted tetrahedral.[24] Force constant data from vibrational spectra indicate the bond strength sequence $Zn_2^{2+} < Cd_2^{2+} < Hg_2^{2-}$.

11.6.2 Mercury(I)

11.6.2.1 Hg_2^{2+} in Solution

The dimercury(I) or "mercurous ion," $Hg_2(H_2O)_2^{2+}$, is readily obtained from Hg^{2+} salts by reduction in aqueous solution; it is stable in acid solution and in the absence of coordinating ligands such as OH^-, F^-, and CN^- that promote disproportionation. The aqueous $Hg^0/Hg_2^{2+}/Hg^{2+}$ system has the following redox potentials:

$$Hg_2^{2+} + 2e = 2Hg^0(l) \qquad E° = 0.7960 \text{ V} \tag{11.18a}$$

$$2Hg^{2+} + 2e = Hg_2^{2+} \qquad E° = 0.9110 \text{ V} \tag{11.18b}$$

$$Hg^{2+} + 2e = Hg^0(l) \qquad E° = 0.8535 \text{ V} \tag{11.18c}$$

Thus the Hg_2^{2+} disproportionation is rapid and reversible:

$$Hg_2^{2+} \rightleftharpoons Hg(l) + Hg^{2+} \qquad\qquad E° = -0.115\ V \qquad\qquad (11.19)$$

These redox potentials indicate that mercury metal can be oxidized to Hg_2^{2+} but not to Hg^{2+} only by oxidizing agents with potentials in the very narrow range from -0.79 V to -0.85 V. Therefore treatment of elemental mercury with an excess of any common oxidizing agent converts it entirely into Hg^{2+}. When free mercury is present in at least 50% excess, however, only Hg_2^{2+} is obtained. Furthermore, any reagents that reduce the activity of Hg^{2+} in a mercury solution, such as ammonia, amines, OH^-, CN^-, SCN^-, sulfide, and acetylacetonate, will promote the disporoportionation of Hg_2^{2+}. Thus the range of stable mercury(I) compounds is rather restricted.

11.6.2.2 Mercury (I) Compounds

Mercury(I) fluoride, Hg_2F_2, is unstable toward water undergoing hydrolysis to HF and mercury(I) hydroxide, which undergoes disproportionation and de-hydration to HgO and elemental mercury. Since the other *mercury(I) halides*, Hg_2X_2 (X = Cl, Br, I), are highly insoluble in water, they undergo neither hydrolysis nor disproportionation. *Mercury(I) nitrate* exists only as the di-hydrate, $Hg_2(NO_3)_2 \cdot 2H_2O$, which contains the diaquodimercury(I) cation, $[H_2O \rightarrow Hg\text{—}Hg \leftarrow OH_2]^{2+}$; the perchlorate behaves analogously. Relatively insoluble mercury(I) derivatives, including not only the halides Hg_2X_2 (X = Cl, Br, I) but also the sulfate, chlorate, bromate, iodate, and acetate, can be pre-cipitated by adding the corresponding anions to an $[H_2O \rightarrow Hg\text{—}Hg \leftarrow OH_2]^{2+}$ solution obtained from mercury(I) nitrate or perchlorate in water. Mercury(I) forms relatively few isolable complexes because most ligands such as cyanide, halides, amines, and alkyl sulfides, form more stable complexes with Hg^{2+}, leading to disproportionation of Hg_2^{2+} upon treatment with such ligands. However, nitrogen ligands of relatively low basicity favour Hg_2^{2+} so that relatively stable mercury(I) complexes of ligands such as aniline and 1,10-phenanthroline can be obtained, for example, $[PhNH_2 \rightarrow Hg\text{—}Hg \leftarrow OH_2]^{2+}$ with aniline.

11.6.3 Mercury Compounds with Formal Oxidation States Below +1: Cationic Mercury Chains

Elemental mercury, like elemental sulfur, selenium, and tellurium (Section 6.2.6), can be oxidized by strong Lewis acid oxidants to polynuclear cluster cations with fractional formal oxidation states.[25] The product depends on the ratio of elemental mercury to the oxidant, for example,

$$2Hg + 3AsF_5 \xrightarrow[\text{SO}_2]{\text{liquid}} Hg_2^{2+}[AsF_6^-]_2 + AsF_3 \qquad \text{(colorless)} \quad (11.20a)$$

$$3Hg + 3AsF_5 \xrightarrow[SO_2]{liquid} Hg_3^{2+}[AsF_6^-]_2 + AsF_3 \qquad \text{(yellow)} \qquad (11.20b)$$

$$4Hg + 3AsF_5 \xrightarrow[SO_2]{liquid} Hg_4^{2+}[AsF_6^-]_2 + AsF_3 \qquad \text{(red)} \qquad (11.20c)$$

The Hg_n^{2+} cations in these products can be characterized by their ^{199}Hg NMR spectra[26] and are shown by single-crystal X-ray diffraction to consist of chains of mercury atoms with approximately linear Hg–Hg–Hg angles. The yellow Hg_3^{2+} cation also can be obtained by dissolving mercury metal in pure fluorosulfuric acid, HSO_3F. Reactions of elemental mercury with more limited amounts of MF_5 oxidant (M = As, Sb) lead to golden metal-like non-stoichiometric solids $Hg_{3-\delta}MF_6$ ($\delta = 0.18$ for M = As and $\delta = 0.1$ for M = Sb), which are shown by single-crystal X-ray diffraction to have structures with two nonintersecting mutually perpendicular chains of mercury atoms. The $Hg_{3-\delta}MF_6$ derivatives with mercury chains are readily transformed into silvery Hg_3MF_6 derivatives, shown by X-ray diffraction to have structures with hexagonal sheets of close-packed mercury atoms separated by sheets of MF_6^- ions.

References

1. Siegel, S. M.; Siegel, B. Z., First Estimate of Annual Mercury Flux at the Kilauea Main Vent, *Nature* (London), 1984, **309**, 146–147.

2. Nielsen, J. W., Baenziger, N. C., The Crystal Structures of NaHg$_2$, NaHg, and Na$_3$Hg$_2$, *Acta Crystallogr.*, 1954, **7**, 277–282.

3. Duwell, E. J.; Baenziger, N. C., The Crystal Structures of KHg and KHg$_2$, *Acta Crystallogr.*, 1955, **8**, 705–710.

4. Corbett, J. D., Metal–Metal Bonding in Square Planar Groups: Na$_3$Hg$_2$ and Other Systems, *Inorg. Nuclear Chem. Lett.*, 1969, **5**, 81–84.

5. Deiseroth, H.-J., Strunck, A., Hg$_8$ ("Mercubane") Clusters in Rb$_{15}$Hg$_{16}$, *Angew. Chem. Int. Ed. Engl.*, 1989, **28**, 1251–1252.

6. Garcia, E.; E.; Cowley, A. H.; Bard, A. J., "Quaternary Ammonium Amalgams" as Zintl Ion Salts and Their Use in the Synthesis of Novel Quaternary Ammonium Salts. *J. Am. Chem. Soc.*, 1986, **108**, 6082–6083.

7. Hirschwald, W. H.; Zinc Oxide: An Outstanding Example of a Binary Compound Semiconductor, *Acc. Chem. Res.*, 1985, **8**, 228–234.

8. Davies, J. I.; Fan, G.; Parrott, M. J.; Williams, J. O., *J. Chem. Soc., Chem. Commun.* 1986, 68–69.

9. Duffy, J. A.; Wood, G. L., *J. Chem. Soc. Dalton Trans.*, 1987, 1485–1488.

10. Müller, A.; Schimanski, J.; Schimanski, U.; Bögge, H., *Z. Naturforsch.*, 1985, **40B**, 1277–1288.

11. Roberts, H. L., Some General Aspects of Mercury Chemistry, *Adv. Inorg. Chem. Radiochem.*, 1968, **11**, 309–339.

12. Davidson, S. R.; Wilsher, C. J., *J. Chem. Soc. Dalton Trans.*, 1981, 833–835.

13. Muetterties, E. L.; Tullock, C. W., in *Preparative Inorganic Reactions*, W. Jolly, Ed., Wiley-Interscience, New York, 1965.

14. Grdenić, D., The Structural Chemistry of Mercury, *Q. Rev.*, 1965, **19**, 303–328.

15. Breitinger, D., Brodersen, K., Development of, and Problems in the Chemistry of Mercury–Nitrogen Compounds, *Angew. Chem. Int. Ed. Engl.*, 1970, **9**, 357–367.

16. King, R. B., Mercury Vertices in Transition Metal Clusters and Alkali Metal Amalgams, *Polyhedron*, 1988, **7**, 1813–1817.

17. Wardell, J. L., Ed., *Organometallic Compounds of Zinc, Cadmium, and Mercury*, Chapman and Hall, London, 1985.

18. Heinze, P. L.; Burton, D. L., *J. Fluorine Chem.* 1985, **29**, 359–361.

19. Rathke, M. W., The Reformatsky Reaction, *Ord. React.*, 1974, **22**, 423.

20. Makarova, L. G.; Nesmeyanov, A. N., *Organic Compounds of Mercury*, North-Holland, Amsterdam, 1967.

21. Seyferth, D., Phenyl(trihalomethyl)mercury Compounds: Exceptionally Versatile Dihalocarbene Precursors, *Acc. Chem. Res.*, 1972, **5**, 65–74.

22. Lagow, R. J.; Eujen, R.; Gerchman, L. L.; Morrison, J. A., *J. Am. Chem. Soc.*, 1978, **100**, 1722–1726.

23. Rabenstein, D., The Aqueous Solution Chemistry of Methylmercury and Its Complexes, *Acc. Chem. Res.*, 1978, **11**, 100–107.

24. Faggiani, R.; Gillespie, R. J.; Vekris, J. E., The Cadmium(I) Ion, Cd_2^{2+}; X-Ray Crystal Structure of $Cd_2(AlCl_4)_2$, *J. Chem. Soc., Chem. Commun.*, 1986, 517–518.

25. Brown, I. D.; Gillespie, R. J.; Morgan, K. R.; Sawyer, J. F.; Schmidt, K. J.; Tun, Z.; Ummat, P. K.; Vekris, J. E., *Inorg. Chem.*, 1987, **26**, 689–693.

26. Gillespie, R. J.; Granger, P.; Morgan, K. R.; Schrobilgen, G. J., *Inorg. Chem.*, 1984, **23**, 887–891.

12

Lanthanides and Actinides

12.1 General Aspects of the Chemistry of the Lanthanides and Actinides

12.1.1 Why Are Lanthanides and Actinides Discussed in a Book on Main Group Elements?

The lanthanides and actinides are characterized by incompletely filled f-shells and thus are actually f-block transition metals rather than main group elements. Since, however, in the case of the lanthanides the f orbitals are *relatively* uninvolved in chemical bonding, lanthanide chemistry is predominantly the chemistry of highly electropositive metals in the $+3$ oxidation state, just as the chemistry of the alkali metals and the alkaline earth metals is the chemistry of highly electropositive metals in the $+1$ and $+2$ oxidation states, respectively (Chapter 10). For this reason, the chemistry of the lanthanides is conveniently discussed in the same book as the chemistry of the alkali and alkaline earth metals. In fact, in many respects the trivalent Ln^{3+} ions (Ln is used as a general designation for any lanthanide) functions like a trivalent version of the heavier alkaline earths, namely strontium and barium.

Discussion of actinide chemistry in a book on main group elements is more questionable since, in contrast to the lanthanides, the actinide f orbitals plays a significant role in actinide covalent bonding.[1] However, comparison of the chemistry of lanthanides and actinides is instructive. Furthermore, since all the other actinides are too radioactive for general laboratory use or most applications, only two of the actinides, namely thorium and uranium, are readily

Table 12.1   The Lanthanides and Their Comparison with Some Other Trivalent Metals

Element	Properties of Ln^{3+}			$E°$(V): $M^{3+} \rightarrow M^0$
	Electronic Configuration	Radius (Å)	Color	
Lanthanides				
Lanthanum (La)	[Xe]	1.17	Colorless	−2.37
Cerium (Ce)	$[Xe]4f^1$	1.15	Colorless	−2.34
Praseodymium (Pr)	$[Xe]4f^2$	1.13	Green	−2.35
Neodymium (Nd)	$[Xe]4f^3$	1.12	Lilac	−2.32
Promethium (Pm)	$[Xe]4f^4$	1.11	Pink	−2.29
Samarium (Sm)	$[Xe]4f^5$	1.10	Yellow	−2.30
Europium (Eu)	$[Xe]4f^6$	1.09	Pink	−1.99
Gadolinium (Gd)	$[Xe]4f^7$	1.08	Colorless	−2.29
Terbium (Tb)	$[Xe]4f^8$	1.06	Pink	−2.30
Dysprosium (Dy)	$[Xe]4f^9$	1.05	Yellow	−2.29
Holmium (Ho)	$[Xe]4f^{10}$	1.04	Pink	−2.33
Erbium (Er)	$[Xe]4f^{11}$	1.03	Lilac	−2.31
Thulium (Tm)	$[Xe]4f^{12}$	1.02	Green	−2.31
Ytterbium (Yb)	$[Xe]4f^{13}$	1.01	Colorless	−2.22
Lutetium (Lu)	$[Xe]4f^{14}$	1.00	Colorless	−2.30
Other trivalent metals				
Yttrium (Y)	[Kr]	1.04	Colorless	−2.37
Scandium (Sc)	[Ar]	0.89	Colorless	−1.88
Aluminum (Al)	[Ne]	0.50	Colorless	−1.66
Gallium (Ga)	$[Ne]3d^{10}$	0.62	Colorless	−0.35

accessible. The chemistry of thorium is largely the chemistry of a single diamagnetic oxidation state with a noble gas configuration, namely +4, like many group elements such as aluminum. The chemistry of uranium is considerably more complicated than that of thorium, but uranium(VI) with the noble gas configuration and uranium(IV), two oxidation units lower, are the two most important oxidation states as in many main group elements such as tin, lead, antimony, and bismuth.

12.1.2 Lanthanides

The 15 elements from lanthanum to lutetium (Table 12.1) with the electronic configurations from $[Xe]4f^0$ to $[Xe]4f^{14}$ are called the *lanthanides* and exhibit such similar chemical and physical properties that separation of individual lanthanides is relatively difficult. Yttrium, Y, forms a similar +3 ion with the krypton electronic configuration (Table 12.1); the chemistry of yttrium resembles that of lanthanides with similar atomic and ionic radii, namely terbium, dysprosium, and holmium. Yttrium is found in nature with the lanthanides. The lanthanides plus yttrium are collectively called the *rare earths*, even though lanthanum, cerium, and neodymium are all more abundant than many "abundant" elements such as lead.

The element above yttrium in the periodic table is scandium, which is surprisingly rare and costly for a light, nonradioactive element. The ionic radius of Sc^{3+} (0.89 Å) is smaller than that of any of the lanthanides and falls between that of Al^{3+} (0.50 Å) and the smallest lanthanide, Lu^{3+} (1.00 Å). For this reason the chemical behaviour of scandium is intermediate between that of aluminum and the lanthanides. In addition, the chemistry of scandium is essentially entirely that of the trivalent oxidation state.

A characteristic feature of the lanthanides is the *lanthanide contraction*, that is, a steady decrease in atomic and ionic size with increasing atomic number. Thus in the series La → Lu the radii of the M^{3+} ions decrease monotonically from 1.17 Å to 1.00 Å (Table 12.1). The lanthanide contraction arises from the imperfect screening of increasing nuclear charge in the La → Lu series by the $4f$ electrons. A consequence of the lanthanide contraction is the nearly same size of the element immediately following the lanthanides, namely hafnium (Table I.1), and its lighter congener zirconium (i.e., atomic radii of 1.45 Å for Zr and 1.44 Å for Hf and ionic radii of 0.86 Å for Zr^{4+} and 0.85 Å for Hf^{4+}). As a result of the lanthanide contraction, the chemical properties of zirconium and hafnium are more nearly similar than those of any pair of similar congeneric elements, and zirconium and hafnium are relatively difficult to separate from each other.

All the lanthanides exhibit the $+3$ oxidation state. The $+4$ and $+2$ oxidation states are stable mainly for lanthanide ions with completely empty half-full, or completely filled f shells, that is, $[Xe]4f^0$, $[Xe]4f^7$, and $[Xe]4f^{14}$ electronic configurations. Thus cerium has a stable $+4$ oxidation state with the $[Xe]4f^0$ electronic configuration. However, Ce(IV) derivatives are good one-electron oxidizing agents with oxidation potentials ranging from $+1.44$ V to $+1.70$ V depending on the anion present. The chemistry of Ce(IV) other than its oxidizing properties resembles the chemistry of other highly electropositive tetravalent metals such as Zr(IV), Hf(IV), and Th(IV).

The most stable $+2$ lanthanide oxidation states are colorless Eu^{2+} with a $[Xe]4f^7$ electronic configuration and yellow Yb^{2+} with a $[Xe]4f^{14}$ electron configuration with M^{3+}/M^{2+} reduction potentials of -0.34 and -1.04 V, respectively. Blood-red Sm^{2+} with a $[Xe]4f^6$ electron configuration is also known, but it is a stronger reducing agent, having a M^{3+}/M^{2+} reduction potential of -1.40 V. The chemistry of the divalent lanthanides aside from their reducing properties resembles the chemistry of the heavier alkaline earths such as strontium and barium.

The lanthanide ions have the maximum possible numbers of unpaired f electrons (up to 7 for Gd^{3+}) and exhibit complicated magnetic behavior. The electronic spectra of the lanthanide ions exhibit extremely sharp f–f transitions similar to free atoms unlike the broad d–d transitions of the d-block transition metals.[2] This is a consequence of the shielding of the lanthanide ion f orbitals from the surroundings of the ions. The color sequence of the Ln^{3+}

ions from La $\rightarrow$ Gd where f electrons are added to a xenon electronic configuration is the same as the color sequence of the Ln^{3+} ions from Lu $\rightarrow$ Gd where f electrons are removed (i.e., holes added) from a filled $4f^{14}$ shell. Thus ions having $[Xe]4f^n$ and $[Xe]4f^{14-n}$ electronic configurations exhibit essentially the same colors (Table 12.1). This corresponds to the same sequence of ion ground states in the La $\rightarrow$ Gd and Lu $\rightarrow$ Gd sequences. The colors of the lanthanide ions arise from $f-f$ transitions and are, therefore, insensitive to the lanthanide ion environment.

Certain lanthanide ions, notably some of the central lanthanides such as europium, terbium, and holmium, exhibit luminescence and fluorescence and thus are used in oxide phosphors for television tubes and related devices. The luminescence of europium can be used as a probe of its environment, thereby providing information on ligand charges, binding constants, site symmetry, and ligand exchange rates.[3]

Several of the paramagnetic lanthanide ions, notably Pr^{3+}, Eu^{3+}, and Yb^{3+}, are used in NMR shift reagents, generally complexed with a bulky β-diketonate such as dipivaloylmethanide, $Bu^tC(O)CH^-C(O)Bu^t$, to provide derivatives soluble in relatively nonpolar organic solvents. The large magnetic moment of the lanthanide ion when complexed to an organic molecule with a donor oxygen or nitrogen atom spreads out the NMR resonances, thereby facilitating interpretation of the resulting spectrum.

The lanthanides are characterized by relatively high coordination numbers, with coordination numbers 8 and 9 being relatively common. The most common coordination polyhedra for seven-, eight-, and nine-coordinate lanthanide complexes are depicted in Figure 12.1. Many compounds initially formulated as hexacoordinate octahedral lanthanide derivatives were later found to have higher coordination numbers with coordinated solvent molecules.

The following are the most important rare earth minerals.

1. *Monazite* and *xenotine*, which are lanthanide phosphates containing appreciable amounts of thorium.
2. *Bastnasite*, which as the approximate composition $LnFCO_3$.

The relative amounts of the rare earths in monazite and bastnasite are similar to those found overall in nature; that is, the early lanthanides predominate. Xenotine is useful as a source of the later (heavier) lanthanides. The scarcest naturally occurring lanthanide, namely thulium, is as abundant as bismuth and more abundant than arsenic, cadmium, mercury, or selenium.

Promethium, unlike the other lanthanides, does not have any stable isotopes and, therefore, is not found in nature with the other lanthanides. Traces of promethium occur in uranium ores arising from the spontaneous fission of ^{238}U. Milligram quantities of $^{147}Pm^{3+}$, a β-emitter with a 2.64-year half-life, can be isolated by ion exchange methods from the fission products of nuclear reactors.

Separation of individual lanthanides presents considerable difficulty because of their close similarity and is avoided for practical applications not requiring

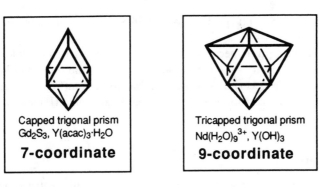

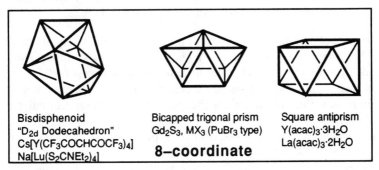

Figure 12.1. Some common polyhedra for coordination numbers 7, 8 and 9 in lanthanide chemistry.

pure individual lanthanides. Thus approximately two-thirds of the commercial uses of lanthanides involve no separation at all. Approximately 15% of commercial lanthanide production is "mischmetall," which is an alloy of lighter lanthanides in the proportions found in ore.

Samples of pure lanthanides used to be extremely rare and valuable because the original procedures for lanthanide separation involved fractional crystallization of suitable derivatives and were extremely tedious, requiring hundreds of crystallizations. The first practical procedures for large-scale lanthanide separation used complexation-enhanced ion exchange processes, which were developed in the 1950s. The lanthanides are eluted from a cation exchange resin in order of increasing size from lutetium to lanthanum. The separation is enhanced by complexing agents such as α-hydroxybutyric acid or ethylenediaminetetracetic acid, which prefer the smaller lanthanide ions. The lanthanides are precipitated as oxalates from the eluates and then ignited to the oxides.

Currently all large-scale commercial lanthanide separation is done by liquid–liquid extraction. Aqueous Ln^{3+} solutions are extracted in a continuous countercurrent process into a nonpolar organic liquid containing tri-*n*-butyl-phosphine oxide or bis(2-ethylhexyl)phosphinic acid. The lanthanides are separated by precipitation as oxalates or fluorides. Typical separation factors are ~ 2.5 per step, and purities of 99–99.9% are routinely achieved in multiple-step processes.

Cerium and europium, because of the existence of a second stable oxidation state (i.e., in addition to $+3$), can be separated from the other lanthanides by redox processes. Thus cerium can be oxidized to cerium(IV) and precipitated as the iodate, whereas europium can be reduced to europium(II) and precipitated as the sulfate.

12.1.3 Actinides

All actinides are radioactive and many are available only in tracer quantities. The availability of the individual actinides is as follows.

1. *Thorium and uranium.* Thorium and uranium[4] are relatively abundant and can be isolated from their ores by normal mining procedures. Their most stable isotopes (^{232}Th and ^{238}U) have half-lives in excess of 10^8 years, and thus their radioactivity poses no special handling problems. Thorium and uranium are the only actinides that can be handled by normal chemical methods.
2. *Protactinium.* The most stable isotope of protactinium is ^{231}Pa, which has a half-life of 3.3×10^5 years. Protactinium is a very minor constituent (0.34 ppm) of uranium ores. Treatment of 60 tons of sludge from the treatment of uranium ores resulted in the isolation of 130 g of protactinium.
3. *Neptunium and plutonium.* Neptunium and plutonium[5,6] are available in multikilogram quantities from the uranium fuel rods of nuclear reactors. The most stable isotope of neptunium is ^{237}Np with a half-life of 2.2×10^6 years. The longer-lived plutonium isotopes include ^{238}Pu, ^{239}Pu, ^{242}Pu, and ^{244}Pu with half-lives of 86.4, 24,360, 3.8×10^5, and 8.3×10^7 years, respectively. The isotope ^{238}Pu is used as a power source for space vehicles, whereas ^{242}Pu is a commonly used "explosive" in nuclear weapons. Since neptunium and plutonium isotopes are *fissionable materials*, their quantities must be kept below the *critical mass* to avoid a nuclear explosion. For a sphere of ^{242}Pu metal, the critical mass is ~ 10 kg. Neptunium and plutonium have the highest known toxicities of any elements.
4. *Americium and curium.* The actinides americium and curium are available in 100 g quantities from nuclear reactors. The most stable isotopes are ^{241}Am, ^{243}Am, and ^{244}Cm, with half lives of 433, 7650, and 18.12 years, respectively. The half-life of ^{244}Cm, the most stable curium isotope, is sufficiently short that curium compounds are highly radioactive.
5. *Actinium.* The most stable actinium isotope, ^{277}Ac, has a half-life of only 21.7 years. It is a trace constituent of uranium minerals, but its recovery from such minerals is not practiced. Actinium can be obtained by neutron irradiation of ^{226}Ra followed by electron capture.
6. *Heavier actinides.* In general, the half-lives of the actinide isotopes decrease with increasing atomic number. The isotopes ^{249}Bk (half-life 325 days), ^{252}Cf

(half-life 2.57 years), and ^{254}Es (half-life 1.52 years) can still be obtained in milligram quantities and ^{257}Fm (half-life 94 days) in microgram quantities from nuclear reactors. The longer-lived berkelium and einsteinium isotopes ^{247}Bk (half-life 10^4 years) and ^{252}Es (half-life 471.7 days) can be obtained only in trace quantities in accelerators, since they do not arise in significant quantities from nuclear reactions in nuclear reactors. Only tracer chemistry is possible with the transfermium elements (elements with atomic numbers greater than 100).

The f orbitals play a much larger role in actinide chemistry than in lanthanide chemistry. Less energy is required for the $5f \rightarrow 6d$ promotion in actinides than for the corresponding $4f \rightarrow 5d$ promotion in lanthanides, with the result that the actinides exhibit higher oxidation states than the lanthanides. In addition, since the $5f$ orbitals of the actinides have greater spatial extension relative to the $7s$ and $7p$ orbitals than the $4f$ orbitals of the lanthanides have relative to the $6s$ and $6p$ orbitals, the f orbitals in actinides can make a significant covalent contribution to their chemical bonding.

Table 12.2 depicts two different sets of the seven f orbitals,[7–9] which are classified by their numbers of major lobes. The major lobes of the f orbitals with six and eight major lobes are directed toward the vertices of a regular hexagon and cube, respectively. The cubic set of f orbitals is used for certain

Table 12.2 The General Shapes of Both the General and Cubic Sets of f Orbitals

Major Lobes	Shape	General Set	Cubic Set
2		z^2	x^3 y^3 z^3
4		xz^2 yz^2	None
6		$x(x^2 - 3y^2)$ $y(3x^2 - y^2)$	None
8		xyz $z(x^2 - y^2)$	xyz $x(z^2 - y^2)$ $y(z^2 - x^2)$ $z(x^2 - y^2)$

highly symmetrical structures—for example, those of cubic (O_h point group), octahedral (also O_h point group), or icosahedral (I_h point group) symmetry— since in the cubic set of f orbitals there are two triply degenerate subsets, namely the x^3, y^3, z^3 subset and the $x(z^2 - y^2), y(z^2 - x^2), z(x^2 - y^2)$ subset. In less symmetrical structures, the generat set of f orbitals is used.

12.2 The Lanthanide and Actinide Metals

12.2.1 Lanthanide Metals

The lanthanide metals can be liberated from their simple binary compounds by treatment with even more electropositive metals at elevated temperatures, for example,

$$2LnCl_3 + 3Ca \xrightarrow{1000°C} 2Ln + 3CaCl_2$$

$$(Ln = La \rightarrow Gd) \qquad (12.1a)$$

$$2LnF_3 + 3Ca \xrightarrow{1000°C} 2LnF_3 + 2Ln + 3CaF_2$$

$$(Ln = Y, Tb \rightarrow Tm) \qquad (12.1b)$$

$$Ln_2O_3 + 2La \xrightarrow{\Delta} La_2O_3 + Ln$$

$$(Ln = Eu, Sm, Yb) \qquad (12.1c)$$

The later lanthanide *chlorides* cannot be used for calcium reduction (equation 12.1a) because their volatilities are too high; hence the corresponding fluorides must be used (equation 12.1b).

The lanthanide metals are silvery white and very reactive. They liberate hydrogen from water upon heating and burn in oxygen to the corresponding oxides. However, yttrium metal, like magnesium and aluminum metals, forms a protective oxide coating that renders bulk yttrium resistant to air. The lanthanide metals react exothermally upon heating with molecular hydrogen to give the hydrides LnH_2 and LnH_3 (Section 1.5.3.1). Europium and ytterbium, which are the two lanthanides exhibiting the most stable $+2$ oxydation states, dissolves in liquid ammonia to form blue solutions like the heavier alkaline earth metals (Section 10.2.3).

12.2.2 Actinide Metals[10]

The actinide metals can be liberated from their fluorides (or occasionally their chlorides or oxides) by treatment of 1100–1400°C with the vapor of a more electropositive metal such as lithium, magnesium, calcium, or barium. In general, actinide metals are very reactive. In addition, pure plutonium metal above its critical mass can initiate a nuclear explosion. Actinium and curium metals are so radioactive that they glow from the heat that is generated by their radioactive decomposition.

12.3 Trivalent Lanthanides

12.3.1 Oxides and Hydroxides

The *lanthanide oxides*, Ln_2O_3, are the most basic known oxides of trivalent metals and resemble the divalent alkaline earth oxides (Section 10.5.1.1) in their behavior. The basicities of the lanthanide oxides decrease in the sequence La → Lu, corresponding to the decrease in the lanthanide ionic radii. The lanthanide oxides absorb CO_2 to form the corresponding carbonates as well as water to form the corresponding *lanthanide hydroxides*, $Ln(OH)_3$. The hydroxides $Ln(OH)_3$ can also be obtained as gelatinous precipitates from aqueous solutions of Ln^{3+} upon treatment with ammonia dilute alkalies and do *not* dissolve in excess alkali; they are, therefore, *not* amphoteric. The hydroxides $Ln(OH)_3$ have hexagonal structures with tricapped trigonal prismatic nine-coordinate lanthanide atoms (Figure 12.1).

12.3.2 Halides[11,12]

The *lanthanide fluorides*, LnF_3, are insoluble in water but dissolve in a nitric acid/boric acid mixture to remove fluoride as BF_4^-. Anhydrous *lanthanide chlorides*, $LnCl_3$, can be made by heating the corresponding lanthanide oxide with solid ammonium chloride[13]:

$$Ln_2O_3 + 6NH_4Cl \xrightarrow{300°C} 2LnCl_3 + 3H_2O + 6NH_3 \qquad (12.2)$$

The anhydrous *lanthanide chlorides*, $LnCl_3$, dissolve in water to form hydrates containing six to seven H_2O molecules. These hydrates *cannot* be dehydrated back to anhydrous $LnCl_3$ because of partial hydrolysis leading to loss of HCl.

12.3.3 Oxo Salts

The lanthanides form numerous double nitrates with stoichiometries such as $2Ln(NO_3)_3 \cdot 3Mg(NO_3)_2 \cdot 24H_2O$ and $Ln(NO_3)_2 \cdot 2NH_4NO_3 \cdot 4H_2O$, Such double nitrates were used in the past for the separation of lanthanides by fractional crystallization procedures. Lanthanide double sulfates include salts with the stoichiometry $Ln_2(SO_4)_3 \cdot 3Na_2SO_4 \cdot 12H_2O$. Such double sulfates of the lanthanides La → Eu are sparingly soluble in aqueous Na_2SO_4, whereas similar salts of the lanthanides Gd → Lu and yttrium are appeciably soluble in aqueous Na_2SO_4, thereby providing a rapid method for separating the lanthanides into two groups, commonly called the cerium and yttrium groups, respectively. Lanthanide ions can be precipitated quantitatively as their oxalates, which can have varying stoichiometries depending on the conditions of precipitation. The oxalates can be ignited to the corresponding oxides, Ln_2O_3, thereby providing a method for the gravimetric determination of the lanthanides.

12.3.4 Lanthanide Complexes of Chelating Oxygen Ligands

Lanthanides form diverse complexes with chelating oxygen ligands. Poly-carboxylic acid anions such as tartrate, citrate, and ethylenediaminetetra-acetic acid form water-soluble complexes useful in the ion exchange separation of lanthanides. The β-diketonates,[14] $Ln(RCOCHCOR)_3$, do not contain octa-hedral six-coordinate lanthanides but usually exhibit higher lanthanide coordi-nation numbers either through polymerization or by solvent coordination. Only when bulky or fluorinated β-diketonates (e.g., dipivaloylmethanide, $Bu^tC(O)$-$CH^-C(O)Bu^t$) are used as ligands can simple octahedral $Ln(RCOCHCOR)_3$ derivatives be obtained; such derivatives are volatile and can be used for the gas chromatographic analysis of the lanthanides.[15] Some alkali metal salts of the eight-coordinate $Ln(\beta\text{-diketonate})_4^-$ anions are sometimes volatile enough to be sublimed without decomposition

12.4 Tetravalent and Divalent Lanthanides

12.4.1 Tetravalent Lanthanides

The $+4$ oxidation state is much more stable for cerium than for any of the other lanthanides, since Ce(IV) has an empty f shell. *Cerium (IV) oxide*, CeO_2, can be simply obtained by heating cerium metal, $Ce(OH)_3$, or cerium(III) salts of oxoacids in air or oxygen. Aqueous solutions of Ce(IV) can be obtained by treatment of Ce(III) solutions with powerful oxidizing agents such as peroxo-disulfate or bismuth(V) in the presence of nitric acid. The aqueous chemistry of cerium(IV) is similar to that of zirconium, hafnium, thorium, or uranium(IV) except for its redox properties.

Cerium can be separated from the other lanthanides by oxidation to the $+4$ oxidation state followed by precipitation of cerium(IV) as a phosphate insoluble in 4 M nitric acid, an iodate insoluble in 6 M nitric acid, or an insoluble oxalate. In addition, cerium(IV) is more readily extracted into organic solvents by tri-*n*-butyl phosphate than the trivalent lanthanide ions because of its higher charge.

Cerium(IV) salts such as $(NH_4)_2[Ce(NO_3)_6]$ are useful strong one-electron oxidizing agents being reduced to cerium(III). The redox potential of cerium(IV) depends significantly on the anion and the concentration, for example,

$$Ce(IV) + e = Ce(III)$$

$$E° = +1.44 \text{ V } (1 \text{ M } H_2SO_4); \quad +1.61 \text{ V } (1 \text{ M } HNO_3);$$

$$1.70 \text{ V } (1 \text{ M } HClO_4) \tag{12.3}$$

This suggests that the relative complexing tendencies of these anions are $SO_4^{2-} > NO_3^- > ClO_4^-$. Cerium(IV) is used to perform a number of useful oxidations including oxidation of aldehydes and ketones at the α-carbon atom, the oxidation of toluene to benzaldehyde, the oxidative cleavage of glycols, and the oxidative liberation of ligands from their metal–carbonyl complexes.[16,17]

Praseodymium(IV) and terbium(IV), with the $[Xe]f^1$ and $[Xe]f^7$ electronic configurations, respectively, are known in solid state oxides and fluoride such as Pr_6O_{11},[18] $NaPrF_5$, Na_2PrF_6, Tb_4O_7, and TbF_4. The oxides Pr_6O_{11} and Tb_4O_7 are examples of mixed oxidation state lanthanide derivatives containing both trivalent and tetravalent lanthanides. However, attempts to generate Pr(IV) and Tb(IV) in aqueous solution by dissolution of these solid tetravalent derivatives in water or aqueous acid result instead of reduction to the trivalent state with oxygen evolution.

12.4.2 Divalent Lanthanides

The most stable divalent lanthanide is Eu(II), which has an $[Xe]f^7$ electronic configuration with a half-filled f shell. Colorless aqueous solutions of Eu^{2+} are readily obtained by reduction of Eu^{3+} with zinc or magnesium metals. Aqueous solutions of Eu^{2+} are stable in water without hydrogen evolution. Less stable divalent lanthanides are blood-red Sm(II) and yellow Yb(II), which can be obtained by sodium amalgam or electrolytic reduction of the corresponding trivalent lanthanides. Aqueous solutions of Sm^{2+} and Yb^{2+} are unstable with respect to hydrogen evolution from water reduction as well as air oxidation. Divalent lanthanides form water-soluble sulfates like the alkaline earths. *Samarium diiodide*, SmI_2, which can be generated from samarium metal and iodine, 1,2-diiodoethane, or HgI_2 in tetrahydrofuran, is useful as a mild one-electron reducing agent in organic chemistry.[19] Samarium(II) is chosen in preference to europium(II) or ytterbium(II) for this purpose because it is a stronger reducing agent.

12.5 Actinide Chemistry

The actinides exhibit a much greater diversity of oxidation states than the lanthanides. Higher oxidation states occur for the actinides than for the lanthanides, with +7 being the maximum actinide oxidation state in well-characterized compounds. The maximum level of oxidation for the early actinides corresponds to the [Rn] noble gas configuration, leading to maximum oxidation states of +3, +4, +5, +6, and +7 for actinium, thorium, protactinium, uranium, and neptunium, respectively. Beyond americium, the +3 oxidation state dominates for the actinides, as it does for all the lanthanides. The +5 and +6 actinide oxidation states, particularly in aqueous media, are dominated by the very stable *actinyl* AnO_2^+ and AnO_2^{2+} (An = actinide) ions,[20] which have a *linear* O—An—O unit and very short actinide–oxide bonds, which can be interpreted as triple bonds with one σ and two orthogonal π components in structures such as $:O^+\!\!\equiv\!\!An\!\!\equiv\!\!O^+:$ for AnO_2^{2+}. The four orthogonal π components in the two linear $An\!\!\equiv\!\!O$ triple bonds can arise from two $d\pi$–$p\pi$

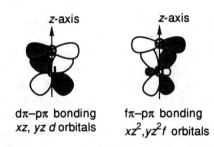

Figure 12.2. Uranium–oxygen $d\pi$–$p\pi$ and $f\pi$–$p\pi$ bonding in UO_2^{2+}.

bonds involving actinide xz and yz d orbitals and oxygen p orbitals and two $f\pi$–$p\pi$ bonds involving actinide xz^2 and yz^2 f orbitals and oxygen p orbitals (Figure 12.2). The oxygen atoms in the AnO_2^{2+} units have unusually low field gradients leading to unusually sharp ^{17}O NMR spectra for UO_2^{2+}, which make it feasible to use ^{17}O NMR data to study the hydrolysis of uranyl ion.[21] The stabilities of AnO_2^{+} and particularly AnO_2^{2+} are so great that these cations survive intact through a variety of chemical transformations.

The properties of the individual actinide oxidation states are as follows.

12.5.1 The +2 Oxidation State

The +2 oxidation state, an unusual one for actinides, occurs mainly in the later actinides (Cf → No), which are generally available only in tracer quantities.

12.5.2 The +3 Oxidation State

The +3 oxidation state is common for the actinides as for the lanthanides. However, Th^{3+} and Pa^{3+} are so strongly reducing that they liberate hydrogen rapidly from water and thus cannot be obtained in aqueous solution. Wine-red aqueous solutions of U^{3+} are not only oxidized rapidly by air but are slowly oxidized by water under ambient anaerobic conditions with hydrogen evolution.[22] Thus the +3 oxidation state is not readily accessible for the two readily available and easily handled actinides, namely thorium and uranium. The trivalent lanthanide and actinide ions are similar in size and solution properties except for the ready oxidation of Th^{3+}, Pa^{3+}, and U^{3+} just noted. Common coordination numbers for trivalent actinides[23] are 6, 8, and 9 in octahedral, bicapped trigonal prismatic, and tricapped trigonal prismatic complexes, respectively, similar to the coordination polyhedra of the trivalent lanthanides (Figure 12.1).

12.5.3 The +4 Oxidation State

The +4 oxidation state is an important oxidation state for the two readily available actinides, namely thorium and uranium, and is also important for a few other actinides such as protactinium and plutonium. The +4 oxidation state is also known for the actinides from americium to californium but is very easily reduced. The *actinide dioxides*, MO_2 ($M = Th \rightarrow Cf$), have the fluorite lattice. The *actinide tetrafluorides*, AnF_4, are isostructural with the two known lanthanide tetrafluorides, namely CeF_4 and TbF_4. Hydrolysis and complexation of the $+4$ actinides can occur in aqueous solution. The coordination number 8 is very common for complexes of the $+4$ actinides and normally exhibits bisdisphenoidal (D_{2d} dodecahedral) or square antiprismatic stereochemistry, although perfect cubic stereochemistry is observed in rare cases (e.g., in $[Et_4N]_4[U(SCN)_8]$).[24]

12.5.4 The $+5$ Oxidation State

The $+5$ oxidation state with the radon configuration is the most stable oxidation state for protactinium. The linear actinyl ions, AnO_2^+, are important in the chemistry of U(V), Np(V), Pu(V), but not Pa(V). Uranium(V) disproportionates readily in water to give uranium(IV) and uranium(VI).[25] Solid actinide(V) derivatives include MF_5 ($M = Pa$, U, Np) as well as binuclear uranium(V) derivatives U_2X_{10} ($X = Cl$, OR, etc.), with octahedral uranium(V) and two μ_2-bridging chloride or alkoxide ligands (**12.1**). These binuclear

12.1

uranium(V) derivatives do *not* have direct uranium–uranium bonds. Penta-coordinate actinide complexes include hexacoordinate octahedral UF_6^- as well as octacoordinate AnF_8^{3-} ($An = Pa$, U, Np) with unusual cubic actinide coordination.

12.5.5 The $+6$ Oxidation State

The $+6$ oxidation state dominates for uranium and is also known for neptunium, plutonium, and americium. In all cases the most important actinide(VI) derivatives contain the actinyl ions AnO_2^{2+} already mentioned, which remain intact in aqueous solution. Some octahedral AnX_6 derivatives ($X = F$; $An = U$, Np, Pu; $X = Cl$, OR; $An = U$) are known; they are readily hydrolyzed to the corresponding AnO_2^{2+} ions. Actinyl complexes are known with octahedral, pentagonal bipyramidal, and hexagonal bipyramidal stereochemistries with four, five, and six ligands, respectively, in the equatorial plane (Figure 12.3).

12.5.6 The $+7$ Oxidation State

For many years the maximum possible oxidation state for actinides was believed to be $+6$. However, the oxoanions $NpO_4(OH)_2^{3-}$, NpO_6^{5-}, and PuO_6^{5-}

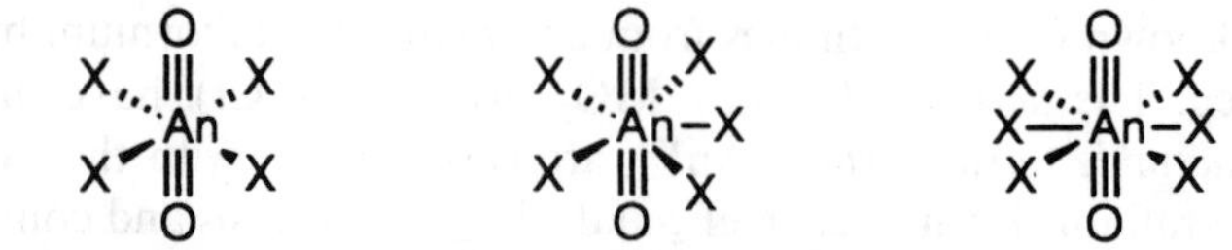

4 equatorial ligands	5 equatorial ligands	6 equatorial ligands
octahedron	pentagonal bipyramid	hexagonal bipyramid
$UO_2Cl_4^{2-}$	$UO_2(H_2O)_5^{2+}$	$UO_2(NO_3)_3^-$

Figure 12.3. Coordination of four to six ligands in the equatorial plane of actinyl derivatives.

containing $+7$ neptunium and plutonium[26] were finally prepared by treatment of the $+6$ derivatives in basic media with strong oxidants such as ozone or by heating the actinide oxide with an alkali metal oxide to $\sim 400°C$ in an oxygen atmosphere. These actinide(VII) oxoanions contain octahedral AnO_6 structural units and are strong oxidizing agents.

The redox chemistry of some of the actinides in aqueous solution (particularly U, Np, Pu, and Am) is rather complicated because of the multiplicity of oxidation states.[27,28] Thus under certain conditions all four oxidation states of plutonium, namely $+3$, $+4$, $+5$, and $+6$, can coexist in aqueous solution in appreciable concentrations. The stable species in aqueous solution for these multiple oxidation state actinides are green U(IV), yellow $U^{VI}O_2^{2+}$, green $Np^VO_2^+$, blue-violet Pu(III), tan Pu(IV), and pink Am(III). The tendency for complex formation for actinides in the oxidation states $+3$ to $+6$ decreases in the sequence $An^{4+} > AnO_2^{2+} > An^{3+} > AnO_2^+$, in approximate agreement with their charge densities. The order of complexing ability is $F^- > NO_3^- > Cl^- > ClO_4^-$ for uninegative anions and $CO_3^{2-} > C_2O_4^{2-} > SO_4^{2-}$ for dinegative anions.

Many of the shorter-lived actinide isotopes decaying by α-emission or spontaneous fission are so radioactive that their aqueous chemistry is complicated by heating and radiolysis effects. Thus the heat outputs from ^{242}Cm, ^{238}Pu, and ^{241}Am with half-lives of 18.12, 86.4m and 433 years, respectively, are 122, 0.5, and 0.1 W/g, respectively. The radiation-induced decomposition of water leads to $H^\cdot$ and $^\cdot OH$ radicals as well as H_2O_2 production, which can result in considerably redox chemistry. For example, the relatively high oxidation states Pu(V), Pu(VI), Am(IV), and Am(VI) are reduced by the $H^\cdot$ and/or H_2O_2 generated by water radiolysis.

12.6 Organometallic Chemistry of Lanthanides and Actinides

All the lanthanides and actinides are highly electropositive metals. For this reason almost all their organometallic compounds are highly sensitive to air and water, like organometallic compounds of the alkali and alkaline earth

metals but unlike many air- and water-stable organometallic derivatives of the
d-block transition metals. This is another important respect in which the
chemistry of the lanthanides and actinides resembles that of the electropositive
main group metals more than that of the d-block transition metals. Strictly
anaerobic and anhydrous conditions are necessary for handling organolanthanide and organoactinide compounds.

12.6.1 Organolanthanide Compounds

Organolanthanide chemistry[29, 30] is almost exclusively the chemistry of anionic
planar hydrocarbon rings such as $C_5H_5^-$, $Me_5C_5^-$, and $C_8H_8^{2-}$ symmetrically
bonding to the lanthanide element with a rather ionic bond. Although lutetium
is one of the rarest lanthanides, it is an attractive lanthanide to use for the
study of organolanthanide chemistry because its diamagnetism leads to sharp
^1H and ^{13}C NMR spectra for product characterization. Most syntheses of
organolanthanide compounds use reactions of an ionic planar hydrocarbon
reagent such as $Na^+C_5H_5^-$, $Na^+C_5Me_5^-$, or $(K^+)_2C_8H_8^{2-}$ (Section 10.3.4) with
a lanthanide halide or similar derivative.

12.6.1.1 Cyclopentadienyllanthanide Derivatives

The *tricyclopentadienyllanthanides*, $Ln(C_5H_5)_3$, can be obtained by reactions of
sodium cyclopentadienide with the corresponding lanthanide trichloride; however, the europium derivative cannot be made by this method because of
reduction to europium(II). The $Ln(C_5H_5)_3$ derivatives form weakly bonded
polymers but can be isolated by vacuum sublimation[31]; they are Lewis acids,
which react with Lewis bases such as phosphines, isocyanides, and tetrahydrofuran to form the corresponding monomeric adducts $L\rightarrow Ln(C_5H_5)_3$. *Dicyclo-
pentadienyllanthanide halides*, $(C_5H_5)_2LnX$, can be made by the following
reactions:

$$LnCl_3 + 2MC_5H_5 \rightarrow (C_5H_5)_2LnCl + 2MCl \tag{12.4a}$$

$$2(C_5H_5)_3Ln + LnCl_3 \rightarrow 3(C_5H_5)_2LnCl \tag{12.4b}$$

$$(C_5H_5)_3Ln + HCl \rightarrow (C_5H_5)_2LnCl + C_5H_6 \tag{12.4c}$$

These halides can be alkylated with organolithium reagents to give the σ-alkyl
derivatives $(C_5H_5)_2LnR$ such as $[(C_5H_5)_2YbCH_3]_2$, which forms a tetrahydrofuran adduct $(C_5H_5)_2Yb(CH_3)(OC_4H_8)$.[32]

12.6.1.2 Pentamethylcyclopentadienyllanthanide Derivatives

The increased bulk of the Me_5C_5 ring relative to the C_5H_5 ring causes major
changes in the structural and chemical properties of their lanthanide derivatives.
Of greatest interest is the ability of Me_5C_5Ln systems to activate C—H bonds
in relatively unreactive molecules such as methane and tetramethylsilane, for

example, (R = H or methyl):

$$(Me_5C_5)_2LuR + Me_4Si \rightarrow (Me_5C_5)_2LuCH_2SiMe_3 + RH\uparrow \qquad (12.5a)$$

$$(Me_5C_5)_2LuR + {}^*CH_4 \rightarrow (Me_5C_5)_2Lu^*CH_3 + RH\uparrow \qquad (12.5b)$$

The methyllutetium derivative $(Me_5C_5)_2LuMe$ is also a very active catalyst for ethylene polymerization.[33]

12.6.1.3 Cyclooctatetraene Derivatives

Lanthanide derivatives of the cyclooctatetraene dianion, $C_8H_8^{2-}$, are known in the formal lanthanide oxidation states $+2$, $+3$, and $+4$. The free lanthanides forming stable divalent compounds (i.e., Eu, Yb) react with cyclooctatetraene in liquid ammonia to form the corresponding MC_8H_8 derivatives; a pyridine adduct $C_8H_8Yb(NC_5H_5)_3$ has been characterized structurally.[34] Reactions of the trivalent lanthanide halides, $LnCl_3$, with $K_2C_8H_8$ can give either the bis(cyclooctatetraene)lanthanide(III) anions, $Ln(C_8H_8)_2^-$, or binuclear derivatives of the type $[C_8H_8Ln(\mu_2\text{-}Cl)(OC_4H_8)_2]_2$ with two bridging chlorine atoms. Oxidation of the cerium(III) derivative, $Ce(C_8H_8)_2^-$, with silver(I) iodide gives the brown-black neutral cerium(IV) derivative $Ce(C_8H_8)_2$, which can also be obtained by reaction of $Ce(OPr^i)_4$ with Et_3Al in the presence of excess cyclooctatetraene.[35] Even this cerium(IV) organometallic derivative is very sensitive toward air oxidation.

12.6.2 Organoactinide Compounds[36]

One or more pentahaptocyclopentadienyl or substituted cyclopentadienyl rings are bonded to the actinide in the largest group of organoactinide compounds. Such compounds are obtained by reactions of actinide halides with an alkali metal or thallium cyclopentadienide or substituted cyclopentadienide. Tetravalent actinide cyclopentadienyl derivatives of the types $(C_5H_5)_4An$ (An = Th, Pa, U, Np), $(C_5H_5)_3AnX$ (An = Th, Pa, U), and $C_5H_5AnX_3$ (An = Th, Pa, U) are known, as well as trivalent actinide cyclopentadienyl derivatives of the types $(C_5H_5)_3An$ (An = Th, U, Pu, Am, Cm, Bk, Cf) and $(C_5H_5)_2AnX$ (An = Tn, U, Bk). Tetravalent unsubstituted cyclopentadienyl derivatives of the type $(C_5H_5)_2AnX_2$ do not appear to be known. The cyclopentadienyl–actinide bond is dative and rather polar with some f-orbital participation.[37] The chlorine atom is labile in $(C_5H_5)_3UCl$ and can be replaced with a variety of other groups by simple metathesis reactions. Thus alkylation of $(C_5H_5)_3UCl$ with alkyllithium compounds gives a variety of alkyl- and aryluranium derivatives of the type $(C_5H_5)_3UR$ in which the uranium atom is σ-bonded to the alkyl or aryl carbon atom.[38] Alkyls of the type $(C_5H_5)_3AnR$ undergo insertion reactions with carbon monoxide and alkyl isocyanides, RNC. Such insertion reactions of $(C_5H_5)_3AnR$ with CO lead to the formation of dihapto-acyl derivatives $(C_5H_5)_3An(COR)$ in which the acyl group is bonded to the actinide through its oxygen as well as a carbon atom, in accord with the high affinity of actinides for oxygen.[39]

Pentamethylcyclopentadienyl actinide derivatives exhibit improved solubility and crystallizability relative to the corresponding unsubstituted derivatives. Furthermore $(Me_5C_5)_2AnCl_2$ (An = Th, U), unlike $(C_5H_5)_2AnCl_2$, is stable and can be used to prepare other $(Me_5C_5)_2AnX_2$ derivatives, including the dialkyls and diaryls $(Me_5C_5)_2AnR_2$ (R = Me, Me_3CCH_2, Me_3SiCH_2, Ph, etc.) with two actinide–carbon σ bonds as well as the two actinide–ring bonds.[40]

The cyclooctatetraene dianion forms actinide complexes of the type $(C_8H_8)_2An$ (An = Th, Pa, U, Np, and Pu), colloquially known as the "actinocenes" in a dubious analogy to ferrocene, $(C_5H_5)_2Fe$. The bis(cyclooctatetraene)actinides are made by reactions of $K_2C_8H_8$ with the corresponding actinide tetrachloride and have "sandwich" structures with parallel C_8H_8 rings and perfect D_{8h} symmetry. Like other organoactinide derivatives, the $(C_8H_8)_2An$ derivatives are sensitive to air and water, although pyrophoric $(C_8H_8)_2U$ is only relatively slowly attacked by air-free water.

Other types of organoactinide derivative include allyls and arene derivatives. The tetraallyl actinides, $(C_3H_5)_4An$ (An = Th, U) are obtained by treatment of the corresponding $AnCl_4$ with an allylmagnesium halide. They have structures in which each of the four allyl groups is bonded to the central actinide through all three of its carbon atoms. The tetraallyl actinides decompose above 0°C and are very sensitive to air and water. Reaction of UCl_4 with hexamethylbenzene in toluene solution at 110°C in the presence of aluminum chloride and aluminum metal gives the maroon hexamethylbenzene–uranium(III) complex $Me_6C_6U(AlCl_4)_3$, shown by X-ray diffraction to have a structure in which the central uranium atom is bonded to all six carbons of the hexamethylbenzene ring as well as two chlorine atoms of each of the three $AlCl_4$ groups.[41] The uranium(IV) is reduced to uranium(III) by the aluminum metal during the course of this reaction.

References

1. King, R. B., Covalent Bonding in Actinide Derivatives, *Inorg. Chem*, 1992, **31**, 1978–1980.

2. Jørgensen, C. K.; Reisfeld, R., Chemistry and Spectroscopy of the Rare Earths, *Top. Curr. Chem.*, 1982, **100**, 127–167.

3. Horrocks, W. de W.; Albin, M., Lanthanide Ion Luminescence in Coordination Chemistry and Biochemistry, *Prog. Inorg. Chem.*, 1984, **31**, 1–104.

4. Cordfunke, E. H. P., *The Chemistry of Uranium*, Elsevier, Amsterdam, 1969.

5. Cleveland, J. M., *The Chemistry of Plutonium*, Gordon and Breach, New York, 1970.

6. Carnall, W. T.; Choppin, G. R., *Plutonium Chemistry*, ACS Symposium Series 216, American Chemical Society, Washington, DC, 1983.

7. Freedman, H. G., Jr.; Choppin, G. R.; Feuerbacher, D. G., The Shapes of the f Orbitals, *J. Chem. Educ.*, 1964, **41**, 354–358.

8. Becker, C., Geometry of the f Orbitals, *J. Chem. Educ.*, 1964, **41**, 358–360.

9. Smith, W.; Clack, D. W., Angular Dependence of Overlap for s, p, d, and f Functions, *Rev. Roum. Chim.*, 1975, **20**, 1243–1252.

10. Spiret, J. C.; Peterson, J. R.; Asprey, L. B., Preparation and Purification of Actinide Metals, *Adv. Inorg. Chem.*, 1987, **31**, 1–41.

11. Brown, D., *Halides of Lanthanides and Actinides*, Wiley, London, 1968.

12. Burgess, J.; Kijowski, J., Lanthanide, Yttrium, and Scandium Trihalides: Preparation of Anhydrous Materials and Solution Thermochemistry, *Adv. Inorg. Chem. Radiochem.*, 1981, **24**, 57.

13. Meyer, G.; Ax, P., An Analysis of the Ammonium Chloride Route to Anhydrous Rare-Earth Metal Chlorides, *Mater. Res. Bull*, 1982, **17**, 1447–1455.

14. Wenzel, T. J.; Williams, E. J.; Haltiwanger, R. C.; Sievers, R. E., *Polyhedron*, 1985, **4**, 369–378.

15. Moshier, R. W.; Sievers, R. E., *Gas Chromatography of Metal Chelates*, Pergamon Press, Oxford, 1965.

16. Richardson, W. H., Ceric Ion Oxidation of Organic Compounds, in *Oxidation in Organic Chemistry*, K. Wiberg, Ed., Academic Press, New York, 1965.

17. Molander, G. A., Application of Lanthanide Reagents in Organic Synthesis, *Chem. Rev.*, 1992, **92**, 29–68.

18. Otsuka, K.; Kunitomi, M.; Saito, T., Oxidation and Reduction of Praseodymium Oxide at Low Temperatures, *Inorg. Chim. Acta*, 1986, **115**, L31–L32.

19. Kagan, H. B.; Namy, J. L., Lanthanides in Organic Synthesis, *Tetrahedron*, 1986, **24**, 6573–6614.

20. Denning, R. G., Electronic Structure and Bonding in Actinyl ions, *Struct. Bonding*, 1992, **79**, 215–276.

21. Jung, W.-S.; Tomiyasu, H.; Fukutomi, H., The First Identification of Hydrolysis Species of Uranyl Ions by ^{17}O NMR Spectroscopy, *J. Chem. Soc., Chem. Commun.*, 1987, 372–373.

22. Drożdżyński, J., Chemistry of Trivalent Uranium, in *Handbook of the Physics and Chemistry of the Actinides*, Vol. 6, A. J. Freeman and C. Keller, Eds., Elsevier, Amsterdam, 1991, pp. 281–336.

23. Bombieri, G.; de Paoli, G., Structural Aspects of Actinide Coordination Chemistry, in *Handbook of the Physics and Chemistry of the Actinides*, Vol. 3, A. J. Freeman and C. Keller, Eds., North Holland, Amsterdam, 1985, pp. 75–141.

24. Countryman, R.; McDonald, W. S., *J. Inorg. Nuclear Chem.*, 1971, **33**, 2213–2220.

25. Ekstrom, A., Kinetics and Mechanism of the Disproportionation of Uranium(V), *Inorg. Chem.*, 1974, **13**, 2237–2241.

26. Keller, C., Heptavalent Actinides, in *Handbook on the Physics and Chemistry of the Actinides*, Vol. 3, A. J. Freeman and C. Keller, Eds., North-Holland, Amsterdam, 1985, pp. 143–184.

27. Choppin, G., Comparison of the Solution Chemistry of the Actinides and Lanthanides, *J. Less-Common Met.*, 1983, **93**, 323–330.

28. Choppin, G., Solution Chemistry of the Actinides, *Radiochim. Acta*, 1983, **32**, 43–53.

29. Evans, W. J., Organometallic Lanthanide Chemistry, *Adv. Organomet. Chem.*, 1985, **24**, 131–177.

30. Marks, T. J., Chemistry and Spectroscopy of f-Element Organometallics, Part I: The Lanthanides, *Prog. Inorg. Chem.*, 1978, **24**, 51–107.

31. Eggers, S. H.; Kopf, J.; Fischer, R. D., *Organometallics*, 1986, **5**, 383–385.

32. Evans, W. J.; Dominguez, R.; Hanusa, T. P., *Organometallics*, 1986, **5**, 263–270.

33. Watson, P. L.; Parshall, G. W., Organolanthanides in Catalysis, *Acc. Chem. Res.*, 1985, **18**, 51–56.

34. Wayda, A. L.; Mukerji, I.; Dye, J. L.; Rogers, R. D., *Organometallics*, 1987, **6**, 1328–1332.

35. Streitwieser, A., Jr.; Kinsley, S. A.; Rigsbee, J. T., *J. Am. Chem. Soc.*, 1985, **107**, 7786–7788.

36. Marks, T. J., Chemistry and Spectroscopy of *f*-Element Organometallics, Part II: The Actinides, *Prog. Inorg. Chem.*, 1979, **25**, 223–333.

37. Bursten, B. E.; Strittmatter, R. J., Cyclopentadienyl–Actinide Complexes: Bonding and Electronic Structure, *Angew. Chem. Int. Ed. Engl.*, 1991, **30**, 1069–1085.

38. Marks, T. J., Organoactinide Compounds with Metal-to-Carbon and Metal-to-Hydrogen Sigma Bonds, in *Handbook of the Physics and Chemistry of the Actinides*, Vol. 4, A. J. Freeman and C. Keller, Eds., Elsevier, Amsterdam, 1986, pp. 491–530.

39. Sonnenberger, D. C.; Mintz, E. A.; Marks, T. J., *J. Am. Chem. Soc.*, 1984, **106**, 3484–3491.

40. Fagan, P. J.; Manriquez, J. M.; Maatta, E. A.; Seyam, A. M.; Marks, T. J., Synthesis and Properties of Bis(pentamethylcyclopentadienyl) Actinide Hydrocarbyls and Hydrides. A New Class of Highly Reactive *f*-Element Organometallic Compounds, *J. Am. Chem. Soc.*, 1981, **103**, 6650–6667.

41. Cotton, F. A.; Schwotzer, W., Synthesis and Structural Comparison of the η^6-Arene Complexes $Sm(C_6Me_6)(AlCl_4)_3$ and $U(C_6Me_6)(AlCl_4)_3$, *Organometallics*, 1987, **6**, 1275–1280.

Index